Ngoc Thanh Nguyen, Radosław Piotr Katarzyniak, and Adam Janiak (Eds.)

New Challenges in Computational Collective Intelligence

Studies in Computational Intelligence, Volume 244

Editor-in-Chief
Prof. Janusz Kacprzyk
Systems Research Institute
Polish Academy of Sciences
ul. Newelska 6
01-447 Warsaw
Poland
E-mail: kacprzyk@ibspan.waw.pl

Further volumes of this series can be found on our homepage:
springer.com

Vol. 221. Tassilo Pellegrini, Sören Auer, Klaus Tochtermann, and
Sebastian Schaffert (Eds.)
Networked Knowledge - Networked Media, 2009
ISBN 978-3-642-02183-1

Vol. 222. Elisabeth Rakus-Andersson, Ronald R. Yager,
Nikhil Ichalkaranje, and Lakhmi C. Jain (Eds.)
Recent Advances in Decision Making, 2009
ISBN 978-3-642-02186-2

Vol. 223. Zbigniew W. Ras and Agnieszka Dardzinska (Eds.)
Advances in Data Management, 2009
ISBN 978-3-642-02189-3

Vol. 224. Amandeep S. Sidhu and Tharam S. Dillon (Eds.)
Biomedical Data and Applications, 2009
ISBN 978-3-642-02192-3

Vol. 225. Danuta Zakrzewska, Ernestina Menasalvas, and Liliana
Byczkowska-Lipinska (Eds.)
Methods and Supporting Technologies for Data Analysis, 2009
ISBN 978-3-642-02195-4

Vol. 228. Lidia Ogiela and Marek R. Ogiela
Cognitive Techniques in Visual Data Interpretation, 2009
ISBN 978-3-642-02692-8

Vol. 229. Giovanna Castellano, Lakhmi C. Jain, and
Anna Maria Fanelli (Eds.)
Web Personalization in Intelligent Environments, 2009
ISBN 978-3-642-02793-2

Vol. 230. Uday K. Chakraborty (Ed.)
*Computational Intelligence in Flow Shop and Job Shop
Scheduling,* 2009
ISBN 978-3-642-02835-9

Vol. 231. Mislav Grgic, Kresimir Delac, and
Mohammed Ghanbari (Eds.)
*Recent Advances in Multimedia Signal Processing and
Communications,* 2009
ISBN 978-3-642-02899-1

Vol. 232. Feng-Hsing Wang, Jeng-Shyang Pan, and
Lakhmi C. Jain
Innovations in Digital Watermarking Techniques, 2009
ISBN 978-3-642-03186-1

Vol. 233. Takayuki Ito, Minjie Zhang, Valentin Robu,
Shaheen Fatima, and Tokuro Matsuo (Eds.)
Advances in Agent-Based Complex Automated Negotiations,
2009
ISBN 978-3-642-03189-2

Vol. 234. Aruna Chakraborty and Amit Konar
Emotional Intelligence, 2009
ISBN 978-3-540-68606-4

Vol. 235. Reiner Onken and Axel Schulte
System-Ergonomic Design of Cognitive Automation, 2009
ISBN 978-3-642-03134-2

Vol. 236. Natalio Krasnogor, Belén Melián-Batista, José A.
Moreno-Pérez, J. Marcos Moreno-Vega, and David Pelta (Eds.)
*Nature Inspired Cooperative Strategies for Optimization (NICSO
2008),* 2009
ISBN 978-3-642-03210-3

Vol. 237. George A. Papadopoulos and Costin Badica (Eds.)
Intelligent Distributed Computing III, 2009
ISBN 978-3-642-03213-4

Vol. 238. Li Niu, Jie Lu, and Guangquan Zhang
Cognition-Driven Decision Support for Business Intelligence,
2009
ISBN 978-3-642-03207-3

Vol. 239. Zong Woo Geem (Ed.)
*Harmony Search Algorithms for Structural Design
Optimization,* 2009
ISBN 978-3-642-03449-7

Vol. 240. Dimitri Plemenos and Georgios Miaoulis (Eds.)
Intelligent Computer Graphics 2009, 2009
ISBN 978-3-642-03451-0

Vol. 241. János Fodor and Janusz Kacprzyk (Eds.)
Aspects of Soft Computing, Intelligent Robotics and Control,
2009
ISBN 978-3-642-03632-3

Vol. 242. Carlos A. Coello Coello, Satchidananda Dehuri, and
Susmita Ghosh (Eds.)
*Swarm Intelligence for Multi-objective Problems in Data
Mining,* 2009
ISBN 978-3-642-03624-8

Vol. 243. Imre J. Rudas, János Fodor, and
Janusz Kacprzyk (Eds.)
Towards Intelligent Engineering and Information Technology,
2009
ISBN 978-3-642-03736-8

Vol. 244. Ngoc Thanh Nguyen, Radosław Piotr Katarzyniak,
and Adam Janiak (Eds.)
New Challenges in Computational Collective Intelligence, 2009
ISBN 978-3-642-03957-7

Ngoc Thanh Nguyen, Radosław Piotr Katarzyniak,
and Adam Janiak (Eds.)

New Challenges in Computational Collective Intelligence

 Springer

Ngoc Thanh Nguyen
Institute of Informatics
Wroclaw University of Technology
Str. Janiszewskiego 11/17
50-370 Wroclaw
Poland
E-mail: thanh@pwr.wroc.pl

Adam Janiak
Institute of Computer Engineering,
Control and Robotics
Wroclaw University of Technology
Wybrzeze Wyspianskiego 27
50-370 Wroclaw
Poland
E-mail: Adam.Janiak@pwr.wroc.pl

Radosław Piotr Katarzyniak
Institute of Informatics
Wroclaw University of Technology
Str. Janiszewskiego 11/17
50-370 Wroclaw
Poland
E-mail: Radoslaw.Katarzyniak@pwr.wroc.pl

ISBN 978-3-642-26928-8   e-ISBN 978-3-642-03958-4

DOI 10.1007/978-3-642-03958-4

Studies in Computational Intelligence   ISSN 1860-949X

© 2009 Springer-Verlag Berlin Heidelberg
Softcover reprint of the hardcover 1st edition 2009

Typeset & Cover Design: Scientific Publishing Services Pvt. Ltd., Chennai, India.

Printed in acid-free paper

9 8 7 6 5 4 3 2 1

springer.com

Preface

Collective intelligence has become one of major research issues studied by today's and future computer science. Computational collective intelligence is understood as this form of group intellectual activity that emerges from collaboration and competition of many artificial individuals. Robotics, artificial intelligence, artificial cognition and group working try to create efficient models for collective intelligence in which it emerges from sets of actions carried out by more or less intelligent individuals. The major methodological, theoretical and practical aspects underlying computational collective intelligence are group decision making, collective action coordination, collective competition and knowledge description, transfer and integration. Obviously, the application of multiple computational technologies such as fuzzy systems, evolutionary computation, neural systems, consensus theory, knowledge representation etc. is necessary to create new forms of computational collective intelligence and support existing ones.

Three subfields of application of computational technologies to support forms of collective intelligence are of special attention to us. The first one is semantic web treated as an advanced tool that increases the collective intelligence in networking environments. The second one covers social networks modeling and analysis, where social networks are this area of in which various forms of computational collective intelligence emerges in a natural way. The third subfield relates us to agent and multi-agent systems understood as this computational and modeling paradigm which is especially tailored to capture the nature of computational collective intelligence in populations of autonomous individuals.

The book consists of 29 chapters which have been selected and invited from the submissions to the 1st *International Conference on Collective Intelligence - Semantic Web, Social Networks & Multiagent Systems* (ICCCI 2009).

All chapters in the book discuss various examples of applications of computational collective intelligence and related technologies to such fields as semantic web, information systems ontologies, social networks, agent and muliagent systems.

Part 1 of the book consists of five chapters in which authors discuss applications of computational intelligence technologies to modeling and usage of semantic web. Semantic web and related technologies are treated as this networking platform with which users (both human and artificial) are able to create mutual content and collective social behavior. Part 2 consists of six chapters devoted to ontology management and applications. Ontologies, understood as formalizations of socially accepted conceptualizations of domains, are fundamental for all forms of social communication and cooperation underlying collective actions. Chapters in Part 2 cover both methodological and theoretical problems related to ontology management, as well as few cases of ontology applications to solving real problems of intelligent collectives. Part 3 of the book is devoted to social networks modeling and analysis. It consists of 5 chapters in which theoretical and practical aspects of social interactions between cooperating individuals are studied and discussed. Part 4 consists of chapters in which few methodological and practical problems related to agent and multiagent systems are presented. In all of them authors discuss issues related directly or indirectly to the concept of collective intelligence. The book completes with Part 5 in which three additional chapters are presented.

The editors hope that the book can be useful for graduate and Ph.D. students in Computer Science, in particular participants to courses on Soft Computing, Multi-Agent Systems and Robotics. This book can also be useful for researchers working on the concept of computational collective intelligence in artificial populations. It is the hope of the editors that readers of this volume can find many inspiring ideas and use them to create new cases intelligent collectives. Many such challenges are suggested by particular approaches and models presented in particular chapters of this book.

We wish to express our great gratitude to Professor Janusz Kacprzyk, the editor of this series, for his interest and encouragement, and to all authors who contributed to the content of this volume. We wish also to express our gratitude to Dr Thomas Ditzinger for his support during preparation of this book.

July 2008   Ngoc Thanh Nguyen
                                                    Radosław Piotr Katarzyniak
                                                             Adam Janiak

Table of Contents

Part III: Social Networks

Part IV: Agent and Multiagent Systems

Part V: Other Applications

Part I
Semantic Web

Web Services Composition Framework with Petri Net Based Schemas

Ewa Ochmańska

Warsaw University of Technology
och@it.pw.edu.pl

Abstract. The paper describes a framework for composite web service based applications, built around domain-specific composition schemas for Petri net based models of service workflows. Those schemas, implemented as a hierarchy of generic and category-specific XSD descriptions, support both user oriented functionality and automatic composing of web services by means of core web technologies. Standard syntactic and semantic descriptions of component web services are referenced by descriptive resources native to the presented framework: XSD schemas, XSLT mappings, XML definitions and automatically generated RDF metadata. Main goals of the proposed solution are: to provide flexible model for composing families of similarly structured domain-specific web service based applications, while exploiting possibilities of automation related to semantic web services, and separating users from advanced formalisms of service ontologies.

1 Introduction

Due to the internet and web expansion, methodological and programmatic efforts in the domain of software engineering driven by such architectural paradigms as CBSA, MDA and SOA, by the end of the previous century have focused on the web service oriented application development [1, 2]. Concurrently evolving ideas and practical experiences of the Semantic Web, with RDF and OWL standard languages [3, 4] for describing ontologies of semantic terms and their interrelations, gave rise to many successful applications in the area of automated information and knowledge retrieval. The semantic concepts were also introduced to the domain of web services (WS), as a support for composition and deployment of complex, value-added applications based on semantic web services (SWS) [5]. Semantic approach promises to automate various tasks related to the web services technology, particularly those to be performed in the dynamic mode. SWS dynamic discovering and composition, still missing effective general solutions, play important role among currently investigated issues. The way of incorporating the vision of complex web applications, being combined "on the fly" of interoperating services, is still far from being obvious, in particular because requirements for such kind of applications are not clear [6].

Since early years of the semantic web it has been well understood that "a crucial aspect of creating the semantic web is to make it possible for a number of different users to create the machine-readable web content without being logic experts" [7].

N.T. Nguyen et al. (Eds.): New Challenges in Compu. Collective Intelligence, SCI 244, pp. 3–14.
springerlink.com   © Springer-Verlag Berlin Heidelberg 2009

In practice, this translates to accompanying acts of introducing new information to the web by automatic generation of semantic metadata. The cited remark is dated on 2001 and concerns machine-readable web content as a means for accessing data and knowledge rather than for developing software. In the meantime, service oriented approach has dominated web based software engineering, and semantic web ideas have been adopted in the area of web service based applications. To popularize semantic support for various advanced aspects of WS, it seems even more important to separate users and - in remarkable extent - also developers from the refined AI and description logic based machinery. Semantic reasoning about advanced WS related tasks should invisibly, automatically work at the background of SWS frameworks.

The paper outlines a concept of self-evolving framework for using and developing composite web service based applications. Service resources of the framework are to be evolved not only by developers but also due to automatically created links to syntactic and semantic descriptions of services being composed and executed, thus gathering knowledge about requirements on composite web services interoperation.

The framework functionality depends on domain-specific composition schemas for Petri net models of service workflows. Petri nets (PN) have been extensively used to model workflows of business processes [8] and, consequently, processes of composite web service execution, e.g. [5]. The proposed schemas, taking form of XML Schema Definition (XSD), delimit the space of available WS compositions according to domain specific needs. On the other hand, XML transformation of such schemas – along with other XML based resources native to the framework – provides an interface to human and software agents making use of services being composed.

The paper has following contents: Section 2 gives background information about composite web services, and motivates the proposed approach. Section 3 introduces a PN based formal model for composite web services, which is then used as a basis for composition schemas described in section 4. Section 5 explains mutual relations between standard and native resources needed for web service composition, and sketches an architectural outline of the framework. Final remarks conclude the paper.

2 Background and Motivation

2.1 Composite Web Services (CWS)

Web Services. The basic features of Web services which distinguish them as a specific kind of software applications are: their identification by URI and description of their public interfaces & bindings by XML [9]. Services execute operations to produce data on behalf of their clients (users, services, applications), typically by means of the basic stack of WS protocols: HTTP for transporting, SOAP for messaging, WSDL for describing, UDDI for publishing. WSDL describes web service by its abstract interface and concrete bindings. An interface specifies any number of operations by their inputs and outputs, and predictable execution faults. Operations can serve as components of complex services, however composition task remains left to users or their software agents. Invoking an operation with proper input and binding returns required output result. Message types of inputs and outputs are carefully defined by means of XSD schema (default format) or by other suitable notation.

Web Service Composition. Automating creation of composite web services is considered as the key issue with regard to development of complex distributed web applications. Tools to enable such automation, e.g. WS-BPEL [10] and WSCI [11], model complex services basing on concepts of business process workflows and service choreography. Petri net based workflow models are also useful in composing web services, and can be translated to BPEL descriptions [12].

Semantic Web Services. In order to enable programmatic reasoning about web services compositional features (necessary for automating several tasks related to complex WS based applications) the vision of Semantic Web was adopted. Frameworks for describing web service semantics by ontologies, like OWL-S [13], SWSO [14] or WSMO [15], address complex web applications based on interoperating services - being discovered, composed, accessed, invoked and monitored in the dynamic and automated mode. Languages expressing web service ontologies and semantic models build on layered concepts of service functional characteristics (OWL-S profile, WSMO capability...), behavioral characteristics (OWL-S process workflows, WSMO choreography & orchestration...) and technical grounding (which typically refers to WSDL bindings). Ontology-based models of web services are intended for automated WS compositions [16].

Semantic Annotations to Syntactic Web Service Description. WSDL-S [17] and SAWSDL [18] standards aim to relate traditional WSDL descriptions with machine-processable semantics. SAWSDL allows to annotate WSDL contents by references to external semantic resources. The specification makes no assumptions about standards used to describe service semantics, neither about standards used to produce mappings between WSDL elements and their semantic representations. Nevertheless, classes and instances of OWL-S ontologies are mentioned as semantic referents, and XSLT - as natural way of transforming XML based WSDL or OWL content to produce desired mappings. Seeking for correspondence between structural elements of WSDL description and basic concepts of OWL-S ontology leads to conclusion that both atomic and complex OWL-S processes should be represented by operations [19].

2.2 Motivation of the Proposed Approach

The paper outlines an attempt to organize a CWS framework by means of core, mature standards of XML technology, while making use - by reference and delegation - of existing SWS resources and software tools. Main assumptions of the proposed solution are: 1° to provide flexible model for composing families of similarly structured domain-specific CWS applications, based on XSD composition schemas; 2° to exploit possibilities of automation related to semantic web services by referring to ontology-based descriptions and by delegating CWS related tasks to middleware services; 3° to separate users from advanced formalisms of service ontologies.

The use of Petri nets in web service composition is motivated in literature by their analytic power to investigate model properties, e.g. to check for absence of deadlocks in a modeled workflow. Contrarily, industrial standards for describing web services [10, 11, 13, 14, 15] lack support for correctness checking [20]. The proposed approach can relax this deficiency by double guard for proper service composition, joining Petri net formal means with XSD validation.

3 Petri Net Based Formal Model for CWS

Petri nets appeared in early sixties of the previous century as a formal model for concurrent systems [21]. Since then, they were broadly used, redefined and extended to suite the wide spectrum of modeling purposes, in particular - those related to various aspects of web service compositions. Examples include their analysis [22], modeling and verifying - also in semantic context [5, 23], and representing WS-BPEL patterns in terms of PN [24]. A new way of exploiting PN modeling power, proposed in this paper, consists in defining PN based WS net model with built-in structural constraints which can be directly implemented by standard XSD schemas.

3.1 Basic Definitions of Petri Nets

Following two definitions form a basis for WS net, introduced in the next subsection.
A classical type of Petri nets, called place/transition net [25], is defined as a tuple:

$$PTN = (P, T, F, M_0) \tag{1}$$

with two disjoint sets of nodes: passive places P and active transitions T, joint by a flow relation (a set of arcs) $F \subseteq P \times T \cup T \times P$, and with an initial marking $M_0 : P \rightarrow N$ assigning nonnegative numbers of tokens to places. We denote sets of input and output places of a transition $t \in T$ as $\bullet t = \{p \in P | (p, t) \in F\}$ and $t \bullet = \{p \in P | (t, p) \in F\}$; sets of input/output transitions of a place $p \in P$ are denoted as $\bullet p$ and $p \bullet$. According to the so-called firing rule, a transition is enabled iff each of its input places contains at least one token. A state of PTN system, described by a marking $M : P \rightarrow N$, i.e. by numbers of tokens contained in its places, can change by activating (firing) an enabled transition which consumes/produces one token per each input/output place.

Among other extension types, colored Petri nets are used to model web service compositions, e.g. in [23]. They extend place-transition type of net systems by adding colors (datatypes) for tokens and some functions influencing system behavior. Colored Petri net is defined [26] as a tuple:

$$CPN = (\Sigma, P, T, A, N, C, G, E, I) \tag{2}$$

with pairwise disjoint sets of: colors Σ, places P, transitions T and arcs A. Arcs join pairs of places and transitions, assigned to them by a node function $N : A \rightarrow P \times T \cup T \times P$. Subsets of acceptable colors (datatypes of tokens) are assigned to places by a color function $C : P \rightarrow 2^\Sigma$; in general, a place may contain a multiset of tokens of different colors (values of various datatypes). Remaining functions control the firing rule for colored Petri nets: Firing of a transition depends on a guard function G which is defined for each transition $t \in T$ as a boolean expression $G(t)$ on typed variables (of types belonging to Σ). The result of transition firing depends on an arc function E which assigns to each arc an expression, returning a set of typed values i.e. tokens to be passed through. An arc expression is defined on a multiset of typed variables, according to a color set assigned to an adjacent place. Finally, initialization function I assigns multisets of colored tokens (values of proper datatypes) to some places.

3.2 Petri Net Model of Composite Web Service Workflow

For modeling web service workflows, WS net – a modification of CPN by structural constraints on classified places and transitions, is introduced and defined as a tuple:

$$WSN = (P, T, F, \Sigma, C_P, \Omega, C_T, G, I) \tag{3}$$

where P, T, F represent disjoint sets of places, transitions and arcs, as defined in (1). Σ is a set of place classes differing by accepted token colors (datatypes), defined as a union of pairwise disjoint subsets IN, OUT, MID corresponding to input data and results of composite service, and to messages passed between its service components:

$$\Sigma = \text{IN} \cup \text{MID} \cup \text{OUT}. \tag{4}$$

$C_P:P \rightarrow \Sigma$ is a place classifying function which assigns one token color to each place.

Ω is a set of transition classes; its elements are defined by pairs of nonempty disjoint color subsets for input and output places. For any $\omega \in \Omega$, we have:

$$\omega = (I_\omega, O_\omega), \quad I_\omega \subseteq \text{IN} \cup \text{MID}, \quad O_\omega \subseteq \text{MID} \cup \text{OUT}. \tag{5}$$

$C_T:T \rightarrow \Omega$ is a transition classifying function; assigning classes to transitions imposes structuring rules for WS nets, by following constraints for each $t \in T$ and $\omega = C_T(t)$:

$$|\bullet t| = |I_\omega| \wedge (\cup p \in \bullet t) \, C_P(p) = I_\omega, \quad |t \bullet| = |O_\omega| \wedge (\cup p \in t \bullet) \, C_P(p) = O_\omega \tag{6}$$

guarantying that classes of input and output places are proper to the transition class. G is a guard function for transitions, analogous to the definition (2).

I is an initialization function assigning one token of color $C_P(p)$ (i.e. value of proper datatype) to each place of the subset $\{p \in P \mid C_P(p) \in \text{IN}\}$; other places remain empty.

Moreover, following constraint holds, for every $p \in P$:

$$C_P(p) \in \text{IN} \Rightarrow \bullet p = \varnothing, \; C_P(p) \in \text{OUT} \Rightarrow p \bullet = \varnothing, \; C_P(p) \in \text{MID} \Rightarrow \bullet p \neq \varnothing \wedge p \bullet \neq \varnothing. \tag{7}$$

Fig. 1 shows sets Σ, Ω of WS net components that can appear in an exemplary excerpt of e-commerce composite web service, concerning goods delivery and payment. WS composition should synchronize and align operations performed on buyer and seller sides, shown as transition classes of Fig. 1.

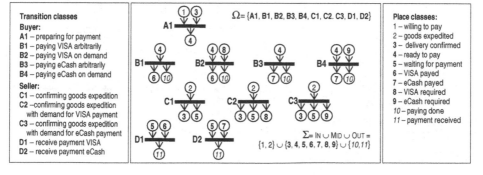

Fig. 1. Exemplary classes of WS net elements

4   Composition Schemas

4.1   Translating WSN Model to XSD Schema

In order to properly compose WS nets, composition schemas will be introduced. Transitions and places in WSN model of CWS refer to their counterparts in syntactic or semantic descriptions of component services, used by a category of applications, (WSDL operations and messages, OWL-S processes and IOPEs etc). Given sets of place and transition classes for application domain, XSD composition schema defines structuring rules for XML based descriptions of WS nets, accordingly to (3)-(7):

✓ Transitions have input and output places proper to their classes (as defined in corresponding service descriptions).
✓ Some input and output places, external to the modelled process, represent input data from CWS users (humans or other applications) and returned results.
✓ Places internal to the process stay on transition outputs (containing results of component services) and inputs (supplying other component services with data).

The above structuring rules for WS nets are forced by references to identity keys in XSD composition schema. Moreover, rules specific for particular application domain can be introduced by XSD constructs for choice groups and occurrence constraints:

1° By grouping of choices, transitions of certain classes may alternatively appear in a workflow (some services have several variants varying by interfaces).
2° Constraining occurrence number of transition or place classes permits their regular, optional or multiple occurrence (some services have to be executed just once in a workflow, other ones may appear not always or more than once).

Fig. 2 a) shows a simplistic outline of composition schema for the exemplary sets of component classes of Fig. 1. Three xsd:choice elements in the example XSD schema define subsets of alternate transition classes, and occurrence number of transitions in WS net is constrained to one per alternative (by default value of XSD standard attributes). The only four WS nets validated by that schema are depicted in Fig. 2 b).

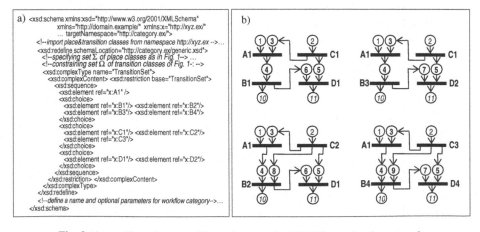

Fig. 2. An outline of composition schema and valid WS nets for the example

Declaring composition rules for a category of CWS applications as constraints in XSD schema supports:

- automated, possibly dynamic composition of service workflows, supported by XSD processing and RDF metadata classifying actual service components
- collecting references to schema elements representing WS net classes, in form of RDF triples, by recording successful (actually executed) service compositions
- transforming schema content and corresponding RDF metadata into contextual creative GUI to formulate requests and composition directives for CWS category.

The concept of schemas for WS based composite applications is general and allows various realizations, conforming to the needs of particular domains and communities. The proposed realization uses the well grounded XML Schema standard with respect to WS nets described in XML format. XSD schema defines allowable structuring of CWS underlying process workflow, according to the WS net definition, extended by additional constraints on service workflow specific for a category of applications.

4.2 Hierarchy of CWS Schemas

Thanks to the inheritance mechanisms built in XSD language, basic constraints on WSN model can be embedded in a base level schema proper for generic model structure. These generic schema is included and semantically précised by higher level category-specific schemas, with types and elements describing classes of transitions and places for component services and their interfaces.

The generic schema (generic.xsd in the example of Fig. 2) defines types and abstract elements for place and transition, general layout of WS net description (3)-(5) in XML format and structural constraints (6)-(7). Abstract elements are substituted and restricted in possibly independent parts, specifying semantic classes of services and of their I/O. The main part of a category specific schema imports needed classes and redefines the generic schema in order to specify sets Σ and Ω, and to introduce additional constraints $1°$, $2°$. Fig. 3 shows how XSD schema for CWS category is organized and how its elements refer to service descriptions distributed in the web.

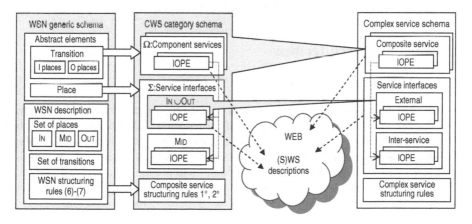

Fig. 3. Layered organization of schemas for composite web services

IOPE shortcut stands for input/output data of a web service and its real world pre-conditions / effects. For conceptual reasons, service ontologies defined in such standard frameworks as OWL-S or WSMO distinguish between data flow and status change related to service execution. However, from WS-based application viewpoint every IOPE instance is perceived as data and can be similarly modeled as token being passed (consumed, produced) by transitions and contained by input/output places.

The generic schema controls CWS structure by means of identity key references, forcing constraint (6) on proper classes of transitions' I/O places and constraint (7) on proper connections of internal and external places. Possible causal / temporal dependencies of component services are described by a schema for CWS category. Elements representing transitions and places of WS net are semantically defined as services and their IOPEs. In an instance XML document defining particular CWS workflow, these elements should refer to actual web services distributed in the web and their interfaces, as is pointed in Fig. 3 by dotted lines. Establishing such references during composition process is based on queries to RDF triples which link known and used service resources to domain-specific ontologies.

The above described structure of a schema for CWS application category can be layered to model more complex configurations, where services are composed of possibly composite services. A "complex service" – composed of CWSs – may then be represented by an upper-layer schema with transition elements referring to corresponding "composite service" lower-layer schemas, as shown in Fig. 3.

Schemas of composite services are nested in a complex service schema by XML Schema import directives and ref attributes. The generic schema constraints force structuring rules of both layers, based on category-specific information about services being composed and about allowable modes of their interoperation. (This two-layered structure of composition schema somehow reflects the scopes of choreography and orchestration, present in the context of (S)WS workflows. However, among several differences, the crucial one is that – on both layers – instead of a single workflow, schemas define families of CWS workflows to be instantiated by XML definitions.)

External IOPEs of nested schemas should be included in IOPE set of the upper-layer schema. Some of them may be external to the complex service being composed, while other ones – called "inter-service" – may pass data / process state changes between interoperating (simple or composite) service components.

5 Framework Architecture

5.1 Standard and Native Resources for CWS

The proposed approach aims to integrate web service resources accessible by standard syntactic or semantic descriptions, supporting composition process by means of core XML based technologies. Several descriptive & programmatic resources are involved in creating and executing CWS applications in the described framework. In addition to standard syntactic and/or semantic descriptions, and implementations of component services, native descriptive resources – in particular, domain specific composition schemas – participate in composing services and requesting their execution.

Descriptive resources native to the framework consist of XSD schemas with corresponding RDF metadata queried by SPARQL [27], XSLT mappings to context dependent user interface, and XML definitions of actual service compositions. Mutual relations between descriptive resources used by the framework are shown in Fig.4.

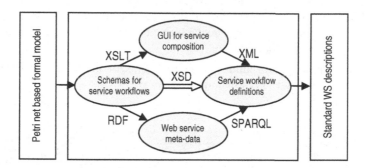

Fig. 4. WS descriptive resources used by the framework

XSD composition schemas are central to the framework. They define families of WS net models for composite web services, delimiting compositions space to some domain specific application frame, in a way permitting to drive user-oriented and resource-evolving functions. System middleware, structurally divided into composition-, execution- and interface managers described in the next subsection, exploits information contained in schemas to perform framework functionality.

Definitions of actual service workflows are created by composition manager as instance descriptions in XML format by means of processing corresponding schemas. XML instance elements of WS net correspond to actual web services, which however are not directly referred by schema classes. Instead, those references are retrieved by SPARQL queries to RDF metadata.

RDF metadata serve as a link to (S)WS space distributed in the web. RDF statements are based on domain/category specific terms, related to class names used by schemas. They define ontologies of terms used to describe CWS category semantics, and record information concerning known relations to actual representatives of particular service classes. Several service features described by RDF triples include proper references and types of published descriptions.

Composing and executing new WS based applications is accompanied by introducing new descriptive resources to the web. Composition manager can automatically create new RDF data, to link schema classes with actual services newly discovered and composed, and gather knowledge about their properties. Thus implicit relations between schemas and actually used service components are recorded to be considered by future composing activities. Ontologies may be extended to capture equivalent terms applied elsewhere to service classes used by the framework.

Schemas are equipped with corresponding XSLT sheets mapping their contents into context dependent user interface to the framework. Structuring rules and sets of component classes defined in a schema are transformed to GUI creators for service requests, and user defined compositions. XML instances of WS net descriptions for actual CWS may be defined or updated following user input.

5.2   Framework Architectural Outline

Presumably, CWS application scope is limited to certain domain and, at the starting point, to certain initial pool of existing and known web services of two types:

- application services as candidate CWS components
- middleware services for particular tasks related to the framework functionality, intended to support (possibly recursive) process of dynamic composing and executing CWS applications.

The framework architecture builds on several kinds of interrelated descriptive and programmatic resources, as is shown in Fig. 5.

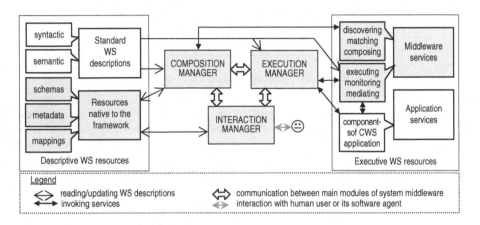

Fig. 5. An architectural outline of the framework

Main parts of the underlying system middleware are dedicated to following functions.

COMPOSITION MANAGER (CM) composes CWS based applications following user input, making use of descriptive resources proper for services being involved. It can delegate composition-related tasks to middleware services. It passes composition results to EM and receives backward information about processes being executed. Moreover, it updates native schemas and metadata by references resulting from newly created compositions, and generates corresponding standard syntactic and semantic descriptions. Thus the self-evolving feature of the framework is implemented by augmenting the pools of descriptive resources with new CWS definitions and with records concerning particular use of their components. CM supplies IM with data about services / service categories available in the framework and consumes user input (service requests and composition directives).

EXECUTION MANAGER (EM) intercepts composition results, referring to proper standard syntactic & semantic descriptions, and delegates execution-related tasks for services components to proper middleware services. It performs global control during executing CWS processes following their descriptions, and notifies CM of subsequent execution stages. It communicates with IM to exchange data corresponding to IOPEs as needed to progress execution of the processes.

INTERACTION MANAGER (IM) provides interactive interface to the framework, separating human user from XML & ontology based formats for service descriptions.

(Currently, human user interface is taken into account; communication with software agents will be considered in the future.) It communicates: with CM to select proper service or service category and to pass user data concerning service requests and compositions; with EM to exchange data during process execution. Service requests and CWS compositions are supported by creative GUI, contextually generated from composition schemas by standard tools for XSLT based processing.

6 Final Remarks

The paper outlines a framework for composite web service based software, built on the core XML-related standards, with some native methods and formats for service description. Central topic concerns using XSD schemas to control Petri net modeling of composite web service workflows, thus bridging mature, well instrumented XML based standards with various technologies of the semantic web. Schemas with related RDF meta-data do not serve as an alternative to standard description languages for composite (semantic) web services, but rather as a tool integrating them in a unified manner, by simpler and well grounded means.

The proposed approach consolidates existing assets for service composition – services published by means of standard syntactic and semantic descriptions: those intended as CWS application components and those exposed as tools for (S)WS processing. The presented architectural conception was elaborated for the purpose of virtual domain-specific web-service based software environment, on the base of earlier work [28], and will be successively implemented to gain better understanding of requirements for complex, automatically composed web service based applications.

There are plenty of problems related to the discussed scope which had to be completely omitted because of space limitations, or because they still remain not fully identified. One of them concerns modeling power of composition schemas, which are suitable to express constraints for relatively simple types of workflows. Extending composition schemas with Petri net based workflow patterns, as defined e.g. in [24], seems an important area for further research.

XSD format in composition schemas is to be replaced by Schema Modeling Language [29], recently published as a new W3C standard concerning composite web services. SML builds on XSD and Schematron [30] to obtain means desirable for its particular purpose, like inter-document references or strong structural constrains. The SML standardization initiative seems to confirm reasonability of organizing CWS frameworks around XML based models, independently of standard languages used to describe syntax and semantics of actual WS components.

References

1. Sheth, A., Miller, J.A.: Web Services: Technical Evolution yet Practical Revolution. IEEE Intelligent Systems (January/February 2003)
2. W3C Web Services Activity, http://www.w3.org/2002/ws/
3. RDF/XML Syntax Specification (Revised). W3C Recommendation 10 February (2004), http://www.w3.org/TR/rdf-syntax-grammar/
4. Web Ontology Language (OWL), http://www.w3.org/2004/OWL/
5. Narayanan, S., Sheila McIlraith, S.: Simulation, Verification and Automated Composition of Web Services. In: WWW 2002, Honolulu, Hawaii, USA, May 7-11 (2002)

6. Brodie, M.L.: Semantic Technologies: Realizing the Services Vision. In: Brodie, M.L., et al. (eds.) Semantic Web Services, Part 1. IEEE Intelligent Systems (September/October 2007)
7. Hendler, J.: Agents and the Semantic Web. IEEE Intelligent Systems Journal (March/April 2001)
8. van der Aalst, W.M.P.: The Application of Petri Nets to Workflow Management. The Journal of Circuits, Systems and Computers 8(1), 21–66 (1998)
9. Dustdar, S., Schreiner, W.: A Survey on Web Services Composition. International Journal on Web and Grid Services 1(1), 1–30 (2005)
10. Web Services Business Process Execution Language Version 2.0, OASIS Standard, April 11 (2007), http://docs.oasis-open.org/wsbpel/2.0/OS/wsbpel-v2.0-OS.pdf
11. Web Service Choreography Interface (WSCI) 1.0. W3C Note 8 August (2002), http://www.w3.org/TR/wsci/
12. Eckleder, A., Freytag, T.: WoPeD 2.0 goes BPEL 2.0. In: AWPN Workshop, Rostock (2008)
13. OWL-S: Semantic Markup for Web Services, http://www.w3.org/Submission/OWL-S/
14. Semantic Web Services Ontology (SWSO), http://www.daml.org/services/swsf/1.0/swso/
15. Web Service Modeling Ontology (WSMO), http://www.w3.org/Submission/WSMO/
16. Majithia, S., et al.: Automated Web Service Composition using Semantic Web Technologies. In: Proc. of the International Conference on Autonomic Computing. IEEE, Los Alamitos (2004)
17. Web Service Semantics - WSDL-S, http://www.w3.org/Submission/WSDL-S/
18. Semantic Annotations for WSDL and XML Schema. W3C Recommendation August 28 (2007), http://www.w3.org/TR/sawsdl/
19. Martin, D., Paolucci, M., Wagner, M.: Toward Semantic Annotations of Web Services: OWL-S from the SAWSDL Perspective. In: OWL-S: Experiences and Directions Workshop, European Semantic Web Conf., Innsbruck (2007)
20. ter Beek, M.H., et al.: Web Service Composition Approaches: From Industrial Standards to Formal Methods. In: ICIW 2007 (2007)
21. Petri, C.A.: Fundamentals of a theory of asynchronous information flow. Information processing. In: Proc. of the IFIP Congress 1962, Munich. North-Holland, Amsterdam (1962)
22. PengCheng, X., et al.: A Petri Net Approach to Analysis and Composition of Web Services. CERCS Tech. Rep. Georgia Institute of Technology (2009), http://hdl.handle.net/1853/27247
23. Cheng, Y., et al.: Modeling and verifying composite semantic Web services based on colored Petri nets. In: ALPIT 2007, Advanced Language Processing and Web Information Technology, pp. 510–514 (2007)
24. van der Aalst, W.M.P., Hofstede, A.H.M.: Workflow Patterns: On the Expressive Power of (Petri-net-based) Workflow Languages. In: Proc. of the Fourth Workshop on the Practical Use of Coloured Petri Nets and CPN Tools (CPN 2002), DAIMI, University of Aarhus, vol. 560 (2002)
25. Reisig, W.: Petri Nets. An Introduction. Springer, Heidelberg (1985)
26. Jensen, K.: Basic concepts, analysis methods and practical use. Springer, Heidelberg (1992)
27. SPARQL Query Language for RDF. W3C Recommendation, January 15 (2008), http://www.w3.org/TR/rdf-sparql-query/
28. Ochmańska, E.: An Open Environment for Compositional Software Development. In: Min, G., Di Martino, B., Yang, L.T., Guo, M., Rünger, G. (eds.) ISPA Workshops 2006. LNCS, vol. 4331, pp. 175–184. Springer, Heidelberg (2006)
29. Service Modeling Language, v 1.1. W3C Recommendation , May 12 (2009), http://www.w3.org/TR/sml/
30. Robertsson, E.: An Introduction to Schematron. O'Reilly Media, Inc., Sebastopol (2003), http://www.xml.com/pub/a/2003/11/12/schematron.html

Proposal of a New Rule-Based Inference Scheme for the Semantic Web Applications*

Grzegorz J. Nalepa and Weronika T. Furmańska

Institute of Automatics,
AGH University of Science and Technology,
Al. Mickiewicza 30, 30-059 Kraków, Poland
gjn@agh.edu.pl, wtf@agh.edu.pl

Abstract. The current challenge of the Semantic Web is the development of an expressive yet effective rule language. In this paper an overview of an integration proposal for Description Logics (DL) and Attributive Logics (ALSV) is presented. These two formalisms stem from the field of Knowledge Representation and Artificial Intelligence, and provide different description and reasoning capabilities. The contribution of the paper consists in introducing a possible transition from ALSV to DL. This opens up possibilities of using well-founded expert systems modeling languages to improve the design of Semantic Web rules.

1 Introduction

The Semantic Web proposal of the next generation Web with rich semantics and automated inference is based on number of formal concepts and practical technologies. This includes Description Logics (DL) [1] as the formalism describing formal ontologies. Currently, the development is focused on providing a flexible rule language for the Web. It should be RIF-compatible [1] on the rule interchange level, and conceptually compatible with ontologies described in OWL with use of Description Logics. The SWRL proposal recently submitted to W3C aims at meeting these requirements.

The Semantic Web initiative is based of previous experiences and research of Knowledge Engineering [2] in the field of Artificial Intelligence [3]. In this field the rule-based expert systems technologies are a prime example of effective reasoning systems based on the rule-based paradigm [4]. The formal description of these systems is based on the propositional calculus, or restricted predicate logic – it is worth considering how the Semantic Web community could benefit from some classic expert systems tools and solutions.

A recent proposal of a new logical calculus for rules aims at extending the expressiveness of the rule language, by introducing an attributive language [5,6]. This solution seems superior to the simple propositional systems, and easier to reason with than the predicate logic. The XTT^2 rule language is based on this solution [6]. It provides visual design and formal analysis methods for decision rules. This would eventually allow to

---

* The paper is supported by the HeKatE Project funded from 2007–2009 resources for science as a research project.
[1] See http://www.w3.org/2005/rules/wiki/RIF_Working_Group.

N.T. Nguyen et al. (Eds.): New Challenges in Compu. Collective Intelligence, SCI 244, pp. 15–26.
© Springer-Verlag Berlin Heidelberg 2009

design rules for the Semantic Web. The current problem with Attributive Logic with Set Values over Finite Domain (ALSV(FD)) is the lack of conceptual compatibility with DL.

In this paper selected important DL concept are discussed in Sect. 2, and a brief introduction to ALSV(FD) is given in Sect. 3. This gives a motivation presented in Sect. 4 for the research aiming at translation from the ALSV(FD), to DL which provides a formalized foundation for ontologies and SWRL. Possible integration approaches of these two calculi are discussed in Sect. 5, using a simple example in Sect. 6. The paper ends with future work in Sect. 7.

2   Description Logics Overview

Description Logics are a family of knowledge representation languages [1]. Historically related to semantic networks and frame languages, they describe the world of interest by means of concepts, individuals and roles. However, contrary to their predecessors, they provide a formal semantics and thus enable for automated reasoning. Basic Description Logics take advantage of their relation to predicate calculus. On one hand they adopt its semantics, which makes them more expressive than a propositional logic. On the other, by restricting the syntax to formulae with maximum two variables, they remain decidable and more human-readable. These features have made Description Logics a popular formalism used for designing ontologies for the Semantic Web. There exist a number of DL languages. They are defined and distinguished by allowed concept descriptions, which influences the languages' expressivity.

The building blocks of vocabulary in DL languages are *concepts*, which denote sets of individuals and *roles* which denote the binary relations between individuals. Elementary descriptions in DL are *atomic concepts* and *atomic roles*. More complex descriptions can be built inductively from them using *concept constructors*. Respective DL languages are distinguished by the constructors they provide. A minimal language of practical interest is the *Attributive Language*.

Definition 1. *Let A denote an atomic concept and R an atomic role. In basic $\mathcal{AL}$ concept descriptions C and D can be formed according to the following rules:*

$$C, D \rightarrow A| \qquad\qquad\qquad\qquad atomic\ concept \qquad (1)$$
$$\top| \qquad\qquad\qquad\qquad universal\ concept \qquad (2)$$
$$\bot| \qquad\qquad\qquad\qquad bottom\ concept \qquad (3)$$
$$\neg A| \qquad\qquad\qquad\qquad atomic\ negation \qquad (4)$$
$$C \sqcap D| \qquad\qquad\qquad\qquad intersection \qquad (5)$$
$$\forall R.C| \qquad\qquad\qquad\qquad value\ restriction \qquad (6)$$
$$\exists R.\top \qquad\qquad limited\ existential\ quantification \qquad (7)$$

In order to define a formal semantics, an *interpretation* $\mathcal{I} = (\Delta^{\mathcal{I}}, \cdot^{\mathcal{I}})$ is considered. The interpretation consists of the *domain of interpretation* which is a non-empty set and an interpretation function, which to every atomic concept A assigns a set $A^{\mathcal{I}} \subseteq \Delta^{\mathcal{I}}$ and for every atomic role R a binary relation $R^{\mathcal{I}} = R^{\mathcal{I}} \subseteq \Delta^{\mathcal{I}} \times \Delta^{\mathcal{I}}$. The interpretation function is extended over concept descriptions by the Definition 2.

Definition 2

$$\top^{\mathcal{I}} = \Delta^{\mathcal{I}} \tag{8}$$

$$\bot^{\mathcal{I}} = \emptyset \tag{9}$$

$$(\neg A)^{\mathcal{I}} = \Delta^{\mathcal{I}} \setminus A^{\mathcal{I}} \tag{10}$$

$$(C \sqcap D)^{\mathcal{I}} = C^{\mathcal{I}} \cap D^{\mathcal{I}} \tag{11}$$

$$(\forall R.C)^{\mathcal{I}} = \{a \in \Delta^{\mathcal{I}} | \forall b, (a, b) \in R^{\mathcal{I}} \rightarrow b \in C^{\mathcal{I}}\} \tag{12}$$

$$(\exists R.\top)^{\mathcal{I}} = \{a \in \Delta^{\mathcal{I}} | \exists b, (a, b) \in R^{\mathcal{I}}\} \tag{13}$$

Basic DL allow only atomic roles (i.e. role names) in role descriptions.

The basic language can be extended by allowing other concept constructors, such as *union* $(\mathcal{U})$, *full negation* $(\mathcal{C})$, *full existential quantification* $(\mathcal{E})$ or *number restriction* $(\mathcal{N})$. Resulting formalisms are called using the letters indicating the allowed constructors, e.g. $\mathcal{ALC}$, $\mathcal{ALCN}$, $\mathcal{ALUE}$ etc. The smallest propositionally closed language is $\mathcal{ALC}$.

Different extensions to basic Description Logics are introduced by allowing *role constructors*. They enable for introducing various constraints and properties of roles, such as *transitive closure*, *intersection*, *composition* and *union*, or *complement* and *inverse* roles. Another kind of extension is obtained by allowing *nominals* in concept definitions and introducing primitive datatypes. These modifications proved to be extremely valuable and important in the context of the Semantic Web and ontologies. However, they are sources of high computational complexity of reasoning in the resulting ontologies.

For expressive DLs the abovementioned naming convention would be too long. Hence, for the basic $\mathcal{ALC}$ language extended with transitive roles, $\mathcal{S}$ is often used. The letter $\mathcal{H}$ is used to represent role hierarchy, $\mathcal{O}$ to indicate nominals in concept descriptions, $\mathcal{I}$ represents inverse roles, $\mathcal{N}$ number restrictions, and $(\mathbf{D})$ indicates the integration of some concrete domain/datatypes. The DL underlying OWL-DL language includes all of those constructs and is therefore called $\mathcal{SHOIN}(\mathbf{D})$.

Description Logics provide tools to build a knowledge base and to reason over it. The knowledge base consists of two parts, namely TBox and ABox.

TBox provides a terminology and contains a taxonomy expressed in a form of set of axioms. The axioms define concepts, specify relations between them and introduce set constraints. Therefore, TBox stores implicit knowledge about sets of individuals in the world of interest. Formally, a terminology $\mathcal{T}$ is a finite set of terminological axioms. If C and D denote concept names, and R and S role names, then the terminological axioms may be in two forms: $C \sqsubseteq D$ ($R \sqsubseteq S$) or $C \equiv D$ ($R \equiv S$). Equalities that have an atomic concept on the left-hand side are called *definitions*. Axioms of the form $C \sqsubseteq D$ are called *specialization* statements. Equalities express necessary and sufficient conditions, whereas specialization statements specify constraints (necessary conditions) only. An interpretation (function) $\mathcal{I}$ maps each concept name to a subset of the domain. The interpretation satisfies an axiom $C \sqsubseteq D$ if: $C^{\mathcal{I}} \subseteq D^{\mathcal{I}}$. It satisfies a concept definition $C \equiv D$ if: $C^{\mathcal{I}} = D^{\mathcal{I}}$. If the interpretation satisfies all the definitions and all axioms in $\mathcal{T}$, it satisfies the terminology $\mathcal{T}$ and is called a *model* of $\mathcal{T}$.

ABox contains explicit assertions about individuals in the conceived world. They represent extensional knowledge about the domain of interest. Statements in ABox may be: concept assertions, e.g. $C(a)$ or role assertions, $R(b, c)$. An interpretation $\mathcal{I}$ maps each individual name to an element in the domain. With regards to terminology $\mathcal{T}$ the interpretation satisfies a concept assertion $C(a)$ if $a^{\mathcal{I}} \in C^{\mathcal{I}}$, and a role assertion $R(b, c)$ if $\langle b^{\mathcal{I}}, c^{\mathcal{I}} \rangle \in R^{\mathcal{I}}$. If it satisfies all assertions in ABox $\mathcal{A}$, then it satisfies $\mathcal{A}$ and $\mathcal{I}$ is a model of $\mathcal{A}$.

Although terminology and world description share the same model-theoretic semantics, it is convenient to distinguish these two parts while designing a knowledge base or stating particular inference tasks.

With regards to terminology $\mathcal{T}$ one can pose a question if a concept is *satisfiable*, if one concept *subsumes* another, if two concepts are *equivalent* or *disjoint*. A concept C is satisfiable with respect to $\mathcal{T}$ if there exists a model (an interpretation) $\mathcal{I}$ of $\mathcal{T}$ such that $C^{\mathcal{I}}$ is not empty. A concept C is subsumed by a concept D w.r.t. $\mathcal{T}$ if $C^{\mathcal{I}} \subseteq D^{\mathcal{I}}$ for every model $\mathcal{I}$ of $\mathcal{T}$. Two concepts C and D are equivalent w.r.t. T if $C^{\mathcal{I}} = D^{\mathcal{I}}$ for every model I of T. Finally, two concepts C and D are disjoint w.r.t. $\mathcal{T}$ if $C^{\mathcal{I}} \cap D^{\mathcal{I}} = \emptyset$ for every model $\mathcal{I}$ of $\mathcal{T}$.

Satisfiability and subsumption checking are the main reasoning tasks for TBox; other can be reduced to them, and either can be reduced to the other.

For ABox there are four main inference tasks: *consistency checking*, *instance checking*, *realization* and *retrieval*. An ABox $\mathcal{A}$ is consistent w.r.t. a TBox $\mathcal{T}$, if there is an interpretation that is a model of both $\mathcal{A}$ and $\mathcal{T}$. Furthermore, we say that an ABox is consistent, if it is consistent w.r.t. the empty TBox. One can acquire an empty TBox by expanding concepts in (acyclic) TBox which means replacing concepts with the right-hand side of their definitions. Instance checking tests if a given assertion is entailed by the ABox. Realization tasks consist in finding the most specific concept for a given individual and retrieval returns individuals which are instances of a given concept.

All these tasks can be reduced to *consistency checking* of the ABox with respect to the TBox.

3 Attributive Logics Concepts

Knowledge representation based on the concept of *attributes* is one of the most common approaches [2]. It is one of the foundations for relational databases, attributive decision tables and trees [7], as well as rule-based systems [5].

A typical way of thinking about attributive logic for knowledge specification may be put as follows. Knowledge is represented by *facts* and *rules*. A fact could be written as $A := d$ or $A(o) := d$, where A is a certain attribute (property of an object), o an object of interest and d is the attribute value. Facts are interpreted as propositional logic atomic formulae. This basic approach is sometimes extended with use of certain syntax modifications [7]. On top of these facts simple decision rules are built, corresponding to conditional statements.

In a recent book [5] the discussion of attributive logic is extended by allowing attributes to take *set values* and providing some formal framework of the *Set Attributive*

Logic (SAL). The basic idea for further discussion is that attributes should be able to take not only *atomic* but *set* values as well, written as $A(o) = V$, where V is a certain set of values.

In [6] the language of SAL has been extended to provide an effective knowledge representation tool for decision rules, where the state description uses finite domains. The proposed calculus *Attribute Logic with Set Values over Finite Domains* (ALSV(FD)) is proposed to simplify the decision rules formulation for classic rule-based expert system, including, business rules, using the so-called XTT[2] rule language. While being a general solution, the language is oriented towards forward chaining intelligent control systems.

Here, an extended notation for ALSV(FD) is introduced, including: 1) the explicit differentiation of equality relation, denoted as = used for comparison in the ALSV(FD) formulas and the fact definition operator, denoted as :=, and 2) the formalization of constants that simplify effective formula notation.

The basic element of the language of *Attribute Logic with Set Values over Finite Domains* (ALSV(FD) for short) are: attribute names and attribute values. **A** – a finite set of attribute names, **D** – a set of possible attribute values (their *domains*), and **C** – a set of constants.

Let $\mathbf{A} = \{A_1, A_2, \ldots, A_n\}$ be all the attributes such that their values define the state of the system under consideration. It is assumed that the overall domain **D** is divided into n sets (disjoint or not), $\mathbf{D} = D_1 \cup D_2 \cup \ldots \cup D_n$, where D_i is a domain related to attribute A_i, $i = 1, 2, \ldots, n$. Any domain D_i is assumed to be a finite (discrete) set.

We also consider a finite set of constants $\mathbf{C} = \{C_1, C_2, \ldots, C_m\}$. Every constant C_j $j = 1, 2, \ldots, m$ is related to a single domain D_i, $i = 1, 2, \ldots, n$ such that $C_j \subset D_i$ There might be multiple constants related to the same domain.

Definition 3. *Attributes. As we consider dynamic systems, the values of attributes can change over time (or state of the system). We consider both simple attributes of the form $A_i: T \rightarrow D_i$ (i.e. taking a single value at any instant of time) and generalized ones of the form $A_i: T \rightarrow 2^{D_i}$ (i.e. taking a set of values at a time); here T denotes the time domain of discourse.*

Let A_i be an attribute of **A** and D_i the domain related to it. Let V_i denote an arbitrary subset of D_i and let $d \in D_i$ be a single element of the domain. V_i can also be a constant related to D_i.

The legal atomic formulae of ALSV along with their semantics are presented in Def. 4 for simple and in Def. 5 for generalized attributes.

Definition 4. *Simple attribute formulas syntax*

$$A_i = d \qquad \text{\textit{the value is precisely defined}} \qquad (14)$$

$$A_i \in V_i \qquad \text{\textit{the current value of } } A_i \text{ \textit{belongs to} } V_i \qquad (15)$$

$$A_i \neq d \qquad \text{\textit{shorthand for} } A_i \in D_i \setminus \{d\} \qquad (16)$$

$$A_i \notin V_i \qquad \text{\textit{is a shorthand for} } A_i \in D_i \setminus V_i \qquad (17)$$

Definition 5. *Generalized attribute formulas syntax*

$$A_i = V_i \qquad \text{equals to } V_i \text{ (and nothing more)} \qquad (18)$$

$$A_i \neq V_i \qquad \text{is different from } V_i \text{ (at at least one element)} \qquad (19)$$

$$A_i \subseteq V_i \qquad \text{is a subset of } V_i \qquad (20)$$

$$A_i \supseteq V_i \qquad \text{is a superset of } V_i \qquad (21)$$

$$A \sim V \qquad \text{has a non-empty intersection with } V_i \qquad (22)$$

$$A_i \not\sim V_i \qquad \text{has an empty intersection with } V_i \qquad (23)$$

In case V_i is an empty set (the attribute takes in fact no value) we shall write $A_i = \{\}$. In case the value of A_i is unspecified we shall write $A_i = \text{NULL}$ (a database convention). If we do not care about the current value of the attribute we shall write $A = \_$ (a PROLOG convention). More complex formulae can be constructed with *conjunction* ($\wedge$) and *disjunction* ($\vee$); both symbols have classical meaning and interpretation.

There is no explicit use of negation. The proposed set of relations is selected for convenience and as such they are not completely independent. Various notational conventions extending the basic notation can be used. For example, in case of domains being ordered sets, relational symbols such as $>$, $>=$, $<$, $=<$ can be used with the straightforward meaning.

The semantics of the proposed language is presented below in an informal way. The semantics of $A = V$ is basically the same as the one of SAL [5]. If $V = \{d_1, d_2, \ldots, d_k\}$ then $A = V$ if the attribute takes all the values specified with V (and nothing more). The semantics of $A \subseteq V$, $A \supseteq V$ and $A \sim V$ is defined as follows:

- $A \subseteq V \equiv A = U$ for some U such that $U \subseteq V$, i.e. A takes *some* of the values from V (and nothing out of V),
- $A \supseteq V \equiv A = W$, for some W such that $V \subseteq W$, i.e. A takes *all* of the values from V (and perhaps some more), and
- $A \sim V \equiv A = X$, for some X such that $V \cap X \neq \emptyset$, i.e. A takes *some* of the values from V (and perhaps some more).

As it can be seen, the semantics of ALSV is defined by means of relaxation of logic to simple set algebra.

To simplify the formula notation *constants* can also be defined as: $C := V$. A knowledge base in ALSV(FD) is composed of simple formulas forming rules and facts defining the current systems state.

The current values of all attributes are specified in the contents of the knowledge-base. From logical point of view the state is represented as a logical formula of the form:

$$(A_1 := S_1) \wedge (A_2 := S_2) \wedge \ldots \wedge (A_n := S_n) \qquad (24)$$

where $S_i := d_i$ ($d_i \in D_i$) for simple attributes and $S_i := V_i$, ($V_i \subseteq D_i$) for complex ones. In order to cover realistic cases an explicit notation for covering unspecified, unknown values is proposed; for example to deal with the data containing the NULL values imported from a database.

Consider a set of n attributes $\mathbf{A} = \{A_1, A_2, \ldots, A_n\}$. Any XTT^2 rule is assumed to be of the form:

$$(A_1 \propto_1 V_1) \wedge (A_2 \propto_2 V_2) \wedge \ldots (A_n \propto_n V_n) \longrightarrow RHS$$

where $\propto_i$ is one of the admissible relational symbols in ALSV(FD) (see legal formulas in Def. 4,5), and RHS is the right-hand side of the rule covering conclusion, including state transition and actions, for details see [5]. In order to fire a rule the preconditions have to be satisfied. The satisfaction of rule preconditions is verified in an algebraic mode, using the rules specified in Fig. 1 The full discussion of ALSV inference for all formulas can be found in [6].

To summarize, main assumptions about the XTT^2 representation are:

- current system *state* is explicitly represented by a set of *facts* based on attributive representation,
- facts are denoted in ALSV(FD) as: $A := V$,
- constants are facts defined as: $C := V$ to simplify compact rule notation,
- system dynamics (state transition) is modelled with *rules*,
- the conditional part of the XTT^2 rule is conjunction built from simple ALSV(FD) formulas,
- rule decision contains state modification – new facts, and optionally external actions execution statements,
- to fire the rule, the condition has to be satisfied, which means that certain *inference task* in ALSV(FD) has to be solved.

$$A = d_i, d_i = d_j \models A = d_j$$
$$A = d_i, d_i \neq d_j \models A \neq d_j$$
$$A = d_i, d_i \in V_j \models A \in V_j$$
$$A = d_i, d_i \notin V_j \models A \notin V_j$$

$$A = V, V = W \models A = W$$
$$A = V, V \neq W \models A \neq W$$
$$A = V, V \subseteq W \models A \subseteq W$$
$$A = V, V \supseteq W \models A \supseteq W$$
$$A = V, V \cap W \neq \emptyset \models A \sim W$$
$$A = V, V \cap W = \emptyset \models A \not\sim W$$

Fig. 1. Inference rules

4 Motivation

Description Logics provide an effective formal foundation for the Semantic Web application based on ontologies described with OWL. They allow for simple inference tasks, e.g. corresponding to concept classification. Currently the main challenge for DL is the rule formulation.

Rules are the next layer in the Semantic Web stack that has to be provided in order to make the Semantic Web applications operate on the knowledge expressed in ontologies. There are several approaches to the integration of rules and ontologies. An example of a homogeneous approach are Description Logic Programs (DLP) [8]. A heterogeneous one can be found in the design of the Semantic Web Rule Language (SWRL) [9]. Other problems, that so-far have rarely been considered in the Semantic Web research include effective design methodologies for large rule bases, as well as knowledge quality issues. To maintain larger knowledge bases is not a trivial task; prone to design errors an logical misformulations. This is why scalable design methods have to be considered.

In the classic AI field of rule-based expert systems numerous solutions for rule formulation and inference, as well as the design and analysis are considered [4,5]. Number of visual knowledge representations equivalent to decision rules are considered, including decision trees and decision tables [2]. They prove to be helpful in the visual design of rules, as well as providing CASE tools for rule-based systems. The theory and practice of rules verification is a well studied field, for details see [10]. Therefore, evaluating the use of mature rule-based systems solutions is a natural choice for the Semantic Web applications.

The XTT^2 [6] rule formulation and design language based on the ALSV(FD) is an example of an expert system design and analysis framework. It offers flexible knowledge representation for forward chaining decision rules, as well as visual design tools. Thanks to the formal description in ALSV(FD) a formalized rule analysis is also possible. As a rule framework XTT^2 provides a subset of functionality that SWRL aims for, at least for the production systems based on decision rules. On the other hand, it has ready design and analysis solutions.

The primary goal of this research is to allow the use of the XTT^2 rule design framework for the Semantic Web rules. This would open up possibility to use the visual design tools for XTT^2 to design Semantic Web rules, as well as exploit the existing verification solutions. The following phases are considered:

1. a transition from ALSV(FD) to a subset of a selected DL language,
2. XTT^2 rules formulation compatible with the above transition procedure,
3. visual design of XTT^2 rules for the Semantic Web,
4. XTT^2 rules translation to SWRL, with possible RIF-only extensions,
5. rule inference on top of OWL-based ontologies.

Ultimately it should be possible to design Semantic Web rules with the XTT^2 visual design tools and provide a formal analysis of rules. XTT^2 could be run with a dedicated XTT^2 engine, or a SWRL runtime (whenever it is available).

In this paper the focus is on the first phase, with some considerations for the 2nd and the 3rd one. In order to provide a transition, a discussion of syntax and semantics of both calculi DL and ALSV(FD) is given in the next section.

5 Syntax and Semantics Analysis

Description Logics enable for complex descriptions of objects in the universe of discourse and the relations between them. The static part of the system is expressed in TBox part of a DL Knowledge Base. The actual state is represented by means of facts asserted in ABox. ABox in DL is limited in terms of its syntax and semantics. There are only simple assertions there, which together with knowledge expressed in TBox lay the ground for inferencing.

The language of DL consists of *concepts*, *roles* and *constants*. The meaning of the symbols is defined by an *interpretation function*, which to each concept assign a set of objects, to each role a binary relation, and to each individual an object in the universe of discourse.

The main goal of ALSV(FD) is to provide an expressive notation for dynamic system state transitions in case of rule-based systems. Thus, the knowledge specification with ALSV(FD) is composed of: state specification with facts, and transition specification with formulas building decision rules.

Both logics describe the universe of discourse by identifying certain entities. In ALSV(FD) they are called attributes and in DL – concepts. Here we will concentrate on simple attributes.

Every attribute in ALSV(FD) has its domain, which constraints the values of the attribute. In DL this kind of specification is done by means of TBox axioms. In order to express a finite domain in DL, a *set of* constructor, denoted by $\mathcal{O}$ is needed (see Table 1).

Table 1. Formulas in AL and terminological axioms in DL – domain definitions

Attributive Logic		Description Logic	
Attribute Name	Attribute domain	Concept Name	Concept constructors
A_i	D_i	A_i	$A_i \equiv D_i$
	$D = \{a_1, a_2, \ldots, a_n\}$		$D \equiv \{a_1, a_2, \ldots, a_n\}$

Once the attributes and concepts are defined, they are used in system rules specification. Legal AL formulas (see Defs. 4,5) specify the constraints that an attribute value has to match in order for a rule to be fired. They constitute a schema, to which states of a system in certain moments of time are matched. ALSV(FD) formulas used in rule preconditions can be represented as terminological axioms in DL as in Fig. 2.

Attributive Logic	Description Logic
Formula	Axiom
$A_i = d$	$A_i \equiv \{d\}$
$A_i \in V_i$	$A_i \equiv V_i$
$A_i \neq d$	$A_i \equiv \neg\{d\}$
$A_i \notin V_i$	$A_i \equiv \neg V_i$

Fig. 2. Simple attributes formulas in AL rules and respective axioms in DL

Attribute values define the state of the system under consideration. State is represented as a logical formula (24) built from a conjunction of formulas specifying the values of respective attributes. A statement in AL that the attribute A_i holds certain value, in DL language corresponds to a statement that a certain object is an instance of concept A_i. Such statements build the ABox in DL systems. State specification is shown in Tab. 2.

At a given moment of time, the state of the system is represented as a conjunction of formulas or concept assertions in DL respectively. In order to check the rules satisfiability, appropriate inference tasks have to be executed. The inference rules in AL

Table 2. State specification in AL and assertions in DL

Attributive Logic	Description Logic
Formula	Assertion in ABox
$A_i := d_i$	$A_i(d_i)$

are presented in Fig. 1. For DL such a task is *consistency checking*. For each rule a *consistency checking* of the state assertions with regards to the rule preconditions is performed. If the consistency holds, the rule can be fired. Every rule is loaded into the DL reasoner together with the actual state of the system. The rule constitues a schema, which is temporary joined with existing TBox (in which concept definitions are stored). The state is a *temporary ABox*. The DL reasoner checks the consistency of the ontology resulting from the TBox and ABox. If it is consistent, then the rule can be fired. The actions specified in the heads of rules may change the system state, which is then represented as a new ABox.

6   An Example of Translation

Let us consider a simple example of a forward chaining rule based system. The goal of the system is to set a thermostat temperature based on the condition, namely the time specification.

Consider a set attributes
$\mathbf{A} = \{day, month, hour, today, season, operation, therm\_setting\}$,
with corresponding domains:
$\mathbf{D} = \{D_{day}, D_{month}, D_{hour}, D_{today}, D_{season}, D_{operation}, D_{therm\_setting}\}$,
defined as $D_{day} = \{sun, mon, tue, wed, thr, fri, sat\}$,
$D_{month} = \{jan, feb, mar, apr, may, jun, jul, aug, sep, oct, nov, dec\}$,
$D_{hour} = \{1 - 24\}$, $D_{today} = \{workday, weekend\}$,
$D_{season} = \{winter, spring, summer, autumn\}$,
$D_{operation} = \{bizhours, notbizhours\}$, $D_{therm\_setting} = \{12 - 30\}$,
In such a system we can consider simple production rules (in Australia!):
$R1 : month \in \{dec, jan, feb\} \longrightarrow season := summer$
$R2 : day \in \{mon, tue, wed, thr, fri\} \longrightarrow today := workday$
$R3 : today \in \{workday\} \wedge hour \in \{9 - 17\} \longrightarrow operation := bizhours$
$R4 : operation = \{bizhours\} \wedge season = \{summer\} \longrightarrow therm\_setting := 24$
In Description Logics language the following concepts are distinguished:
$Day, Month, Hour, Today, Season, Operation, Therm\_setting$.
The definition of the concepts is as follows:
$Day \equiv \{sun, mon, tue, wed, thr, fri, sat\}$,
$Month \equiv \{jan, feb, mar, apr, may, jun, jul, aug, sep, oct, nov, dec\}$,
$Hour \equiv \{1, 2, \ldots, 24\}$,
$Today \equiv \{workday, weekend\}$,
$Season \equiv \{winter, spring, summer, autumn\}$,
$Operation \equiv \{bizhours, notbizhours\}$,
$Therm\_setting \equiv \{12, 13, \ldots, 30\}$

To simplify the notation one can introduce the following concepts:
$SummerMonths \equiv \{dec, jan, feb\}$,
$Bizhours \equiv \{9, 10, \ldots, 17\}$,
$WorkingDays \equiv \{mon, tue, wed, thr, fri\}$.
and using the transition specified in Fig. 2 write the rules as follows:
$R1 : (Month \equiv SummerMonths) \rightarrow Season(summer)$,
$R2 : (Day \equiv WorkingDays) \rightarrow Today(workday)$,
$R3 : (Today \equiv \{workday\}) \wedge (Hour \equiv BizHours) \rightarrow Operation(bizhours)$,
$R4 : (Operation \equiv \{bizhours\}) \wedge (Season \equiv \{summer\}) \rightarrow Therm\_setting(24)$
    Let the state in moment 0, be represented as $ABox_0$ as follows:

$$Month(jan).Day(mon).Hour(11). \tag{25}$$

The inference process is as follows: The state $ABox_0$ (25) and the preconditions of rule R1 are loaded into a DL reasoner. The DL reasoner performs the consistency checking of the state with respect to rule preconditions. Because the ontology built from the state assertions and R1 precondition formulas is consistent ($Month(jan)$ is consistent w.r.t. $Month \equiv SummerMonths$ ($SummerMonths \equiv \{dec, jan, feb\}$)) the rule is fired. The conclusion of the rule generates a new assertion in the state formula. The new $ABox_1$ replaces the old one ($ABox_0$). The new state $ABox_1$ is as follows:

$$Month(jan).Day(mon).Hour(11).Season(summer). \tag{26}$$

The state $ABox_1$ (26) and the preconditions of rule R2 are loaded into the DL reasoner. The DL reasoner performs the consistency check of the state with respect to rule preconditions. Because this time the ontology built from the state assertions and R2 precondition formulas again is consistent ($Day(mon)$ is consistent w.r.t. $Day \equiv WorkingDays$ ($WorkingDays \equiv \{mon, tue, wed, thr, fri\}$) the rule is fired. The conclusion of the rule generates a new assertion. The reasoning continues with new state and the next rules.

7   Future Work

The research presented in the paper affects integration of the selected expert system rule design method and the Semantic Web. Several design tools have been implemented for the design and verification of XTT^2-based systems. These could be used for building such systems in the context of the Semantic Web.

    To use DL modelling and reasoning capabilities to a larger extent, the ontology of the example used above and other systems could be built in a different manner. Some of the rules could be included into subsumption relations such that those rules would be executed as an intrinsic DL inference.

    The XTT^2 toolset (see `hekate.ia.agh.edu.pl`) also provides a custom inference engine able to execute the XTT^2 rules. The engine has an extensible architecture and is implemented in Prolog. Future work includes developing an interface able to use some dedicated DL reasoners such as Pellet. This would allow to reason with XTT^2 rules on top of existing OWL ontologies.

Acknowledgements. The authors wish to thank Claudia Obermaier for her valuable remarks concerning the final version of the paper.

References

1. Baader, F., Calvanese, D., McGuinness, D.L., Nardi, D., Patel-Schneider, P.F. (eds.): The Description Logic Handbook: Theory, Implementation, and Applications. Cambridge University Press, Cambridge (2003)
2. van Harmelen, F., Lifschitz, V., Porter, B. (eds.): Handbook of Knowledge Representation. Elsevier Science, Amsterdam (2007)
3. Russell, S., Norvig, P.: Artificial Intelligence: A Modern Approach, 2nd edn. Prentice-Hall, Englewood Cliffs (2003)
4. Giarratano, J., Riley, G.: Expert Systems. Principles and Programming, 4th edn. Thomson Course Technology, Boston (2005)
5. Ligęza, A.: Logical Foundations for Rule-Based Systems. Springer, Heidelberg (2006)
6. Nalepa, G.J., Ligęza, A.: Xtt+ rule design using the alsv(fd). In: Giurca, A., Analyti, A., Wagner, G. (eds.) ECAI 2008: 18th European Conference on Artificial Intelligence: 2nd East European Workshop on Rule-based applications, RuleApps 2008, July 22, 2008, pp. 11–15. University of Patras, Patras (2008)
7. Klösgen, W., Żytkow, J.M. (eds.): Handbook of Data Mining and Knowledge Discovery. Oxford University Press, New York (2002)
8. Grosof, B.N., Horrocks, I., Volz, R., Decker, S.: Description logic programs: combining logic programs with description logic. In: Proceedings of the Twelfth International World Wide Web Conference, WWW 2003, pp. 48–57 (2003)
9. Horrocks, I., Patel-Schneider, P.F., Boley, H., Tabet, S., Grosof, B., Dean, M.: Swrl: A semantic web rule language combining owl and ruleml, w3c member submission 21 May 2004. Technical report, W3C (2004)
10. Ligęza, A., Nalepa, G.J.: Logical Representation and Verification of Rules. In: Handbook of Research on Emerging Rule-Based Languages and Technologies: Open Solutions and Approaches. Information Science Reference (2009)

Applying Caching Capabilities to Inference Applications Based on Semantic Web

Alejandro Rodríguez[1], Enrique Jimenez[1], Mateusz Radzimski[1], Juan Miguel Gómez[1], Giner Alor[2], Rubén Posada-Gomez[2], and Jose E. Labra Gayo[3]

[1] Department of Computer Science, Universidad Carlos III de Madrid, Avenida de la Universidad 30, 28911, Leganés, Madrid
{alejandro.rodriguez,enrique.jimenez,mateusz.radzimski,
juanmiguel.gomez}@uc3m.es
[2] Division of Research and Postgraduate Studies, Instituto Tecnológico de Orizaba, Mexico
{galor,rposada}@itorizaba.edu.mx
[3] Department of Computer Science, Universidad de Oviedo
labra@uniovi.es

Abstract. Nowadays there is a large number of Expert Systems available to users requiring the extraction of data relevant to specific domains, many of which are based on reasoning and inference. However, many of these tools offer slow execution time, resulting in delayed response times to the queries made by users. The strategy of caching to define specific patterns of results enables such systems to eliminate the requirement to repeat the same queries, speeding up the response time and eliminating redundancy. This paper proposes a caching strategy for an Expert System based on Semantic Web and reasoning and inference techniques. Caching strategies have previously been applied to simple XML queries. Performance has been evaluated using an existing system for medical diagnosis, which demonstrates the increased efficiency of the system.

Keywords: caching, semantic web, ontologies, inference, rule, pattern recognition.

1 Introduction

Caching techniques play a very important role in optimizing systems performance [1]. They have been widely adopted in many different areas for years, therefore a great deal of effective strategies [2] and algorithms [3] were invented. Advanced use of caching techniques using frequent patterns recognition [4] [5] and query caching [6] have been broadly discussed in the literature. Many works focus on applying semantics to cache tables [7], or retrieving data based on already performed, different queries [8]. Even though this paper incorporates some of those ideas, it proposes a new approach for boosting performance of semantic queries containing frequently used subparts. Previous caching strategies have been principally applied to simple XML

N.T. Nguyen et al. (Eds.): New Challenges in Compu. Collective Intelligence, SCI 244, pp. 27–37.

queries; however, our work proposes the application of caching to queries based on logical inference.

The paper is structured as follows. The following section details the principal features of the system, including the pattern caching strategy (PCS), the preconditions of the system, the main components which comprise the system, the cache driver and the rules processing agent. Section 3 describes the architecture, followed by a use case example in Section 4. Section 5 describes the evaluation made. Section 6 outlines future work and conclusions.

2   System

2.1   Pattern Caching Strategy (PCS)

The strategy of pattern-based caching consists of searching within the table of already cached elements, with the aim of extracting determined patterns from the input data. This process is carried out systematically by an agent, which is responsible for the search and retrieval of these patterns. Once one or various patterns have been established, the retrieved information is processed, with the objective of establishing whether the patterns are valid or adequate to enable an increase in the efficiency of the system.

The objective of this strategy is to increase the efficiency of the system. The initial step is an inference process, which is supported by a knowledge base represented by an ontology, and another base which is a set of rules, both related by means of the inference engine. This inference process has generally a non-linear time or cost, which increases as the size of the knowledge base increases, as well as the knowledge base, which represents the rules (the number of rules).

Once the PCS has been applied, some patterns of inputs, which are used with a given frequency, can be found. Therefore, the goal is to create new sets of rules based on these patterns, in such a way that when the inference engine receives a new input query, the engine verifies whether this query matches any previously found patterns prior to performing the inference process. If a matching is found between the input and the stored patterns, the system automatically assigns this query to the corresponding knowledge base of rules, increasing the performance of the system given that it is not required to verify the input consulting the global set of rules, which is of greater size and whose inference time is much longer.

2.2   Preconditions of the System

Some basic preconditions for the efficient use of these types of caching strategies can be outlined. The most important feature is the following: based on the output of the inference system, it should be possible to obtain new inference rules, which generate the same results. This feature is required, because once an inference pattern has been discovered, all of the possible results with this pattern need to be determined. Based on these results, a new set of rules is generated which is smaller than the initial set. These rules lead to the outcome, which has generated the result just obtained. That is, to establish:

Set of rules for the pattern $\subset$ Set of total rules

The set established above should always be smaller than the total set. It is just this feature what enables an increased speed and efficiency of the system. When the inference engine is run with a smaller and more easy to use set of rules.

Therefore, it is very important to establish that the outputs of the system enable the generation of inference rules, that is, they allow the determination of the premises that lead to the conclusion. Another issue that should be considered is that this solution can only be used in systems where the number of inputs directly affects the number of output results. More specifically, if the system receives X inputs and returns Y outputs, in the case that it receives these X inputs a second time with an additional input element, the number of outputs in this case should be less than Y, and additionally, they should be elements of Y. This is because the new inputs that have been added to X have redefined the result such that it is a more closed set than the previous set:

X *(inputs)* $\rightarrow$ *infers* $\rightarrow Y$ *(outputs)*
$X+T$ *(inputs)* $\rightarrow$ *infers* $\rightarrow Z$ *(outputs)*
Where: $Z \subset Y$

Although in this paper, we have used Jena Rules as the rule language and the inference engine is the inference engine provided by Jena, our proposal could be applied to other kinds of rule languages and systems.

3 System Architecture

The basis of the architecture of our proposed system is that of a traditional Expert System: a Knowledge Base, a set of inference rules and an inference engine. The novel feature is the inclusion of a caching system for the creation of new subsets and rules, in the form of minor modifications in the rules container, which is composed of two repositories, one global which stores all of the initial rules of the system, and another which is responsible for storing the distinct subsets of rules which will be generated by the Caching System.

The figure below shows the logical architecture of the system, where it can be viewed how the distinct subsystems communicate, as well as the flow of messages that are interchanged in order to realize the inference process. In the following sections , the internal functioning of each of these elements will be described, placing special emphasis on the Rules Manager and Caching System, the components which represent the basis of this article.

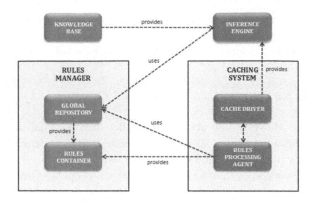

Fig. 1. System architecture

3.1 Knowledge Base

This component contains the Knowledge Base [9] of the Expert System that is used to carry out the reasoning process. Specifically, it is the ontology manager [10] of the system. It provides all of the data that is stored in the repository, so that the inference engine can perform with explicit knowledge about the domain in question. Thus, it receives new facts from the engine, which are the result of constant interaction and reasoning that this base orchestrates.

3.2 Inference Engine

The inference engine is the kernel of the Expert System [11], which carries out the entire reasoning and inference process to generate a response based on the experience and knowledge it contains about the query, always working using the Knowledge Base and the rules that are provided to it.

At the same time, it generates hash codes for each input which it receives, with the aim that they are stored in the cache and can be later consulted if it is possible to obtain a subset of rules which is capable of solving the problem.

3.3 Rules Manager

The Rules Manager transfers the rules that provide the restrictions for the inference process to the inference engine, working on the Knowledge Base. It is divided into two subsystems, the first one being the global repository, which is comprised of the total set of rules of the system, while the Rules Container systematically stores the distinct subsets of rules that are generated via the caching system.

3.3.1 Global Repository

This repository stores the global inference rules, which is the repository of the rules that affect the entire problem domain, and which may increase as new facts are introduced or as the application domain of the system increases. It is managed by the expert, and can only be modified if the expert extends or reduces the magnitude of the problem domain.

3.3.2 Rules Container

The Rules Container is a repository where distinct subsets of rules are stored, generated by interaction with the Caching System module. The Agent, based on the Caching System, analyses the total set of rules, searching for patterns that are repeated in the cache table, and which are determined based on a concrete group of rules. In this way, the rules that are not necessary for this step may be eliminated, reducing the size of the problem domain, thus decreasing the response time of the system.

3.4 Caching System

The goal of this module is to reduce the size of the problem at hand, decreasing the number of rules, which the inference engine has to process, and notably limiting the magnitude of the domain in question.

3.4.1 Cache Driver

The Cache Driver is responsible for managing the cache and all of its related objects: the database, content, etc. It ensures that the number of elements is adequate, and manages the policy of replacement of the elements: Least Recent Used (LRU), Most Recent Used (MRU), etc. It mainly works as a container interacting with the rest of the systems, such as the processing of the rules or the inference engine.

The principal working is based on its interaction with the inference engine. The engine is responsible for verifying with the Cache Driver if the input has a pattern that can match with it in the cache. In the case that they are inside the cache, this indicates that a search can be performed for a subset of rules, which is not the global set, therefore obtaining the knowledge in a more efficient time span.

Additionally, this component is in charge of verifying that the size of the cache does not exceed a particular threshold, eliminating tuples based on specific criteria. Generally, the criteria to be taken into account are that the tuples that contain more "hits" are those more likely to survive. The tuples that are consulted with greater frequency are characterized by a greater number of hits.

3.4.2 Rules Processing Agent

This module is responsible for carrying out the processing of the global set of rules to generate new subsets that can solve the problem based on the entries stored in the cache, as well as the management of their storage. It is implemented by means of an Intelligent Autonomous Agent [12][13], which, at systematic moments in time, carries out a map of the cache in search of new patterns, freeing space within it in the case that it is required. The method selected for this process is LFU (*Least Frequently Used*) [14], given that it abolishes the least common combinations and those that have a lower probability of occurrence. Therefore, they have a higher or lower impact on the optimization of the system. The agent will be responsible for searching combinations within the Global Repository that can match the pattern in question, storing these new resulting subsets in the Rules Container [15]. At the same time, in the case of carrying out deletions in the cache table, it will also have as a goal to abolish those subsets that correspond to the deleted patterns.

Its fundamental objective is to obtain the patterns of inputs that are stored in the cache and generate new sets of rules that will be synchronized according to the global repository. This process can be divided into a number of steps:

1. Search for new patterns.
2. Updating the use of old patterns.
3. Reorganization of the cache (deletion of old patterns and addition of new patterns).
4. Reasoning (inference) of results based on the new patterns.
5. Computation of the results of the inference in step 4 to generate new sets of rules.
6. Updating the repository of rules, adding the new sets generated based on new patterns, and deleting the sets of patterns that have been eliminated.

This process can be done depending on the duration of the process over every X units of time. Depending on the latency of this process, the units of time change so that parameter X can decrement or increment to achieve an improvement in the global performance of the system. The following section describes the pattern recognition algorithm.

The algorithm that searches new patterns developed to test the caching system consists of the following set of steps:

1. Obtaining all the hashes from the Database. Suppose this 6 hashes:

    55f5a7216d1c107c62508bc690d78392
    e4673bd7788f8a1b55266f3f77464b1f
    444096cb35814b0857df5fbf6d3372d2
    335c8c67e7bda3abe46b5a0dbd645bbd
    67323fa32b004ea1013dbf444fe976cf
    9c975b4518210e6fecb874559e3027ec

2. For each hash, generate an object (Caching Consult) that will encapsulate the hash and the corresponding code associated with it. In our example (medical system), it will be symptoms, so we will have:

    55f5a7216d1c107c62508bc690d78392 - *sA*
    e4673bd7788f8a1b55266f3f77464b1f - *sA,sB,sC*
    444096cb35814b0857df5fbf6d3372d2 - *sA,sB,sD*
    335c8c67e7bda3abe46b5a0dbd645bbd - *sA,sB,sD,sF*
    67323fa32b004ea1013dbf444fe976cf - *sG,sH,sI*
    9c975b4518210e6fecb874559e3027ec - *sG,sH,sI,sL*

    In this case, we are able to identify some patterns. The pattern depends on the length that we establish that should have it. In this case, we establish at least 3 elements to build a pattern.

3. Now, we create a string of lists that contains only the codes of the previous objects. So, we will have:

    sA
    sA,sB,sC
    sA,sB,sD

sA,sB,sD,sF
sG,sH,sI
sG,sH,sI,sL

4. Now, we take every previous string and compare each one with the rest with some restrictions, trying to search matches. So, the trace of this part of the algorithm could be:

sA,sB,sC / sA,sB,sD -> (Match: 2 / Discard. Minimum: 3)
sA,sB,sC / sA,sB,sD,sF -> (Match: 2 / Discard. Minimum: 3)
sA,sB,sC / sG,sH,sI -> (Match: 0 / Discard. Minimum: 3)
sA,sB,sC / sG,sH,sI,sL -> (Match: 0 / Discard. Minimum: 3)

sA,sB,sD / sA,sB,sD,sF -> (Match: 3 / Pattern found: **sA,sB,sD**)
sA,sB,sD / sG,sH,sI -> (Match: 0 / Discard. Minimum: 3)
sA,sB,sD / sG,sH,sI,sL -> (Match: 0 / Discard. Minimum: 3)

sA,sB,sD,sF / sG,sH,sI -> (Match: 0 / Discard. Minimum: 3)
sA,sB,sD,sF / sG,sH,sI,sL -> (Match: 0 / Discard. Minimum: 3)

sG,sH,sI / sG,sH,sI,sL -> (Match: 3 / Pattern found: **sG,sH,sI**)

The algorithm discards automatically the elements that did not have enough elements to be a pattern. For example "sA" is a single element and the algorithm establishes that at least three elements are needed. Furthermore, it also does not check elements that have already been checked. For example: sA,sB,sC / sA,sB,sD is the same as sA,sB,sD / sA,sB,sC.

The step 4 makes a canonicalization of the input strings. This canonicalization takes the inputs as a unique string. For example: "sA,sB,sC". We divide this String as an array that contains all the elements separated by coma: data[] = { "sA", "sB", "sC" }, and finally we make an alphabetic sort of the array. With this operation we always have the inputs in the same format, so it's more easily and efficient make the comparisons.

4 Use Example

In the current section, an example of how the system works will be outlined, particularly given that there are certain features that should be adapted as a function of the problem in question for which a solution is required.

We use a knowledge base of a medical diagnosis domain [16], which is linked to another knowledge base, with rules to infer results for the domain. In the case that the rules container contains a very high number of rules, it is highly possible that the speed or performance of the system is affected, because of this elevated number. The advantage is provided by the possibility to work with a reduced set of rules, which means that the performance of the system is not notably degraded as frequently as it occurs in other cases.

The inputs are strings that represent the code of the symptom that the system receives to make the diagnosis. Each input will uniquely be represented by a single string. Strings will be separated by comas. In this case, for simplification purposes, the outputs will also be strings, in such a way that each string is an output.

Bearing the example in mind, Table 1 below is a table that represents an example of the cache of the current system:

Note: SYM_A = sA, SYM_B = sB, etc. (To clarify with previous notation).

Table 1. Example of the data in the caching system

Hash	Input	Output	Times	Date
2b0dc568e588	SYM_A	*DIS_M,DIS_N,DIS_P,DIS_T,DIS_X,DIS_Z*	356	10/01/08
e86410fa2d6e	SYM_A,SYM_B,SYM_D,SYM_F	*DIS_M*	245	10/01/08
8827a41122a5	SYM_G,SYM_H,SYM_I,SYM_L	*DIS_W*	240	09/01/08
ab5d511f23bd	SYM_A,SYM_B,SYM_C,SYM_E	*DIS_N*	145	09/01/08
3c6f5dee2378	**SYM_A,SYM_B,SYM_C**	*DIS_N,DIS_X*	98	08/01/08
6b642164604d	**SYM_G,SYM_H,SYM_I**	*DIS_V,DIS_W*	98	08/01/08
8c328f9a099b6	**SYM_A,SYM_B,SYM_D**	*DIS_M,DIS_Z*	98	08/01/08

The above table represents an example of the approximated content of an individual caching strategy. Notice that a number of patterns has been found in the inputs and are indicated in bold.

The system will carry out the strategy previously described with this set of patterns. In this case, the system will search possible common patterns within all of the inputs, and it would find the patterns as displayed in the current example. Once the system has extracted all of the new patterns (if a pattern is extracted which is already present in the system, it will be omitted, as it already exists), the system will communicate with the inference engine to attempt to obtain all of the possible outputs which it (or the pattern) would provide. In the current example, the system will make a query to the inference engine so that it returns all of the outputs based on the set of inputs of *{SYM_A, SYM_B, SYM_C}, {SYM_G, SYM_H, SYM_I}* and *{SYM_A, SYM_B, SYM_D}*.

The system, based on this inputs, returns the following:

SYM_A, SYM_B, SYM_C → DIS_X, DIS_N
SYM_G, SYM_H, SYM_I → DIS_V, DIS_W
SYM_A, SYM_B, SYM_D → DIS_Z, DIS_M

Looking at the outputs that the system has just returned, it can be observed that these are also present in all of the elements that have been output by the inference engine. It is evident that more elements exist than those that are in the table, but it is important that all of the elements that are displayed in the table are present. This situation is the result of one of the preconditions discussed in Section 3.1.

Once this step has been performed, the goal of the system is to generate all of the possible inference rules that lead to this output based on the output that it has just obtained.

5   Evaluation

We have evaluated the system taking into account the improvement of the temporal performance of the application when the developed caching system is working. To do that, we developed two kinds of tests to contrast results. The first one obtains the execution time of the entire system, checking the times applying the caching system and without applying it and the results are analyzed. The second one only takes into account the inference time, because it is the time that the system tries to improve.

Table 2 shows the inference times for one single execution running the entire system while table 3 shows the same time in 100 executions.

Table 2. Results for entire system. One execution.

Test	Time Caching Disabled	Time Caching Enabled
ABDF	219 ms	234 ms
GHIL	188 ms	188 ms
ABCE	156 ms	156 ms

Table 3. Results for entire system. 100 executions.

Test	Time Caching Disabled	Time Caching Enabled
ABDF	7 ms	6 ms
GHIL	4 ms	4 ms
ABCE	5 ms	5 ms

As table 2 shows, the execution of the entire system in one single execution does not improve the inference times, and even in one case, the caching system deteriorates the time. This is because in one execution of the inference system it takes more time to load all the data and to compare it against the ontology. If we add also the time to consume the pattern search algorithm trying to match the given input with a pattern of the caching we obtain that the times for one execution are increased.

However, the times shown in table 3 for 100 executions, are the times per execution, so, for example, in the case of ABDF test with Time Caching Disabled, the total time for the 100 executions was 700 ms.

Furthermore, it must be taken into account that this time is for the entire system, and not for the inference engine, which is the system whose performance we are trying to improve.

Table 4 shows the time of the inference engine in one execution while table 5 shows the time for 100 executions.

Table 4. Results for inference system. One execution.

Test	Time Caching Disabled	Time Caching Enabled
ABDF	16 ms	15 ms
GHIL	32 ms	16 ms
ABCE	< 1 ms	< 1 ms

Table 5. Results for inference system. 100 executions.

Test	Time Caching Disabled	Time Caching Enabled
ABDF	1 ms	< 1 ms
GHIL	< 1 ms	< 1 ms
ABCE	1 ms	< 1 ms

Now, we can see the real performance of the caching system. In the first case, we can see that in two cases the times are improved. In one of them, with a reduction of one millisecond of the inference time and in the other with a reduction of 50% (16 ms).

With this data, we can conclude that the caching system obtains an increment of the temporal efficiency for small data sets. We have done this evaluation with these types of tests because the system where the caching system was integrated and tested uses them. It would be difficult to introduce biggest datasets to try to make an evaluation with them.

6   Future Work and Conclusions

The most important advantage of this kind of applications is its efficiency. The time taken for the completion of a query decreases and it will facilitate the management of the system by the user through reduced response time.

Besides that, the architectural design will ease the adaptation of the system to many and diverse fields, and it will easily be adaptable to modifications or changes.
Therefore, we present the system as a general solution to productivity boosting, without carrying large computational or implementation costs.

Future work will include the interesting task of optimizing the pattern search algorithm. However, it is not an easy task because, as stated previously, in every case it will be necessary to traverse the entire set of tuples that comprise the cache.

Acknowledgements

This work is supported by the Spanish Ministry of Industry, Tourism, and Commerce under the project SONAR (TSI-340000-2007-212), GODO2 (TSI-020100-2008-564) and SONAR2 (TSI-020100-2008-665), under the PIBES project of the Spanish Committee of Education & Science (TEC2006-12365-C02-01) and the MID-CBR project of the Spanish Committee of Education & Science (TIN2006-15140-C03-02).

References

[1] Franklin, M.J.: Client Data Caching: A Foundation for High Performance Object Database Systems. Kluwer Academic Publishers, Dordrecht (1996)
[2] Alonso, R., Barbara, D., Garcia-Molina, H.: Data caching issues in an information retrieval system. ACM Transactions on Database Systems 15(3), 359–384 (1990)

[3] Johnson, T., Shasha, D.: 2Q: A Low Overhead High Performance Buffer Management Replacement Algorithm. In: Proceedings of the 20th International Conference on Very Large Data Bases, pp. 439–450 (1994)

[4] Bei, Y., Chen, G., Hu, T., Dong, J.: Caching System for XML Queries Using Frequent Query Patterns. In: Shen, W., Yong, J., Yang, Y., Barthès, J.-P.A., Luo, J. (eds.) CSCWD 2007. LNCS, vol. 5236. Springer, Heidelberg (2007)

[5] Yang, L.H., Lee, M.L., Hsu, W., Huang, D., Wong, L.: Efficient mining of frequent XML query patterns with repeating-siblings. Inf. Softw. Technol. 50(5), 375–389 (2008)

[6] Adali, S., Candan, S., Papakonstantinou, Y., Subrahmanyan, V.: Query Caching and Optimization in Mediator Systems. Technical Report. Stanford University (1995)

[7] Ren, Q., Dunham, M.: Semantic caching and query processing. Technical Report 98-CSE-4. Southern Methodist University (May 1998)

[8] Godfrey, P., Gryz, J.: Semantic query caching for heterogeneous databases. In: Proceedings of 4th KRDB Workshop at VLDB (1997)

[9] Gruber, T.R.: Toward Principles for the Design of Ontologies used for Knowledge Sharing. International Journal of Human-Computer Studies 43, 907–928 (1995)

[10] Guarino, N.: Formal Ontology in Information Systems. In: Proceedings of the 1st International Conference on Formal Ontologies in Information Systems FOIS, pp. 3–15. IOS Press, Amsterdam (1998)

[11] Hayes-Roth, F., Waterman, D.A., Lenat, D.B.: Building expert systems (1983)

[12] Girardi, R., Faria, C., Marinho, L.: Ontology-based Domain Modeling of Multi-Agent Systems. In: Gonzalez-Perez, C. (ed.) Proceedings of the Third International Workshop on Agent-Oriented Methodologies at International Conference on Object-Oriented Programming, Systems, Languages and Applications (OOPSLA 2004), Vancouver, Canada, pp. 51–62 (2004)

[13] Luke, S., Spector, L., Rager, D., Hendler, J.: Ontology-based Web Agents. International Conference on Autonomous Agents. Marina del Rey, California, United States (1997)

[14] Ruay-Shiung, C., Hui-Ping, C., Yun-Ting, W.: A Dynamic Weighted Data Replication Strategy in Data Grids. In: IEEE/ACS International Conference on Computer Systems and Applications, AICCSA 2008 (2008)

[15] Hanli, W., Kwong, S., Yaochu, J., Wei, W., Kim-Fung, M.: Agent-based evolutionary approach for interpretable rule-based knowledge extraction. Systems, Man, and Cybernetics, Part C 35(2), 143–155 (2005)

[16] Rodriguez, A., Mencke, M., Alor Hernandez, G., Posada Gomez, R., Gomez, J.M.: Medboli: Medical Diagnosis Based on Ontologies and Logical Inference. In: The Third International Conference on Digital Society, ICDS 2009, Cancun, Mexico, February 1 - 7 (2009)

Semantic Web System for Automatic Plan Scheduling

Piotr Czerpak, Paweł Drozda, and Krzysztof Sopyła

Faculty of Mathematics and Computer Science
University of Warmia and Mazury in Olsztyn, Poland
piotrczerpak@matman.uwm.edu.pl,
pdrozda@matman.uwm.edu.pl,
ksopyla@uwm.edu.pl

Abstract. In this paper we describe an ontology-based scheduling system which generates classes plans for university departments in an automatic way. Ontologies are proposed as structures which contain required data about courses, students, teachers, classrooms, programs, etc. In this way we can benefit from the existing ontologies, which describe parts of university needed for planning. FOAF ontology [9] is used to provide information about teachers and group of students and iCalendar ontology [5], [6] describes availability of teachers and rooms. For all other information, we develop University Ontology. For the scheduling the poly-optimization algorithm is implemented which utilizes gathered information from the above mentioned ontologies.

Keywords: Multi-Agent Systems, Planning and Scheduling, Semantic Web, Ontologies.

1 Introduction

The problem of timetabling exists almost as long as universities exist. Creation of a timetable which satisfies all interested individuals is a very hard and time consuming process. In the past as well as at present many departments at universities create plans of classes without any software support, which often causes overlapping of courses, lack of laboratories in a specific time etc. For this reason many researches started to investigate possibility of automatic planning (first system was developed in 1967 by E. Burke [4]) which would facilitate the solution of timetabling task. In the literature that problem is considered in many aspects. For example, it can be considered from student's point of view when he wants to take courses – these courses should not overlap. On the other hand, the schedule should satisfy all teachers, for instance long breaks between lessons should be avoided. Furthermore, timetable is necessary for many organizations like primary schools, high schools, universities etc.

The main goal of this work is an application of Semantic Web idea in scheduling. The concept of Semantic Web was introduced by Tim Berners-Lee [3]. Its aim is to allow to process knowledge, gathered in the Web, by human agents as well as by

N.T. Nguyen et al. (Eds.): New Challenges in Compu. Collective Intelligence, SCI 244, pp. 39–50.
springerlink.com
© Springer-Verlag Berlin Heidelberg 2009

computer agents. It is achieved by annotating current web pages in such way, that each concept has significance and is understood by every application or user in the same way. To ensure common understanding, annotated knowledge is structured in *ontologies*. The ontology is defined as "a specification of a conceptualization" [11]. Intuitively, ontology means the description of a part of the world in one of formalisms understandable by machines. One of these formalisms is OWL [2] which was chosen for ontology construction in our system. The Semantic Web was chosen for the system presented in this work since it is considered as "a New Web" and it is possible that in the near future major part of WWW pages will be encoded in one of Semantic Web formalisms. Moreover, there are plenty of already created ontologies useful for scheduling from which we can benefit and make use of. For example, our system utilizes two ontologies: FOAF ontology [9] and iCalendar ontology [5], [6].

For all required information not included in mentioned ontologies, we develop University Ontology necessary for scheduling. Besides of ontologies our system contains an application – Planner System. The major task of this module is automatic generation of timetable using all necessary information obtained from ontologies. Planner System does utilize the algorithm developed in [7].

The rest of the paper is structured as follows: Section 2 presents related work in the fields of timetabling and the Semantic Web. In Section 3 an architecture of the system with scheduling algorithm is briefly explained. Description of system usage is in Section 4. Finally, conclusions and future directions are given in the last section.

2 Related Work

There are plenty of systems which deal with the timetabling problem, yet scheduling is specific, dependent on an organization, which makes it impossible to create one application being able to generate plans in the universal manner.

The first type of systems, summarized in [18], proposes a solution of timetabling problem imitating human behavior. In particular, the lectures are added to schedule one by one until all of them are inserted into the plan. One of these systems called SHOLA is described in [12]. Author introduces three strategies for timetabling. The first begins scheduling with the most restricted lectures. The second verifies if there exists any place in the schedule in which only one lesson can be inserted. If so, mentioned lesson is planned. The last deals with situations when in particular point of algorithm the lecture cannot be added to the plan. In such cases one of the planned lesson is moved into other appropriate place in the schedule.

Other solution of that problem is proposed in [16], [20]. Authors apply graph theory for scheduling.

Moreover, there are known methods, which take advantage of genetic algorithms [17], local search including Tabu Search [10], Simulated Annealing [14], Ant Colony Optimization [8].

All of propositions mentioned above employ only one criterion for plan evaluation or transform criterions into constraints which makes solution almost impossible to find. The algorithm applied in our system, described in [7], utilizes multi-criterion optimization.

In our opinion, approach which utilizes only one aggregated criterion is less effective and flexible than poly-optimization. Implementation of our approach ensures simple adjustment of benchmarks as well as easy replacement of preference model. It makes adaptation of different requirements of departments or universities uncomplicated.

The next factor which differentiates our system from already created is the application of Semantic Web technology. To the authors' knowledge, no implementations of Semantic Web into the timetabling problem were proposed before. The main aim of introducing Semantic Web is possibility of usage of already created ontologies. Moreover, the information published in the Web are being changed by addition of semantic and in the next future we can expect that sites of public organizations including universities will be encoded with semantic formalism. Therefore, the system described in this work will be prepared to adaptation to major part of universities and will be able to create lesson plans in automatic way without grant modifications. Finally, Semantic Web enables utilization of published schedules by other agents.

In our system we take advantage of two ontologies, FOAF ontology and iCalendar ontology. The first one is utilized for identification of students as well as academic teachers working in the department. It ensures gaining essential information for time tabling in automatic way. The second one stores personal calendars of the teachers.

Moreover, we considered ontologies, which describe university, yet none of them was suitable for requirements of our system. First of them was introduced in [13]. The ontology gathers concepts related to IT education and materials. The main goal of the application is to make possible reuse of the results of other research.

Similar ontology is described in [21]. The materials useful for academic teachers are ordered in the hierarchy which facilitates teachers to organize the sequence of a lesson plan.

Next ontology introduced by [19] describes university domain. It contains, encoded in RDF, concepts and properties regarding lessons lectured at universities as well as information about software, devices, documentation, etc. It gives the viewpoint of what and how people teach and learn. The main aim of ontology creation is the ability of information sharing between different universities.

None of mentioned ontologies is included in the described system since they generally contain specific information about courses. They do not contain information about programme, classrooms, assignments subjects to the teachers which are crucial for presented system.

3 Scheduling System Architecture

The basic architecture and coordination schema of the system described in this paper is shown in Figure 1.

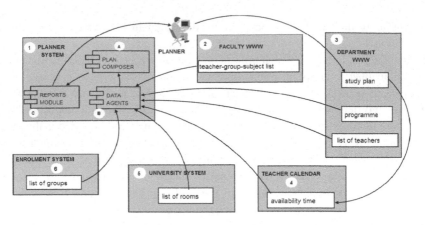

Fig. 1. System Architecture

The system was developed in order to generate optimal lesson plan for the department. To perform this task, our system consists of collaborating components.

First of them, named Planner System, is the most important part of the system. It contains three modules. The first module, Plan Composer uses algorithm proposed in [7] for creating lesson plan. The general description of the algorithm is provided below. It avails from data managed by Data Agent module. The main goal of Data Agent is collecting essential information from ontologies regarding lesson plan, such as names and availability of academic teachers, study programme or assignment of lectures. Last module, named Reports, ensures conversion of schedule generated by Plan Composer. It creates plans for each teacher, each group of students, each laboratory, as well as summary documents for a deanery. All other components illustrated in Figure 1 contain necessary data for plan generation.

3.1 Description of the Algorithm

The main goal of the scheduling algorithm is automatic generation of course plan, which maximizes evaluation functions proposed by planner. The algorithm begins scheduling from course which is the most difficult to plan (according to evaluation functions described in Lesson Choice Procedure subsection). Afterwards, the chosen course is planned in the best allotment according to evaluation functions described in Ordering of Course Allotments subsection. When the considered course is planned the iteration of the algorithm is repeated for the next course. Algorithm terminates when all courses are scheduled.

Due to the specific nature of universities assumptions, constraints, input data and data structures utilized in the algorithm are described below.

Assumptions and Scheduling Constraints
Each organization such as an university generates its timetables in the individual manner. The way the organization operates implies the assumptions and constrains for

the timetable generation task (for example the timetable can be created for each week, for two weeks, for entire semester). For our purpose we make the following simplistic assumptions and constraints:

- lessons are planned in week periods,
- a day is divided into fifteen minutes time intervals,
- at the end the fifteen minutes break is added to each lesson,
- Teachers determine hours and days in which they are available. Availability time of teachers is stored in the matrix DW:

$$DW = [c_{wh}] \begin{pmatrix} w = 1, ..., W \\ h = 0, ..., 239 \end{pmatrix}$$

where w – index of teacher, h – index of fifteen minutes interval.

The elements of matrix is equal to 1 if the teacher w is available in the interval h and 0 otherwise.

- Each subject can take place only in the appropriate type of classroom. Possibilities of the classroom usage for specific subject are encoded in the matrix MS:

$$MS = [d_{ps}] \begin{pmatrix} p = 1, ..., P \\ s = 1, ..., S \end{pmatrix}$$

where P is the number of all subjects and S is the number of all classrooms.

The elements of matrix is equal to 1 if the classroom s is appropriate for the subject p and 0 otherwise.

- The group of students and teacher cannot have more than one lesson in the same time,
- In classroom can be held only one lesson in specific time,
- The duration of lessons is set a priori.

Input Data

All data utilized by algorithm are stored in three ontologies: FOAF, iCalendar and University Ontology, which are described in the next subsection. In particular the following data are essential for algorithm execution:

programme represented by ordered triples (teacher, subject, group of students),

list of teachers, list of students' groups, list of rooms, availability time of teachers (matrix DW), for each subject the set of appropriate classrooms (matrix MS) and duration of lessons.

Data Structures Employed by the Algorithm

Authors propose three data structures which facilitate algorithm execution. In particular, the following structures are:

- Courses Queue,

The queue contains the ordered triples (teacher, subject, group of students). At the beginning of the queue, the lessons which are the most difficult to plan are placed.

The difficulty is calculated by evaluation function which is defined in the Lesson Choice Procedure subsection.

- Courses Stack,

The stack contains lessons already scheduled. The order in which lessons are placed on the stack is the same as the order in which lessons are planned.

- Allotments Queue,

The queue consists of sets – the classroom and the interval in which the course can be held. Allotment queue is formed for each lesson when the lesson stays at the beginning of the course queue. The order of the allotment queue is specified by evaluation function described in the Ordering of Course Allotments subsection and at the beginning of the queue stays "the best" allotment according to evaluation function.

Workflow of Algorithm

The Figure 2 illustrates the main steps of algorithm.

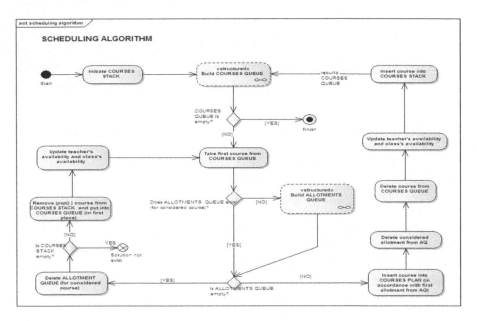

Fig. 2. Scheduling Algorithm

At the beginning of the algorithm the courses stack is initiated and the courses queue is created. Successively, the first course p_i is removed from the courses queue. If there does not exist allotment queue for course p_i, the queue is built. After the course p_i is inserted into the best allotment according to evaluation function – criterions, which evaluate plan, are set on minimal values. It means that the considered allotment will not be considered in the next steps of algorithm. It should be noted that allotments in the allotments queue fulfils constraints included in matrixes DW and MS as well as constraints from vector T. Subsequently, the course p_i is moved from the courses queue to courses stack. Then, the tables of teacher and classroom availability time are updated. Next step of algorithm ensures conversion of courses queue. In particular, the available time of teacher, classroom and group of students are changed and have impact on planning remaining courses. Then, the next iteration of algorithm is performed in the same way as first.

If course p_i is impossible to plan (allotment queue is empty for course p_i) then allotment queue is deleted and the last planned course is moved from the course stack to the beginning of the courses queue. In addition, the available times of appropriate teacher and classroom are updated. For the first course in the courses queue new allotment is taken from the beginning of allotment queue. The algorithm terminates when all courses are scheduled.

Poly-optimization

The main contribution of algorithm is application of poly-optimization for automatic scheduling. In particular, difficulty of lesson choice for planning and determination of appropriate lesson allotment are based on poly-optimization. The most important advantage of poly-optimization usage in automatic scheduling is the possibility of changing evaluations functions in easy way. It ensures adaptability to similar organizations and makes algorithm independent from the specific requirements of organization.

The algorithm consists of two optimization objectives: course choice procedure and ordering of course allotments procedure, which are described in the rest of this subsection.

Lesson Choice Procedure

The algorithm of lesson choice procedure is based on poly-optimization. Sequence of a lesson settings determinates the complexity of scheduling. In particular, there can be a case when the schedule creation is impossible. Thus, lectures which are the most difficult to plan (according to fixed criterions) will be first in the queue. It should be noted that the lesson choice procedure is called not only at the beginning of the algorithm but also each time when the lecture is put in to plan. When the mentioned modification occurs the difficulty of scheduling queued lessons changes which motivates necessity of queue rebuilding.

For the lesson choice procedure the following optimization objective is proposed:

$$(X, F, M(P))$$

where:

X – set of lessons, which remains for addition to lesson plan. The elements of X are triples (teacher, subject, group) which determinate each lesson,
$F: X \rightarrow N^3$ - evaluation function described below, where $F(x) = (F_1(x), F_2(x), F_3(x))$,
$M(P)$ - preference model described below.

The choice of criterions, number of them and preference model has decisive impact on ordering lessons in the queue. The adequate selection of elements mentioned above is not evident neither simple. The presented approach enables easy modification of criterions as well as the change of whole preference model. It makes the algorithm independent from processed data.

Authors specified three valuation functions:

- F_1- the number of possible allotments for each lecture,
- F_2- indicator of difficulty in lesson planning for given teacher,
- F_3- indicator of difficulty in lesson planning for given group.

Allotment is defined as classroom, triple (teacher, subject, group) and time in which teacher and the group are available. Indicator F_2 informs about the level of difficulty of lesson planning for a teacher. It determinates the number of fifteen minutes intervals in reserve which corresponds to each lecture. The bigger number means higher flexibility during scheduling. When the indicator equals to negative number for specific teacher, the complete lesson plan is impossible to create.

Following equation presents formula for calculation the indicator of difficulty in lesson planning for given teacher:

$$F_2 = \frac{(b_3 - b_2)}{b_1}$$

where:

b_1 − number of lessons to plan for given teacher,
b_2 − sum of fifteen minutes intervals corresponding to lectures had by given teacher,
b_3 − number of fifteen minutes intervals in which teacher is available.

Indicator F_3 describes difficulty of lesson planning for given group of students. It determinates the number of fifteen minutes intervals in reserve which corresponds to each lecture for given group. Following formula calculates mentioned indicator:

$$F_3 = \frac{(c_3 - c_2)}{c_1},$$

where:

c_1 − number of lessons to plan for given group of students,
c_2 − sum of fifteen minutes intervals corresponding to lectures attended by given group,
c_3 − number of fifteen minutes intervals in which group is available.

Preference model is the last part of optimization objective. The most often preference model is expressed in the form of domination relation $(\mathcal{M}(\mathcal{P}) = R)$ [1], which is defined as:

$R \subset Y \times Y$, such as

If $(y, w) \in R$ then y is "better" than w and $y \neq w$.

For purposes of the system authors introduce relation based on hierarchical approach and PARETO optimization [1]. First the lessons are compared in relation to first evaluation function (F_1). If the result is the same for two individuals the PARETO relation for other evaluation functions (F_2, F_3) is employed.

It can be formalized, as:

let us assume, that:

$$y = (y_1, y_2, y_3), z = (z_1, z_2, z_3) \ y, z\epsilon X;$$

then:

$$R = \{(y, z) \ \epsilon \ X \times X; y_1 < z_1 \lor (y_1 = z_1 \land y_2 \le z_2 \land y_3 \le z_3)\}.$$

Ordering of Course Allotments

The ordering of allotments algorithm is based at poly-optimization as well. For given lesson there should be specified allotment which will have the best influence in relation to definite criterions. The choice of specific allotment determinates not only the schedule but it can also precise if lesson plan can be created. If the lessons in the first free period are inserted without any criterions there might appear unnecessary intervals for given group of students or teacher. Each of such interval decreases availability time of particular resources, and as consequence it increases meaningfully constraints for not inserted lessons.

The following optimization objective for ordering of course allotments for course $z = (w, p, g)$ is proposed:

$$(X, F, M(P)),$$

where:

X — set of allotments (set of acceptable solutions), where allotment is a triple *(s, h, n)*
s — index of classroom
h — the time of lesson beginning
n — the lesson duration time

$$F(x) = (F_1(x), F_2(x), F_3(x), F_4(x), F_5(x)) - evaluation \ functions$$
$M(P)$ — preference model (relation R).

The following criterions are taken into account for allotment valuation:

F_1 — sum of time intervals in lesson plan of teacher w after insertion of allotment,
F_2 — sum of time intervals in lesson plan of group g after insertion of allotment,
F_3 — sum of time intervals of given allotment after four pm,
F_4 — sum of time intervals in lesson plan of classroom s in which no lesson can be scheduled,
F_5 — the time of lesson beginning — allotments with earlier hour of beginning are preferred.

The following preference model for ordering of lesson allotments is proposed:

Let us assume, that:

$$F: X \to Y, \qquad Y \in R^5$$
$$x_1 = (s_1, h_1, n), x_2 = (s_2, h_2, n) - \text{allotments for lessons } (w, p, g)$$
$$y = (y_1, y_2, y_3, y_4, y_5), z = (z_1, z_2, z_3, z_4, z_5) \ y, z\epsilon Y.$$

Then, the relation R can be formalized:

$$
R = \left\{ (y,z) \epsilon Y \times Y; \left| \begin{array}{l} (n * y_1 + m * y_2 < n * z_1 + m * z_2) \vee \\ \left(\begin{array}{l} n * y_1 + m * y_2 = n * z_1 + m * z_2 \wedge \\ y_3 < z_3 \end{array} \right) \vee \\ \left(\begin{array}{l} n * y_1 + m * y_2 = n * z_1 + m * z_2 \wedge \\ y_3 = z_3 \wedge \\ y_4 < z_4 \end{array} \right) \vee \\ \left(\begin{array}{l} n * y_1 + m * y_2 = n * z_1 + m * z_2 \wedge \\ y_3 = z_3 \wedge \\ y_4 = z_4 \wedge \\ y_5 < z_5 \end{array} \right) \end{array} \right. \right\},
$$

where:

n, m −coefficients, which define the importance of evaluation functions F_1 and F_2. First, the relation R takes into account results of evaluation functions F_1 and F_2 (considering coefficients n, m). If results are different then allotment with superior value is "better". In other case the function F_3 is calculated. If for F_3 results are equal the F_4 is taken into account. Finally, if results for F_4 are equal the values for F_5 are compared.

3.2 Description of Ontologies

The most important information (according to scheduling) about academic teachers is available on department web site encoded in RDF ontology. In particular, the FOAF ontology is proposed as a repository of teachers' data.

The following concepts and properties defined in FOAF ontology are utilized:

Person (data about specific person), Title, Mbox (e-mail), First_name, Second_name, Homepage, Group (container with all members of specific group).

Additionally, the tag with reference to calendar is added, which stores the address to personal calendar. For the personal calendars iCalendar ontology is used. The most important concept for purposes of our system is Vevent. It contains information about all events in which owner of the calendar is involved.

At department web site information about programme and study plan are also included. Thus, for purposes of our system university ontology is created. It contains not only information about programme and study plan but also assignment of lectures encoded in triples (teacher, group, lesson). Therefore, System Planner as well as department web site take advantage of university ontology. The mentioned ontology contains the rest of information which are encoded in the following concepts:

Programme, Subject, Subject Description, Specialization, Courses and Classroom.

All above mentioned concepts are crucial for automatic scheduling. Part of information gathered in our ontology are also included in university ontologies described in section 2, yet none of them is complete. The listing of the part of source code of university ontology is presented in listing 3. The version created in Protégé can be downloaded from [22].

Listing 3 Part of University Ontology

```
<owl:Class rdf:ID="Subject">
    <rdfs:comment
rdf:datatype="http://www.w3.org/2001/XMLSchema#string">
```

```
      Subject to be taught</rdfs:comment>
      <rdfs:subClassOf>
        <owl:Restriction>
          <owl:onProperty>
            <owl:FunctionalProperty
 rdf:ID="hasSubjectDescription"/>
          </owl:onProperty>
 <owl:cardinalityrdf:datatype=http://www.w3.org/2001/XML
 Schema#int>1</owl:cardinality>
        </owl:Restriction>
      </rdfs:subClassOf>
      ...
</owl:Class>
```

The listing presents class – Subject. For concept Subject the property hasSubjectDescription with restriction minCardinality equal to one is defined.

The last data crucial for plan generation concerns students. These data from enrolment system are collected. In order to ensure automatic processing during scheduling information about students and groups of students are encoded in FOAF ontology, which contains teachers' features and availability as well.

Concepts which describe teachers, groups of students, programme, subjects, courses, classrooms are utilized during plan generation.

4 Description of System Usage

The execution of the system modules is performed in following steps:

a) The Data Agent collects the data for automatic plan generation:

- o The list of academic teachers and programme from faculty web site;
- o Calendars of academic teachers; if the calendars for part of teachers are not available agent informs planner to contact these persons;
- o The lists of students with group division from enrolment system;
- o The list of classrooms from university system;

b) The gathered data are converted and passed to Plan Composer;
c) The plan is generated in automatic manner;
d) The plan to Reports Module is passed;
e) Reports Module generates specific schedules for each academic teacher, each group of students and each classroom in OWL file;
f) The lesson plan in OWL file is placed on faculty web site, in order to make possible data update in students' and teachers' calendars.

5 Conclusions and Future Work

In this paper the Semantic Web system for automatic scheduling is described. The main goal of the system is automatic generation of timetable. The system is based on the combination of university ontology proposed by authors and other ontologies developed in the past. So far, the skeleton of the system was implemented and first algorithm tests based on exemplary data set were performed. The results are promising and in the next future we intend to perform following tasks:

development of testing environment, implementation of interface for data entry, execution of algorithm tests for real data, development of Data Agents and integration of all components into complete system.

References

[1] Ameljańczyk, A.: Optymalizacja Wielokryterialna, WAT (1986)
[2] Antoniou, G., van Harmelen, F.: Web Ontology Language: OWL, Handbook on Ontologies. Springer, Heidelberg (2004)
[3] Berners-Lee, T., Fischetti, M.: Weaving the web (1999)
[4] Burke, E.: Home Page, http://www.cs.nott.ac.uk/~ekb/
[5] Calendar ontology iCal, http://www.w3.org/2002/12/cal/ical
[6] Calendar RDF - an application of the Resource Description Framework to iCalendar Data, http://www.w3.org/TR/2005/NOTE-rdfcal-20050929/
[7] Czerpak, P., Sopyła, K.: Optimization of Courses Plan for University Department (in polish), Master Thesis, University of Warmia and Mazury in Olsztyn (2007)
[8] Dorigo, M., Di Caro, G., Gambardella, L.M.: Ant algorithms for discrete optimization. Artificial Life 5(2), 137–172 (1999)
[9] FOAF Vocabulary Specification 0.91, http://xmlns.com/foaf/spec/
[10] Glover, F.: Tabu search. part I. ORSA Journal of Computing 1, 190–206 (1989)
[11] Gruber, T.: Toward principles for the design of ontologies used for knowledge sharing. International Journal of Human-Computer Studies, 907–928 (1994)
[12] Junginger, W.: Timetabling in Germany – a survey. Interfaces 16, 66–74 (1986)
[13] Kasai, T., Yamaguchi, H., Nagano, K., Mizoguchi, R.: A Semantic Web System for Helping Teachers Plan Lessons Using Ontology Alignment. In: Proceedings of the International Workshops of Applications of Semantic Web Technologies for E-Learning (SW-EL 2005) at AIED 2005, pp. 9–17 (2005)
[14] Kirkpatrick, S., Gerlatt Jr., C.D., Vecchi, M.P.: Optimization by simulated annealing. Science 220, 671–680 (1983)
[15] Legierski, W.: Automated Timetabling via Constraint Programming, PhD Thesis, Silesian University of Technology Gliwice (2005)
[16] Neufeld, G.A., Tartar, J.: Graph coloring conditions for the existence of solutions to the timetable problem (1974)
[17] Rich, D.C.: A smart genetic algorithm for university timetabling. In: Burke, E.K., Ross, P. (eds.) PATAT 1995. LNCS, vol. 1153, pp. 181–197. Springer, Heidelberg (1996)
[18] Schaerf, A.: A Survey of Automated Timetabling. Artificial Intelligence Review 13(2), 87–127 (1999)
[19] University Ontology, http://www.patrickgmj.net/project/university-ontology
[20] Werra, D.: An introduction to timetabling (1985)
[21] Yang, T.D., Lin, T., Wu, K.: An Agent-Based Recommender System for Lesson Plan Sequencing. In: Proceedings of 2nd ICALT, Kazan, Russia (September 2002)
[22] http://matman.uwm.edu.pl/~pdrozda/ont/university.owl

Knowledge and Data Processing
in a Process of Website Quality Evaluation

Janusz Sobecki and Dmitrij Żatuchin

Wroclaw University of Technology
Wyb. Wyspiańskiego 27, 50-370 Wrocław, Poland
{Janusz.Sobecki, Dmitrij.Zatuchin}@pwr.wroc.pl

Abstract. Different data are present in every branch of computer science, statistics, mathematics, and such, often treated as soft techniques, as human-computer interaction, usability evaluation and ergonomics. Using standard quality evaluation methods the received results obtained empirically or analytically, are not systemized and that increases the cost of output production. Website quality tests are usually characterized by a repeatability of operations, similar to the tasks defined for the offline tests with the participants. Because every person, an object of valuation, is different and his or her answers, actions and behavior during user experience tests are hard to predict. The role of analytic is to select and classify properly transcribed answers of such users and input them to a knowledge database system. We believe that it is possible to standardize and translate all data given as input and received as output during such tests into one consistent model.

Some of artificial computational intelligence methods may help and be effective to increase productiveness of processing the knowledge in a right way basing on the previously received data, by making i.e. feedback to next website analysis. In that case we propose to look closer at uncertainty management, frames and finally propose the idea of expert system enhanced by data mining and inferencing mechanism which will make possible to decide about website quality.

In this paper we generally describe quality of website interface, chosen methods of website usability evaluation, the data obtained during evaluation process and prototypes for organizing data in frames for further use in an expert system.

Keywords: computational intelligence, data, frames, knowledge, testing, usability, websites.

1 Introduction

Quality of system's service should be understood as general set of properties and features of service which decide about ability of product to fulfill needs. Interface is a shell via which users have access to the services of i.e. banking or educational system. It has also its own quality which is described as value dependent of usability and

N.T. Nguyen et al. (Eds.): New Challenges in Compu. Collective Intelligence, SCI 244, pp. 51–61.
springerlink.com   © Springer-Verlag Berlin Heidelberg 2009

accessibility of interface. Usability in the forward will be understood as ease of use during process of acquirement of system's service through its interface.

Usability is also the key element of the Human-Centered-Design [9] process. Context of use prejudges the characteristics of the system, and the tasks that will be performed during receiving system's service have a significant influence on the architecture of the product during software production process. Figure 1 presents the life cycle of the system design in Human-Centered-Design approach and attention should be paid at the evaluation domain. The system and in particular interface are linked with tasks, classes of users and contexts of use. All of that results in using standards and experts' opinions as a source of correct knowledge of interface design process. For example, such information is given at ISO 9241, ISO 9001 or ISO 13407 standards, W3C WAI (Web Accessibility Initiative) [10], or design guidelines such as the Sun Java Look & Feel, Gnome 2.0, Apple or MS Windows Guidelines.

Website interface quality valuation in evaluation domain is used to verify the following elements of system visible to users:

- structure of the page (column layout, type of layout - float, static, amount of whitespace, number of links etc.);
- naming of sections, links and controls;
- distribution of information;
- logical grouping of elements (similar in a one group);
- order of correct and expected events, in a way as state machine does.

By providing the interface for user tests, designer have to validate assumptions, requirements and needs finishing with the quality rank of interface and list of errors.

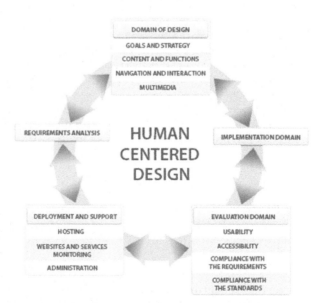

Fig. 1. Human-Centered-Design process

Website evaluation should have goals to detect such problems as: 'Why users do not visit "that" section of a website?', 'Is it easy for users to go through a registration process?' or 'Does designed website meets the needs of users?'.

After testing process a lot of different-type data is collected and then it has to be intelligently processed – by set of human experts or i.e. by expert systems with knowledge data base and rules of inference.

2 Data in a Website Quality Evaluation

2.1 Website Evaluation Methods

Website evaluation methods could be classified into: empirical testing, user oriented methods, and analytical testing. Then the empirical testing could be divided into following methods: focus groups, surveys, interviews, user tests, measurements. The user oriented methods encompass: paper prototype evaluation, experimental tests, observations and free exploration. Finally the analytical testing comprises the following methods: GOMS [7], cognitive walkthrough, heuristics verification [5] and expert analysis, simulations (referred to the automated testing with the scenarios through website's sitemap) and non-intrusive automated testing (movements, clicks and action tracking) or gaze tracking testing.

2.2 Data

Before, during and after testing process different data collections are received. They require further processing to obtain the final report [4] and quality rank.

After its digitalization we can group it by files:

- text (.txt, .doc, .xls files),
- graphics (screenshots, photos, scanned prototypes and wireframes),
- multimedia (video or audio record)

or by source:

- profile of system user:
  - o data about user (i.e. age, gender, country, used system etc.),
  - o data about interface usage (i.e. track of mouse, points of gaze fixation, spent time etc.),
- profile of interface (i.e. number of columns, position of search form, length of website, number of module),
- data about quality of system (measurements, opinions, comments, surveys).

Depending on the applied evaluation methods there is obtain the following kind of data:

- observer's notes from the evaluation with users;
- checklist's answers;

- structured data set with finite number of fields, supplemented during evaluation in a database or spreadsheet by the researcher or observer,
- pre- and post surveys' results;
- data collected automatically by hardware and/or software;
- expert analysis in form of heuristics adequacy ranks or comparisons to specimen interface;
- specified metric values (i.e. time of completing the task, number of operations done to receive service).

One of the most important area of the interest in this paper are types of data collected automatically by some hardware for example Eyetrack 6000 [1] or software i.e. GTAnaly, GoogleAnalytics [11], ergoBrowser [12], CrazyEgg [13], Noldus The Observer [14], Stat24 [15] which could have the following interpretation:

- gaze trails;
- look zones;
- contours;
- heat-maps;
- user actions;
- click and cursor maps, which present: cursor position, time stamps, the average time of completing the task, distribution of users that completed the task and reconstructed track of visit;
- parameters of the user framework (browser name, resolution, GeoIP location, source of entry and exit from website etc.).

Moreover, during automated testing with the usage of clicktracking, eyetracking or validations (i.e. TawIt, W3C), usually the response returns big portions of information, saved as the log files (Tab. 1), entries in separate databases, or cache in a browser session (client-side) or system's cache (server-side).

Table 1. Data received through gaze tracking and click tracking tools

Simplo Tracker	592	95	998	opera	1	1200958709735	82.143.157.23
	1122	238	1257	firefox	1	1200958981668	62.21.16.7
Eyetrack6000	Mozilla Firefox, F, 58, 206, 604, 20.718, 20.858, 0.140, 0,
	Mozilla Firefox, F, 60, 822, 603, 21.098, 21.358, 0.260, 0,

As above shows, the data obtained during various testing methodologies are numerous and very different, they need a special approach for its organization.

One of the possible solutions for this problem is creation of four major databases, combined with the central expert system, supporting the interface quality evaluation process [3]. These databases contain the following data: the basic specification, the test results, the potential tester candidates and users and tools to assist the evaluation. However this is not the best solution to the problem of Website evaluation and quality ranking. Building a database of web interfaces, associated user groups, multimedia models, tests and speeches associated by the relationship of many-to-many, could be more effective. The perfect system should recommend the similar reports, future us-ers' profile for new evaluation or if possible the set of errors defined on the base of previous analysis of other interfaces.

3 Framing Collected Data

By "framing" previously mentioned data we understand the decision process, which will be used to create frames with finite properties represented as objects with attrib-utes in a database and connect them with the modules of expert system used for automated interface quality evaluation method. The main task is to define criteria of framing and choosing right object properties and that is thing to be done.

On Figure 2 there is demonstration of how such expert system could be constructed.

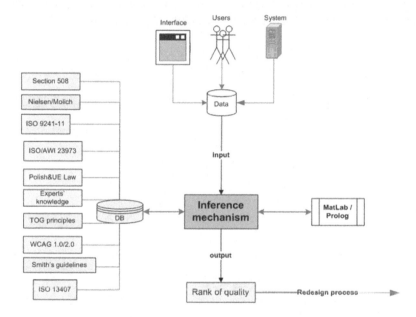

Fig. 2. Expert system for quality valuation

In *"DB"* part knowledge is gathered and processed, in *"Data"* part frames such as *Website* interface or *Person* profile are described and in *Inference mechanism* such parameters are analyzed and classified with f.e. functions of membership.

3.1 First Frame Model: Target of Usability Evaluation

To start design of computational system with the centralized database that will sup-port the quality evaluation process of the web interface, it should be defined an object model of the test itself and its participants. In such tests may participate: users, observers, experts, moderators – persons. Hence we define a class on an abstract level - *Person*, which may be inherited by a particular type of experiment sample. A *Person* may have the most general attributes of the survey participant (demographical such as personal data or age, historical such as experience of working with computer / Internet or websites which were most used in the last 24 months) and the type of test (intrusive, remote or local, in the native environment, with restrictions) or methods of evaluation which are used to collect data.

3.2 The "Person" Objects

Person object is an advanced construction and it covers a wide range of people that are involved in quality evaluation process. On Figure 3 there is shown a general model of object-relational database system to assist the quality evaluation of web pages. We distinguish the key features such as the type of person, specify the roles such as moderator, observer, tester, and an expert user, and inform about the type of cooperation - a remote, on-site with or without a computer. For the needs of example, there are listed two types - *RemoteUser_t* and *OfflineUser_t* (offline test participant).

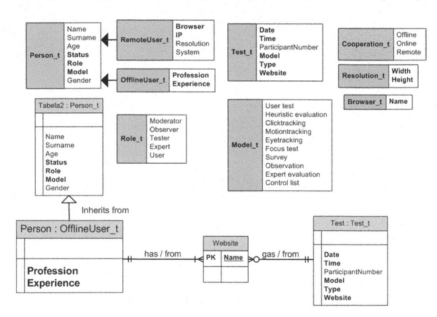

Fig. 3. Scheme represents data types and relation between basic objects

Every entity of *Model_t* will represent appropriate method of evaluation applied to a given Web interface and should be defined separately. It requires establishing a connection with other entities, including *Result* object and *Data* object, which contain data collected during the evaluation.

Test of a *Test_t* type is a general definition of model for each test, which takes place at a set time, with the certain finite number of participants assigned to one or several models of *Model_t* type and associated with the Website, which object has been defined in the next section.

3.3 The "Website" Frame

Website is a key frame in the whole model of general studies, as it is the target point of the testing process. The decision process of choosing the proper method of evaluation depends on the input characteristics of the evaluated Web service and requirements of external parties. Figure 4 presents the object-relational model of a *Web page* frame. *InternalAttributes* entity is appropriate to the architecture and characteristics of information describing the page. *ExternalAttributes* entity contains information such as results of validation of the service in terms of accessibility, code compliance with the standards and visibility in search engine agents. It can be noted in a *InternalAttributes* the column named *Architecture*, which in tests is connected to the statistics of clicks on each page or interface elements or i.e. entity of the opinion survey on distribution of information on the page.

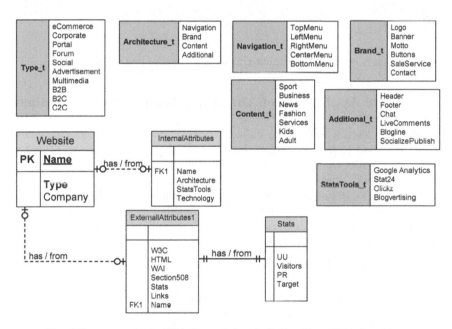

Fig. 4. Data types in the Website model and relationship of Website object

Concluding, there should be created another entity – *OtherWebsites*, which is the list of indexes of websites with the short distance of similarity function to the evaluated one.

3.4 The "Person-Test-Website" Relationship

On the Figure 5 there is presented the model of survey evaluation method, where data such as user information, user satisfaction response etc. are kept and processed. It illustrates relationship between test, participant and a website.

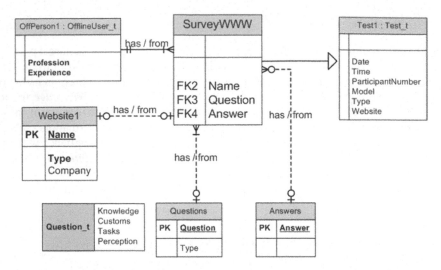

Fig. 5. Model of website survey evaluation method (SurveyWWW)

4 Proposal of Application of Automated Data Collection Methods

We propose the special tool that is called *SimploActions*, which is intermediate between the user and tested interface. It base on methods described forward and was implemented with technologies: PHP, AJAX, JavaScript and SOAP, so there is no need to interfere into users' environment. During the tests, data are collected transparently for the target users of interface, in a real time on the server side, put directly into the database, log files or cache, so the users' computers does not perform any additional operations. The collected data are available remotely to usability engineers. The *SimploActions* aggregates the following information:

- time between taking by user individual action (totally idle time spent with the interface);
- time stamps and registry of actions (clicks, moves on a website, transitions through forms, writing on the keyboard);
- navigational data (metric conversion page, entrances and exits from the page) [6].

During tests using automatic tracking or eye tracking user action are returned as the large amount of data recorded in the form of log files.

To analyze the collected data it is required to create a model of users' behavior and its possible ways of data representation in avoid unnecessary and burdensome data processing. This model may contain the following elements, previously processed by data mining methods:

- reports summarizing the statistics of visits and page impressions, the most clickable links and buttons, the time spent at the whole website and each individual subpage,
- user's actions imposed on the website screen-shots;

- video Replays;
- heat Maps;
- click Maps;
- noop points Maps.

4.1 Mouse Tracking

Remote usability tracking with the mouse is an alternative for the eye-tracking method. Usually it requires usage of sophisticated devices and also sometimes some physical interference with the human body, which could be at least inconvenient for the users. One possible solution to eye-tracking is using for example ASL devices such as head mounted optics (HMO) or Pan/Tilt (P/T) [1,2] and Head Tracker (HT). These devices together with appropriate hardware such as EyeTrack 5000 or EyeTrack 6000 and GT Analy software are able to track several parameters concerning user interaction with computer application or web interface, such as: gaze fixation points coordinates, pupil size and head position together with time stamps, cursor movements and keyboard actions. They have however two major disadvantages – they are quite expensive and more complicated than standard participation user tests.

We can intercept somehow the position of the user attention by using mouse tracking. Mouse traces are easy to obtain, they are an alternative to focus mapping and may help in creating a recommended path for user. On other site, it can be stated, that such traces may serve as recommendation itself for item and product placement on a website.

It is, however, noted that the eye-tracking is able to capture fast subconscious processes, and the mouse tracking software registers only points where users' attention was given consciously what may result in incorrect data received from the evaluation. That's why simple mouse tracking is not enough to show users' behavior on a website.

4.2 Click Tracking

Click tracking is a method inherited from the action tracking that enables tracking of all users clicks with the pointing device (mouse, touchpad, trackball, etc.). The clicking may result in the following web browser operations:

- entering a webpage;
- switching between bookmarks or other elements of a website;
- the movements of pointer in a single window;
- clicking and filling-in a form;
- clicking on a button, link or another clickable or not clickable page element;
- selecting, copying and pasting, editing;
- dragging and dropping.

Click tracking is used to track users' activity on the website. It's also used as a tool for A/B split testing. The A/B testing allows comparing the two alternative versions of some solution to verify the functionality of two competing prototypes of the system. Click tracking allows users to create a relationship of map-to-link on the website. It can reproduce the number of clicks on each link, or any other element on the page.

In this way unpopular content and locations are detected, what may bring very valuable information for interaction designers and web-architects. Click tracking could be very useful in e-commerce applications. It helps to solve the problem of interface elements placing (i.e. advertisements, product, special offers boxes etc.) and analyze the effectiveness of shopping cart.

One of possible ways of presenting data given by click tracking is a click map (single points) and heat map (cold to hot areas). Click map is a two-dimension image, which consists of two layers – the screenshot of tested system, and imposed layer of information about the clicks. Heat map [16] is a graphic representation of points compiled in some areas, remote from each other by defined constant (radius). So the area could be formed within the radius must be found an appropriate minimum number of clicks. All clicks on a given area are presented as a thermal image, the place where the highest coefficient of clicks takes the warm color, or in the case of a small number of clicks – cold.

4.3 Motion Tracking

Motion tracking is a method for recording movements and actions performed by the user on the interface. This method belongs to the class of action tracking methods and uses the mouse tracking. Recording is carried out discreetly and is carried out in very short intervals (i.e. less than 150ms).

Motion tracking can be used also for: identification of similar users, by finding those with close distance of behavior on the webpage, to identify the most common webpage behaviors and then to modify the page prototype according to these users needs.

5 Summary

The aim of this work was to show the potential for increasing the efficiency of conducting quality testing of website interfaces, especially usability tests through the use of standardized frames for object-relational database model and additional computational methods for data processing in usability evaluation.

For each method of website usability testing there is required proper establishment of a relationship of described objects. Preliminary work shows that the number of objects, their types and relationships can be very large and a database, technically, may work in complications to the process of studies, especially when users will have a little or none experience with DBMS (abbr. Database Management Systems). On the other hand, such approach will systematize the knowledge generated in the course of such evaluation, and create a knowledge repository of the files described in the database giving additional possibilities for mining of collected data. There should be an intuitive expert system designed in accordance with the methodology of human-centered design (Fig. 1), which will use such a database, and thus will facilitate the work of designers, testers and user experience experts.

We also believe that using modern gaze-tracking devices such as ASL or TOBI installed on some web kiosks with click tracking utilities and integrating it via interfaces or API such as SOAP or REST with the central usability DB system will make possible to do a step forward into automation of quality evaluation process.

References

1. ASL (Applied Science Laboratories), Eye Tracking System Instructions ASL Eye-Trac 6000 Pan/Tilt Optics, EyeTracPanTiltManual.pdf, ASL Version 1.04 01/17/2006
2. ASL (Applied Science Laboratories), Eye Tracking System Instruction Manual Model 501 Head Mounted Optics. In: ASL 2001 (2001)
3. Butler, M., Lahti, E.: Lotus Notes Database Support for Usability Testing, Lotus Development Corporation (2003)
4. Kuniavsky, M.: Observing the User Experience – A Practitioner's Guide to User Research. Morgan Kaufmann Publishing, San Francisco (2003)
5. Nielsen, J.: Alterbox: Current issues in Web Usability (2003-2007), http://www.useit.com/alertbox, ISSN 1548-5552
6. Norman, K.L.: Levels of Automation and User Participation in Usability Testing (2004)
7. Raskin, J.: The Human Interface. New Directions for Designing Interactive Systems. Addison-Wesley, Boston (2000)
8. Traub, P.: Optimising human factors integration in system design. Engineering Management Journal Publication 6, 93–98 (1996)
9. New Standards in Usability Downloaded, Userfocus ltd. (downloaded, January 2009), http://www.userfocus.co.uk/articles/ISO9241_update.html
10. W3C Recommendation, WCAG 2.0 Guidelines, W3C Organization (December 11, 2008), http://www.w3.org/TR/WCAG20/ (downloaded January 2009)
11. Google Analytics Software, Google Inc., http://www.google.com/analytics/
12. ergoBrowser, ergosoft laboratories, http://www.ergolabs.com/resources.htm
13. Crazyegg software, Crazy Egg Inc., http://crazyegg.com/demo
14. The Observer XT, Noldus Information Technology, http://www.noldus.com/human-behavior-research/products/the-observer-xt
15. Clickmap tool, stat24, http://stat24.com/en/statistics/clickmap/
16. gemius HeatMap Explained, GEMIUS SA (downloaded, June 2009), http://www.gemius.pl/pl/products_heatmap_about
17. Welcome to Eyetrack III, A project of the Poynter Institute, http://www.poynterextra.org/eyetrack2004/about.htm
18. ISO 13407, Human Centered Design, UsabilityNet (downloaded, May 2009), http://www.usabilitynet.org/tools/13407stds.htm

Part II
Ontology Management and Applications

Syntactic Modular Decomposition of Large Ontologies with Relational Database

Pawel Kaplanski

Gdansk University of Technology, Samsung Poland R&D Center

Abstract. Support for modularity allows complex ontologies to be separated into smaller pieces (modules) that are easier to maintain and compute. Instead of considering the entire complex ontology, users may benefit more by starting from a problem-specific set of concepts (signature of problem) from the ontology and exploring its surrounding logical modules. Additionally, an ontology modularization mechanism allows for the splitting up of ontologies into modules that can be processed by isolated separate instances of the inference engine. The relational algorithm, described in this paper, makes it possible to construct an inference engine that can run in a highly scalable cloud computing environment, or on a computer grid.

1 Introduction

Gaining knowledge in the world around us starts with specifying its elements (objects, beings, instances). Objects can be ascribed some properties, analyzed in terms of relationships and grouped in classes/sets with certain properties. Description logic (DL) [6,8,9,7] facilitates modeling of such relationships because it represents the knowledge of its domain – "world" – by defining given concepts within one domain (its terminology) so that later, by using these concepts it can describe object properties – concept instances – occurring in the world description. Formally DL is a well behaved fragment of first order logic (FOL) equipped with decidable reasoning tasks. The domain of interest is modeled by means of concepts (which denote classes of objects) and relationships. The main components, that forms DL are: a description language, a knowledge specification mechanism and a set of automated reasoning tasks.

The description language specifies how to construct complex concept and relational expressions, by starting from a set of atomic symbols and by applying suitable constructors. The difference between description logic languages concerns the grammar constructors that are allowed.

The knowledge specification mechanism specifies how to construct ontology, in which properties of concepts and relationships are asserted. Ontology, described by means of description logic is divided into three parts: TBox (Terminology Box) - describing concept terminology, ABox (Assertion Box) describing assumptions about named instances and RBox (Relationship Box) – describing terminology of relations.

The set of automated reasoning tasks provided by the DL implement algorithms of automatic knowledge discovery which were proved to be computable (they always complete the task). One of the possible inferences is concept classification, that is, a

N.T. Nguyen et al. (Eds.): New Challenges in Compu. Collective Intelligence, SCI 244, pp. 65–72.
springerlink.com

hierarchical arrangement of concepts within the notion of subsumption operator. Another one is classification of instances to certain concepts.

As the size and complexity of ontologies continues to grow in the current fractured manner, methods of breaking them down into smaller pieces become more and more important. Support for modularity allows complex ontologies to be separated into smaller pieces (modules) that are easier to maintain and compute. Instead of considering the entire complex ontology, users may benefit more by starting from a problem-specific set of concepts (signature of problem) from the ontology and exploring its surrounding logical modules. Additionally, an ontology modularization mechanism allows one to split up ontologies into modules that can be processed by isolated separate instances of the inference engine.

In this paper I first introduce the syntactic modularity algorithm introduced by Bernardo Cuenca Grau et al.[1], then I introduce a version of the algorithm developed by myself, that can be applied to a relational database. A relational algorithm makes it possible to construct a inference engine that can run in a highly scalable cloud computing environment, or on a computer grid.

2 Related Work

In opposition to the idea of knowledge sharing that initiated the utilization of ontologies, maintenance and usage of ontologies is completely centralized. Large ontologies are engineered by teams using tools (i.e. Protege [11]) with no support for modularization apart from the import construct that copies all axioms of one ontology into another. There are tools for ontology development and a couple of reasoners [12,13] available but approaches to modular ontologies are still in the fledgling stages. However syntactic locality approach was introduced in most important description logic dialects $\mathcal{SHOIQ}$ [1,2] and $\mathcal{EL}+$ [3] there is still a lack of system tools, for very large ontologies, which can deal with such dialects and make use from the syntactic modularity concept. This paper tries to bridge this gap by introducing a version of syntanctic locality based algorithm which can be implemented on a relational database.

3 Preliminaries

The notion of syntactic locality [1], which can be checked in polynomial time for $\mathcal{SHOIQ}$ ontologies is the core idea for a modularization algorithm.

Definition 1. *(Syntactic Locality for $\mathcal{SHOIQ}$).*

Let $\mathcal{S}$ be a DL signature (signature is the disjointed union of the sets of concept names, instance names and role names). The following grammar recursively defines two sets of concepts $\mathcal{C}_{\mathcal{S}}^{\top}$, $\mathcal{C}_{\mathcal{S}}^{\perp}$ and set of roles $\mathcal{R}_{\mathcal{S}}^{\perp}$ for a signature $\mathcal{S}$:

$$\mathcal{C}_{\mathcal{S}}^{\perp} ::= A_{C}^{\perp}|(\neg C^{\top})|(C \sqcap C^{\perp})|(\exists R^{\perp}.C)|(\exists R.C^{\perp})|(\geqslant nR^{\perp}.C)|(\geqslant nR.C^{\perp})$$
$$\mathcal{C}_{\mathcal{S}}^{\top} ::= (\neg C^{\perp})|(C_{1}^{\top} \sqcap C_{2}^{\top})$$
$$\mathcal{R}_{\mathcal{S}}^{\perp} ::= A_{R}^{\perp}|(R^{\perp})^{-}$$

where R is a role, and C is a concept, $A_C^\perp \notin S$ is a atomic concept, $A_R^\perp \notin S$ is a atomic role, $C^\perp \in C_S^\perp$, $C_1^\top, C_2^\top \in C_S^\top$ and $R^\perp \in R_S^\perp$.

An axiom is syntactically local w.r.t. S if it is of one of the following forms:

$$\text{role subsumption} \longrightarrow R^\perp \sqsubseteq R \tag{1}$$
$$\text{transitive role} \longrightarrow R^\perp \circ R^\perp \sqsubseteq R^\perp \tag{2}$$
$$\text{functional role} \longrightarrow \top \sqsubseteq \leqslant 1 R^\perp . \top \tag{3}$$
$$\text{concept subsumption} \longrightarrow C^\perp \sqsubseteq C \tag{4}$$
$$\text{concept subsumption} \longrightarrow C \sqsubseteq C^\top \tag{5}$$
$$\text{instance of concept} \longrightarrow C^\top(a) \tag{6}$$

A $\mathcal{SHOIQ}$ ontology $\mathcal{O}$ is syntactically local w.r.t. S if all its axioms are syntactically local w.r.t. S.

A module of an ontology $\mathcal{O}$ is a subset $\mathcal{O}' \subseteq \mathcal{O}$ that preserves an axiom of interest or the axioms over a signature of interest. Using Locality Condition we can define a module as follows [1]:

Definition 2. *(Modules based on Locality Condition).*

Given an ontology $\mathcal{Q}$ and a signature S, we say that $\mathcal{Q}_S^{loc} \subseteq \mathcal{Q}$ is a locality-based S-module in $\mathcal{Q}$ if $\mathcal{Q} \setminus \mathcal{Q}_S^{loc}$ is local w.r.t $S \cup Sig(\mathcal{Q}_S^{loc})$.

4   The Algorithm

The modularization algorithm presented in this paper deals with ontologies stored in a relational database and therefore can be implemented in an effective way on any database system that is able to handle SQL queries. It can be proven that this algorithm is equivalent to a modularization algorithm based on syntactic locality condition. Instead of counting the syntactic locality condition for each DL expression, the algorithm uses a Relational locality function.

Definition 3. *(Relational locality function for $\mathcal{SHOIQ}$)*

The following grammar recursively defines two functions of value from relations of atomic concepts $G_\perp$ and $G_\top$ with domain of $\mathcal{SHOIQ}$ expression and value of one-column relation [5] of sets (eventually empty) of atomic concepts $\begin{bmatrix} s \\ \{A_1, A_2, ...\} \\ \{B_1, B_2, ...\} \\ ... \end{bmatrix}$:

$$G_\perp(A^C) = \begin{bmatrix} s \\ \{A^C\} \end{bmatrix} \tag{7}$$
$$G_\perp(\neg C) = G_\top(C) \tag{8}$$
$$G_\perp(C \sqcap D) = G_\perp(C) \cup G_\perp(D) \tag{9}$$
$$G_\perp(\exists R.C) = G_\perp(R) \cup G_\perp(C) \tag{10}$$
$$G_\perp(\geqslant n R.C) = G_\perp(R) \cup G_\perp(C) \tag{11}$$
$$G_\top(\neg C) = G_\perp(C) \tag{12}$$

$$G_\top(C \sqcap D) = \delta_{p \cup q \to s} \pi_{p \cup q} \left(\delta_{s \to p} G_\top(C)\right) \bowtie \left(\delta_{s \to q} G_\top(D)\right) \tag{13}$$

$$G_\perp(R^-) = G_\perp(R) \tag{14}$$

$$G_\perp(A^R) = \begin{bmatrix} s \\ \{A^R\} \end{bmatrix} \tag{15}$$

$$G_\top(otherwise) = G_\perp(otherwise) = \emptyset \tag{16}$$

where R is a role, and C, D are a concepts, A^C is a atomic concept and A^R is a atomic role. Relational locality function $L(.)$ is defined as:

$$L(R \sqsubseteq S) = G_\perp(R)$$
$$L(R \circ R \sqsubseteq R) = G_\perp(R)$$
$$L(\top \sqsubseteq\, \leqslant 1R.\top) = G_\perp(R)$$
$$L(C \sqsubseteq D) = G_\perp(C) \cup G_\top(D)$$
$$L(C(a)) = G_\top(C)$$
$$L(otherwise) = \emptyset$$

Example 1. For $\mathcal{SHOIQ}$ expression $A \sqcap B \sqsubseteq \neg C \sqcap \neg D$ we have:

$$L(A \sqcap B \sqsubseteq \neg C \sqcap \neg D) = G_\perp(A \sqcap B) \cup G_\top(\neg C \sqcap \neg D) =$$

$$= G_\perp(A) \cup G_\perp(B) \cup \delta_{p \cup q \to s} \pi_{p \cup q} \left(\delta_{s \to p} G_\top(\neg C)\right) \bowtie \left(\delta_{s \to q} G_\top(\neg D)\right) =$$

$$= \begin{bmatrix} s \\ \{A\} \end{bmatrix} \cup \begin{bmatrix} s \\ \{B\} \end{bmatrix} \cup \delta_{p \cup q \to s} \pi_{p \cup q} \left(\delta_{s \to p} \begin{bmatrix} s \\ \{C\} \end{bmatrix}\right) \bowtie \left(\delta_{s \to q} \begin{bmatrix} s \\ \{D\} \end{bmatrix}\right) =$$

$$= \begin{bmatrix} s \\ \{A\} \\ \{B\} \end{bmatrix} \cup \delta_{p \cup q \to s} \pi_{p \cup q} \begin{bmatrix} p \\ \{C\} \end{bmatrix} \bowtie \begin{bmatrix} q \\ \{D\} \end{bmatrix} = \begin{bmatrix} s \\ \{A\} \\ \{B\} \end{bmatrix} \cup \delta_{p \cup q \to s} \pi_{p \cup q} \begin{bmatrix} p & q \\ \{C\} & \{D\} \end{bmatrix} =$$

$$= \begin{bmatrix} s \\ \{A\} \\ \{B\} \end{bmatrix} \cup \delta_{p \cup q \to s} \begin{bmatrix} p \cup q \\ \{C, D\} \end{bmatrix} = \begin{bmatrix} s \\ \{A\} \\ \{B\} \end{bmatrix} \cup \begin{bmatrix} s \\ \{C, D\} \end{bmatrix} = \begin{bmatrix} s \\ \{A\} \\ \{B\} \\ \{C, D\} \end{bmatrix}$$

Proposition 1. *(Locality relation function and syntactic locality)*

An axiom X is syntactically local w.r.t. $\mathcal{S} \iff \exists_{s \in L(X)} s \cap \mathcal{S} = \emptyset$.

Proof. Directly from properties of relations:

$$\exists_{s \in X \cup Y} s \cap \mathcal{S} = \emptyset \iff \exists_{s \in X} s \cap \mathcal{S} = \emptyset \vee \exists_{s \in Y} s \cap \mathcal{S} = \emptyset$$

$$\exists_{s \in \delta_{p \cup q \to s} \pi_{p \cup q}(\delta_{s \to p} X) \bowtie (\delta_{s \to q} Y)} s \cap \mathcal{S} = \emptyset \iff \exists_{s \in X} s \cap \mathcal{S} = \emptyset \wedge \exists_{s \in Y} s \cap \mathcal{S} = \emptyset$$

and from definitions1,3.

Proposition 2. *(Modules based on Relational locality function)*

$\mathcal{Q}_1 \subseteq \mathcal{Q}$ is a locality-based $\mathcal{S}$-module in $\mathcal{Q}$ if

$$\forall_{X \in \mathcal{Q} \setminus \mathcal{Q}_\mathcal{S}^{loc}} \exists_{s \in L(X)} s \cap \left(\mathcal{S} \cup Sig(\mathcal{Q}_\mathcal{S}^{loc})\right) = \emptyset$$

Proof. Directly from proposition 1 and definition 2.

Proposition 3. *(Algorithm for syntactic modularization of $\mathcal{SHOIQ}$ ontologies based on Relational locality function)*

Algorithm is based on relations $\mathbb{C}$, $\mathbb{T}$ and $\mathbb{E}$. The relationship between expression and symbols it contains $sort(\mathbb{C}) = \{e, a\}$, the relationship between the expression and components of its Relational locality function $sort(\mathbb{T}) = \{e, s\}$ and the relationship between components of the Relational locality function and contained symbols $sort(\mathbb{E}) = \{s, a\}$:

Algorithm 1. GetModuleFromSignature

Input:
   $\mathcal{Q}$: ontology
   $\mathcal{S}$: signature
Output:
   $\mathcal{Q}_\mathcal{S}^{loc}$: a locality based $\mathcal{S}$-module in $\mathcal{Q}$

1: $S_1 = \bigcup_{a \in S} \{a\}$
2: $\mathbb{C} = \bigcup_{e \in \mathcal{Q}} \bigcup_{a \in Sig(e)} \{e, a\}$
3: $\mathbb{T} = \bigcup_{e \in \mathcal{Q}} \bigcup_{s \in L(e)} \{e, s\}$
4: $\mathbb{E} = \bigcup_{e \in \mathcal{Q}} \bigcup_{s \in L(e)} \bigcup_{a \in s} \{s, a\}$
5: $\mathcal{Q}_\mathcal{S}^{loc} = \bigcup_{e \in \mathcal{Q} \wedge L(e) = \emptyset} \{e\}$
6: $I_1 = \delta_{count(e) \to t} \pi_{e;count(e)} \mathbb{T}$
7: **loop**
8:    $S_2 = S_1 \cup \pi_a(\mathbb{C} \bowtie \mathcal{Q}_\mathcal{S}^{loc})$
9:    $T_1 = \pi_{e,s} \mathbb{T} \bowtie \mathbb{E} \bowtie S_2$
10:    $I_2 = \delta_{count(e) \to f} \pi_{e;count(e)} T_1$
11:    $I_3 = \pi_e \sigma_{f=t}(I_1 \bowtie I_2)$
12:    $I_4 = I_3 - \mathcal{Q}_\mathcal{S}^{loc}$
13:    **if** $I_4 = \emptyset$ **then**
14:       **return** $\mathcal{Q}_1$
15:    **end if**
16:    $\mathcal{Q}_\mathcal{S}^{loc} := \mathcal{Q}_\mathcal{S}^{loc} \cup I_4$
17: **end loop**

The algorithm applies to any ontology $\mathcal{Q}$ with a notion of entailment of signature $\mathcal{S}$ and locality-based $\mathcal{S}$-module $\mathcal{Q}_\mathcal{S}^{loc}$.

The program (line 1) converts a signature into corresponding relation [5] S_1 (line 1). There is a relationship (line 2) between each expression in ontology $\mathcal{Q}$ and its signature. This relationship is stored in relation $\mathbb{C}$. In other words, $\mathbb{C}$ is the table with two columns e and a, and each row denotes occurrence of specific atomic concept in expression e.

A relation $\mathbb{T}$ (line 3) ia given between expression e and locality relation L - given by a set comprised of atomic symbols. Expressions are chosen for each row where $s = L(e)$ set exists into table containing two columns e and s. It is worth noting that, domain of s is set of atomic concepts.

Relation $\mathbb{E}$ (line 4) establishes connection between relation sets s and contained atomic concepts a. Output ontology module $\mathcal{Q}_\mathcal{S}^{loc}$ is initially filled (at line 5) with all expressions that have global property ($L(e) = \emptyset$).

Relation I_1 is given by aggregated projection π and renaming δ operators applied to relation $\mathbb{T}$. Corresponding pseudo-SQL syntax can be written as:

```
I1 := select distinct e, count(e) as t from T group by e
```

The main loop of the algorithm starts (line 8) with computation of $S \cup Sig(Q_S^{loc})$ placed in S_2 using the natural join between Q_S^{loc} and $\mathbb{C}$ and then projection on a column and then summing it with signature (computed in line 1). Corresponding SQL:

```
S2 := S1 union (
          select distinct a
              from C
              join QlocS on C.e=QlocS.e
       )
```

Next step (line 9) is inner join of $\mathbb{T}$ and $\mathbb{E}$ tables with S_2.

```
T1 := select distinct e, s
          from T
          join E on T.s=E.s
          join S2 on E.a=S2.a
```

Then second time but on T_1

```
I2 := select distinct e, count(e) as t from T1 group by e
```

Finally (line 11) I_1 is joined with I_2 and only rows are selected that have $\forall_{e \in I_3} |L(e)| = |s \in L(e) : s \cap (S \cup Sig(Q_S^{loc}))|$ what is possible iff $\forall_{e \in I_3} \forall_{s \in L(e)} s \cap (S \cup Sig(Q_S^{loc})) \neq \emptyset$ so assuming closed world assumption we obtain $\forall_{e \in Q/I_3} \exists_{s \in L(e)} s \cap (S \cup Sig(Q_S^{loc})) = \emptyset$.

```
I3 := select distinct e
          from I1
          join I2 on I1.e=I2.e
          where I1.t=I2.f
```

```
I4 := select e from I3 where e not in QlocS
```

Iff $I_3 = Q_S^{loc}$ algorithm stops satisfying $\forall_{e \in Q/Q_S^{loc}} \exists_{s \in L(e)} s \cap (S \cup Sig(Q_S^{loc})) = \emptyset$ so Q_S^{loc} is locality-based S-module according to the definition.

The locality-based S-module Q_S^{loc} grows in each loop or the algorithm stops (in pessimistic scenario until $Q_S^{loc} = Q$).

5 Discussion

However, the algorithm is easy to understand and implement, although special situations do arise where it might be desirable to implement a modified version of the algorithm rather than work with a standard version.

The simple modification of the algorithm resulting in the modification of the L(.) function can be achieved by extending it with cases:

$$L(C(a)) = G_\perp(a) \cup G_\top(C)$$
$$L(R(a,b)) = \delta_{p\cup q \to s}\pi_{p\cup q}\left(\delta_{s\to p}G_\perp(a)\right) \bowtie \left(\delta_{s\to q}G_\perp(b)\right)$$
$$G_\perp(A^I) = \begin{bmatrix} s \\ \{A^I\} \end{bmatrix}$$

(where A^I is instance name). It is sufficient to ensure better granularity of modularization especially for ontologies with a large number of instances. This modification corresponds to extension of syntactic locality criterion (definition 1) by cases for instance of concept $\longrightarrow C(a_\perp)$ and relation between instances $\longrightarrow R(a_\perp, b_\perp)$ where $a_\perp, b_\perp \notin S$ are instance names. Unfortunately such extension results in wrong modularization due to the fact that instance of concept has a global property:

Proposition 4. *(Assertions Cannot be Local) [2]*

For every assertion $C(a)$ there exists a syntactically local TBox $\mathcal{T}$ such that $\mathcal{T} \cup \{a\}$ is inconsistent.

In other words: instances cannot be threaded as subconcepts because they imply implicit facts that can have global property [2].

6 Conclusion and Future Work

The algorithm described in this article has been implemented, by author of this article, as a part of Ontorion Knowledge Server [10]. Ontorion Knowledge Server is a project that tries to bridge the gap in modern ontology tools, by introducing a implementation of syntanctic locality based algorithm which described in this paper. This implementation uses a relational database. It enables inspecting, browsing, codifying and modifying ontologies and therefore supports the ontology development and maintenance tasks. The current implementation promises to build a scalable knowledge inference technology for the Semantic Web based on cloud computing or a computer grid.

References

1. Grau, B.C., Horrocks, I., Kazakov, Y., Sattler, U.: Modular Reuse of Ontologies: Theory and Practice. Journal of Artificial Intelligence Research 31, 273–318 (2008)
2. Grau, B.C., Horrocks, I., Kazakov, Y., Sattler, U.: A Logical Framework for Modularity of Ontologies
3. Suntisrivaraporn, B.: Module Extraction and Incremental Classification: A Pragmatic Approach for EL+ Ontologies (2008)
4. Grau, B., Parsia, B., Sirin, E., Kalyanpur, A.: Automatic Partitioning of OWL Ontologies Using E-Connections (2005)
5. Abiteboul, S., Hull, R., Vianu, V.: Foundations of Databases. Addison-Wesley Publishing, Reading (1995)
6. Baader, F., Calvanese, D., McGuinness, D., Nardi, D., Patel-Schneider, P.: The Description Logic Handbook: Theory, implementation, and applications. Cambridge University Press, Cambridge (2003)
7. Horrocks, I., Sattler, U.: Decidability of SHIQ with Complex Role Inclusion (2003)

8. Horrocks, I., Kutz, O., Sattler, U.: The even more irresistible SROIQ (2006)
9. Horrocks, I., Kutz, O., Sattler, U.: The Even More Irresistible SROIQ
10. Ontorion Knowledge Server, http://ontorion.com
11. Noy, N., Sintek, M., Decker, S., Crubezy, M., Fergerson, R., Musen, M.: Creating semantic web contents with protege-2000 (2001)
12. Tsarkov, D., Horrocks, I.: FaCT++ Description Logic Reasoner: System Description (2006), http://www.comlab.ox.ac.uk/people/ian.horrocks/Publications/download/2006/TsHo06a.pdf
13. Baader, F., Lutz, C., Suntisrivaraporn, B.: CEL—A Polynomial-time Reasoner for Life Science Ontologies (2006)

Relational Database as an Ontology Framework

Andrzej Macioł

Faculty of Management, AGH University of Science and Technology,
ul. Gramatyka 10, 30-067 Kraków, Poland
amaciol@zarz.agh.edu.pl

Abstract. In this paper we discuss application of relational data model and attributive logic as an alternative for knowledge modelling tools which are based on
descriptive logic. Our experiences indicate that the most effective solution, capable of direct cooperation with majority of industrial information systems, which
simultaneously provides decidability, is a combination of relational model with
inference system, which utilizes attributive logic. Such solution, named Inference
with Queries (IwQ) is realized in accordance with principles of Variable Set Attributive Logic (VSAL). The use of IwQ method for formulation of knowledge
requires creation of data metamodel as well as increase in flexibility of inference
engine by introduction of so called variable formula mechanism. Such framework
is very useful, in the sense that: (i) direct access to information sources stored in
relational databases is possible, (ii) well-known and reliable SQL language can
be utilized, (iii) introduction of explicit knowledge structure and inference engine
control is possible.

Keywords: ontologies, relational databases, rule based systems, attributive logic.

1 Introduction

It is widely believed that relational databases have several disadvantages. Thus, they
are not suitable for formulation of ontologies which are required if information system
is designed not only for data storage, but for knowledge representation as well. In our
opinion above-mentioned statement is untrue. We claim that it is worthy to consider
usefulness of concept based on set theory and attributive logic for formulating universal
description of reality. There is no evidence to substantiate the claim that description
logic (OWL) or object data model (not grounded on any logic) can perfectly describe
information systems, whereas relational database model cannot.

Codd's concept, which found application in relational databases, was designed to
collate and represent data. Its rigorous assumptions and connection with mathematical
principles resulted in high data integrity. However, they limited possibility of storing
such knowledge, which reaches beyond the set of known facts. Query, regardless of
its formulation, cannot find answer to questions if there are no appropriate examples.
There is no possibility to obtain such replies as: 'I don't know' or 'There is no example, but relationship you queried for is not possible'. Moreover, there is no direct

N.T. Nguyen et al. (Eds.): New Challenges in Compu. Collective Intelligence, SCI 244, pp. 73–84.
springerlink.com   © Springer-Verlag Berlin Heidelberg 2009

querying which would allow to detect new assertions on the basis of known examples and/or their negations. Formulation of complex queries requires *a priori* knowledge on structure of relations, which limits possibility of using open tools for data (knowledge) exploration. Therefore it is necessary to store intensional data (knowledge) model by means of programming languages, which are external to database management system. Object-oriented databases and querying languages consistent with this concept were supposed to be solution for this inconvenience. Simultaneously standards of relational databases were evolving - reducing primary limitations and introducing features characteristic for object-oriented solutions (e.g. declaration of types and introduction of multidimensional structures as elements of relations). Unfortunately, presently there are no fully useful implementations of object-oriented concept. Extended standards of querying languages for relational databases are not applied in commercial management systems and do not resolve every mentioned drawback.

Designers of new concepts of knowledge management are focused on creating completely new tools for knowledge storage. These tools are grounded on description logic and create new possibilities for flexible modification of reality description. Difference between these solutions and databases (either relational or object-oriented) is not in addressing of atomic objects (in RDF by URI, in relational databases by key, and in object-oriented databases by identifier) but in the manner of saving relations (roles) between objects. As a consequence of an open approach for representation of relations, which is offered by solutions based on description logic, we gain new possibilities of data mining. Still no satisfactorily efficient tools have been created so far, and they cannot be a competition to database management systems in commercial applications. In our opinion, there is possibility to overcome deficiencies of relational databases and preserve integrity with their primary concept. Our solution is based on application of data metamodel, stored in form of relational database, as specific type of ontology. We consider usage of relational data model and attributive logic as an alternative for knowledge modelling tools based on description logic.

2   Problem Formulation

In recent years search for new features offered by computer technology became common trend. Main spur to this search is undoubtedly popularization of access to knowledge stored in digital form, accessible mainly by the Internet. Also, expectations from corporate information systems have rapidly grown. It should be connected with the concept of knowledge management. Users of information systems expect them not only to collect and present data, but manage knowledge as well. In fact we want our computer, having it connected to the Internet, to solve problems which previously required our intellectual effort. As result of these expectations two concepts were developed: Web Semantics and Business Rules Management. Seemingly irrelevant concepts can be expressed by single expectation: not only pure data but also knowledge must be available to any user of the system. At this point we should answer a question: what is knowledge and whether can it be 'freed' from human by transferring it into artificial system such as information system. The Internet is source of knowledge for typical computer user, and the only tool to access web resources are browsers. Yet, in spite of continuous development efforts, web browsers do not satisfy expectations of users. Without knowledge and experience, user cannot obtain desired information. Questions asked in natural language result in most cases with acquisition of many excess answers, often inconsistent with user's intentions. User may take advantage of built-in features, but in case of information (knowledge) which was not included, he must either ask IT specialist for help or construct more complex queries (e.g. in SQL language), with an assuming that he has appropriate skills and is authorized for such action. Hence, a question is posed: is possible to build information system which would allow users to communicate with the machine in language resembling natural one? In other words, can modern information systems pass the Turing test? Unfortunately not, and there are no signs that in the nearest future any artificial system could pass such test. Obviously, it does not mean that development works related to introduction of 'intelligence' mechanisms in web and corporate solutions are groundless. However, we should be aware of the fact that independently on established concept of knowledge modelling, there are boundaries which should not be pushed. In our opinion, introduction of ontology concept as a new quality in realization of information systems is a mistake. Tools based on descriptive logic have some advantages with regard to traditional solutions, but are not a breakthrough and should be treated as an alternative, not as the only correct method of knowledge systems modelling.

Independently on concept of knowledge modelling, efficient 'intelligent' system must consist of the following three subsystems:

- static data container,
- information search mechanism,
- mechanism of inference and inference control.

In case of the first subsystem, we can consider storage of data in form of relational database (no wide scope of application of object-oriented databases has been found) or in form of XML-related languages. In relational databases the best known tool which allows realization of the second subsystem is SQL, whereas in case of XML or languages such as RDF, RDFS or OWL-D1 the case is more complex. The simplest solutions like XQL, Xpath, Xlink and XQuery provide searching resources over the Web, yet they have many restrictions. Hybrid systems have thus been proposed, which are constituted of two or more subsystems, each of which deals with a distinct portion of the knowledge base and uses specific representation formalisms and reasoning procedures. The improvement in the deductive power of hybrid systems is in terms of both the inferences the system is able to make, and the efficiency of the reasoning process, since any subsystem can take advantage of the inferential power of the other subsystems, whereas the use of specialized reasoning procedures. Example of such solution could be r-hybrid KB [1] which has a structural component (ontology) and a rule component. Among other similar solutions, some may be listed: the first formal proposal for the integration of Description Logics and rules - AL-log [2] or Conceptual Logic Programming (CLP) [3], an extension of answer set programming (i.e., Datalog) towards infinite domains. All these solutions are classified between querying sub-system and inference subsystem. That gave a rise to their advantage in contrary to solutions founded on relational databases and SQL.

In the case when SQL is used as querying mechanism, creation of separate inference subsystem is necessary. It results from declarativity of querying languages, i.e. their orientation towards direct formulation of search target rather than means leading to the result. In classical programming languages programmer formulates target by introducing sequence of computer commands, whose execution is supposed to return the result. Querying languages differ from such idea - processing target is formulated directly by the query, and computer commands are chosen automatically by querying processor. Concept of querying languages assumed (by definition) lack of algorithmic universality. Since such universality is required in database-grounded applications, an assumption was made that querying languages will become 'sub-languages' of software design environment. Hence, such environment would be developed by means of standard programming language. It entails integration of querying language with programming language in such a manner, that [4]:

- queries could be used inside applications,
- queries could be parameterized (dynamically, in any manner) by values of variables of programming language,
- query results could be processed by applications.

It turned out that differences in concepts of various languages cause significant technical difficulties in realization of such connection. However, they were not cause for negative attitude of many specialists and programmers towards the project. The main disadvantage was substantial degradation of software development environment. This degradation manifested itself in a variety of ways, which were eventually called 'impedance mismatch'. The term defines a set of adverse features accompanying formal connection between querying language (in particular SQL) and universal programming language such as C, C++, Pascal or Java. In attempt to remove this inconvenience, a concept of uniform theory known as stack-based approach was developed. This approach is based on assumption that querying languages are types of programming languages. Hence, we should employ notions and concepts which are known and effectively applied in these standard languages. Stack-based approach enables development of universal theory which is independent on specific data model. It can be applied in relational, object-oriented and object-relational databases as well as in XML repositories. Works conducted in this field resulted in development of SBQL (Stack-Based Query Language) standard [4].

Regrettably no significant, practical applications have been found yet for solutions grounded on Datalog language and SBQL standard. At the same time, vast majority of information systems used in economy, including Web solutions, are continually built with usage of relational databases. It induces some researchers to search for methods which could unify ontology (within the scope of description logic), with relational model [5], [6], [7]. It does not seem to be prospective solution.

Our experiences indicate that the most effective solution, capable of direct cooperation with majority of industrial information systems, which simultaneously provides decidability, is a combination of relational model with inference system which utilizes attributive logic. Such solution, named Inference with Queries (IwQ) [8] was developed for its application in Business Rules Management Systems. It may also constitute a basis for formulation of ontology grounded on relational database. Tool which is result of our research is a processor of rules created according to IF...THEN...ELSE formula, in which preconditions are attribute values and conclusions operate on attributes. Purpose of this paper is to present capabilities of this method and show changes which should be made in order to obtain universal solution, which would allow to solve complex decisive problems.

3 Proposed Solution

Above-mentioned IwQ tool has been developed as knowledge model and inference engine for formulation of Business Rules Management Systems. Knowledge storage is realized in accordance with principles of Variable Set Attributive Logic (VSAL) [9]. Basic statement in attributive calculus is generally of the form: $\langle attribute \rangle \, (\langle object \rangle) = \langle value \rangle$ or $A_i(o) = v$ and it has the meaning that attribute A for object o takes value v. For example, $Debt(client) = "high"$ means that the debt of a specific client is high. In case of VSAL, atomic formula definition is expanded: If $o \in O$ is a certain object, $A_i \in A$ is an attribute and $X \in V$ is a certain variable, then any expression of the form:

$$A_i(o) = X \tag{1}$$

and

$$A_i(o) \in X \tag{2}$$

is an atomic formula of VSAL where

- O - a set of object name symbols,
- A - a set of attribute names,
- V - a set of variables.

Attribute A_i is a function of the form:

$$A_i : O \rightarrow 2^{D_i} \tag{3}$$

where D is a set of attribute values names (the domains).

Constant values belonging to domain D denote values of specific attributes for given objects. Variables are used to denote the same elements in case the precise name of an element is currently unknown, unimportant, or a class of elements is to be represented.

Our language allows the use of other than equality relational symbols, i.e. $=$, $>$, $<$, etc.

The generic form of a rule in our inference engine can be presented as follows:

$rule(i) : F_1 \wedge F_2 \wedge ... \wedge F_n$
$\rightarrow$
$\quad G_1 \wedge G_2 \wedge ...G_g$
$\quad P_1 \wedge P_2 \wedge ...P_p$

$next(i)$
$else$
  $H_1 \wedge H_2 \wedge ...H_h$
  $Q_1 \wedge Q_2 \wedge ...Q_q$
  $else(j)$

where

1. F_i, G_i, H_i are in the form of basic statements:

$$A_i(o) = v \qquad (4)$$

or

$$A_i(o) = X \qquad (5)$$

in case of preconditions, $satisfaction$ of the formula is verified and in case of conclusions the $satisfaction$ is stated

2. P_i, Q_i are in form

$$X = v \qquad (6)$$

which means that value v is assigned to a certain variable X

3. $next(j)$, $else(k)$ are the specifications of control; the $next(i)$ part specifies which rule should be examined immediately after successful execution of rule i and $else(j)$ part specifies which rule should be tried in case of failure.

Variables are of great importance in attributive logic. They may serve two main purposes. Firstly, they are undefined, not known *a priori*, yet identifiable values. Secondly, variables allow to transfer values of attributes, even if these values were initially unknown. It follows from the fact that variable name is valid within the whole theory boundaries.

It is widely known that this attributive logic is omnipresent in various applications in them in relational databases. In relational databases formulae of attributive logic are used to define selection criteria for information retrieval and for the so-called $\theta - join$ operation for joining tables. An assumption was made on the basis of above statement, that in rule-based decisive system attributes as well as variables will be stored in form of relations.

Unfortunately, there is no notion of object in relational databases (similarly as in decision tables) which constitutes a problem, especially in case when we want to decide on conditions truthfulness on the base of conclusions logical value. In this case, if we confirm $satisfaction$ of rule's conclusion, we may assume that all premises of such rule are $satisfied$. If inference process had to answer what type of object such formulation concerns, without knowledge of the key we would not be able to determine it. It may be presented on such example:

we search for the most favourable material supplier for production of particular product, which may be understood as follows: find object s in set of suppliers S, which attribute $current\_level(s) = "high"$ *we apply the rule*

$rule(i) : offered\_price(s) = "low" \wedge offered\_quality(s) = "high"$
$\rightarrow$

$\quad current\_level(s) = "high"$
$\quad next(i)$

We know that neither quality, nor price always identify supplier. Therefore it may happen, that in relation or join of relations there might be more than one tuple satisfying given condition. A question arises: what type of an action should inference system take? In our solution we have made an assumption, that inference engine will examine every hypothesis which arises in such case. That behaviour is possible owing to introduction of control specifications into the rules. In a case when several objects are identified by single formula we are dealing with collection (some sort of a SQL cursor), hence it was also necessary to introduce index which points to the consecutive elements of this cursor.

Our experience shows that such formulated model, as well as application which was developed on its basis, fits perfectly for solving elementary decision problems, characteristic of Expert Systems or Business Rules Engines. By introduction of control specifications our solution has some features of programming language in logics, thereby it may find its application in agent-based simulation. In every application which has been tested until now, entire knowledge required for problem solving was stored in set of rules, relations which describe objects, attributes and variables. This model is a kind of ontology, yet limited to very narrow range of applications. Structure of the model is evidently connected with the purpose for which it was developed. Whereas when we talk about knowledge representation by means of ontology, we usually think of formal description of notions. Such description can be used in many, not *a priori* defined purposes. Obviously it is possible to store all current knowledge on given domain in form of interconnected relations, but formulation of some rules would be required in order to allow data mining. This is however not possible without knowing the purposes for which the knowledge is used. In our opinion, solution for this problem is on one hand development of data metamodel and on the other increase in flexibility of inference engine by introducing mechanism of so called variable formulas.

Concept of data metamodel comes down to partial transfer of intensional database structure to its extensional part. In primary version of IwQ method, knowledge was stored in set of relations which described attributes. These attributes were grouped in accordance with general principles so as to simplify formulation of the rules. It should be noted, that there is a possibility to store all values of attributes in one relation. Simultaneously, sets of attribute names and their relationships can be stored in two connected relations. Scheme of such ontology model is depicted on Fig. 1. Ontology, owing to its specific structure, may be presented in form of directed graph, where vertices represent formulas of attributive logic, and edges depict relations between formulas.

Essence of the structure might be analyzed on the following example which shows knowledge required for early-warning system for predicting bankruptcy [10]. Exemplary rules used for inferencing on bankruptcy warning can be written in the following form:

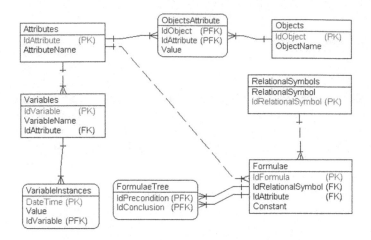

Fig. 1. Entity-Relationship Diagram of ontology model

Rule: 1
*if capital structure is bad **and** profitability is medium **and** debt is high **then** bankruptcy probability is high*

Rule: 2
*if the share of profit in incomes is between 9% and 12% **and** profitability of the assets is between 10% and 13% **then** profitability is medium*

Rule: 3
*if the share of own capital in assets is less then 10% **and** the share of short-term capital is grater then 50% **then** capital structure is bad*

Rule: 4
*if debt in years is grater then 12 **then** the debt is high*

These rules will be recorded in the following tuples:

Attributes	
IdAttribute	**AttributeName**
1	capital structure
2	profitability
3	debt
4	bankruptcy probability
5	share of profit in incomes
6	profitability of the assets
7	share of own capital in assets
8	share of short-term capital
9	debt in years

Formulae

IdFormula	IdAttribute	RelationalSymbol	Constant
1	1	=	bad
2	2	=	medium
3	3	=	high
4	4	=	high
5	5	>=	9%
6	5	<=	12%
7	6	>=	10%
8	6	<=	13%
9	7	<	10%
10	8	>	50%
11	9	>	12%

FormulaeTree	
IdPrecondition	IdConclusion
1	4
2	4
3	4
5	2
6	2
7	2
8	2
9	1
10	1
11	3

Presented above relationships can be shown in form of graph (Fig. 2).

It should be noted that these formulas can be also used for completely different purposes. It is possible owing to the fact, that association between formulas is characterized by many-to-many relationship. Thus, the same formula may become a premise for multiple conclusions.

As mentioned before, redevelopment of inference engine was necessary in order to allow exploration of such ontology. Variable formula was used for formulation of rules. This formula takes value of basic statement, which is determined on the basis of relation representing ontology. Therefore an additional operation is included in the conclusion block, which assigns value to variable formula. Simultaneously an additional element is introduced into inference engine - stack of not verified formulae.

Inference procedure might be depicted on the scheme (Fig. 3).

It should be stressed here, that knowledge on $satisfaction$ of basic statement can be either introduced by user or obtained from external database. Then, values of attributes and variables (if only they are present in formula) are determined by means of SQL query. Formulation and execution of such queries is made in accordance with principles of IwQ method.

Presently, a backward reasoning mechanism is being developed. Nonetheless, usage of forward reasoning is possible, if mechanisms similar to those above-mentioned are implemented.

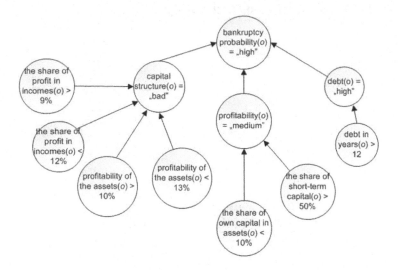

Fig. 2. Graph of knowledge (*o* is an object

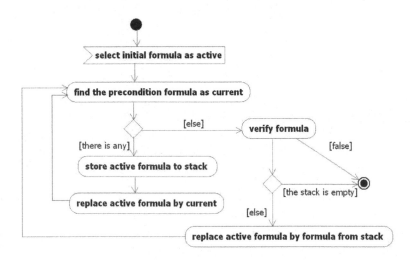

Fig. 3. Inference procedure scheme

4   Conclusions

In recent years, research in the field of knowledge management systems is focused on resolving inconveniences which are related to representation of knowledge by a single language. The trend is to search for tools which provide integrity of data and rules. Main stream of these research activities is connected with application of description logics for that purpose. Difficulties of this approach arise from the fact, that the great majority of information kept in information systems of organizations is stored in form of databases (usually relational). Thus, there is a necessity to transform such knowledge. It becomes an additional load for computer systems and may lead to data loss and/or its misrepresentation.

In this paper we have proposed formulation of relational data metamodel, which could become a rival solution to ontologies developed on the basis of description logics. We have additionally presented inference engine which is an expansion of IwQ method. Such mechanism facilitates flexible exploration of knowledge which is stored in relational database.

Among other advantages our solution provides:

- direct access to information sources stored in relational databases,
- utilization of well-known and reliable SQL solutions,
- transactional access to the ontology provided "for free" by most relational database,
- possibility to introduce explicit knowledge structure and inference engine control which significantly facilitates inference process.

Such features were achieved owing to application of mechanisms grounded on principles of attributive logic.

The aim of our further research is to develop tools which will allow to realize concept of Business Rules Management Systems. Our previous experiences indicate, that concepts elaborated within the research of that field can be effectively applied in other solutions related to knowledge management as well as formulation and utilization of ontologies.

Acknowledgments

This research has been partially supported by the Innovative Economy Operational Programme EU-founded project (UDA-POIG.01.03.01-12-163/08-00).

References

1. Rosati, R.: On the decidability and complexity of integrating ontologies and rules. J. Web. Semant. 3, 61–73 (2005)
2. Donini, F.M., Lenzerini, M., Nardi, D., Schaerf, A.: AL-log: Integrating Datalog and description logics, J. Intell. Inform. Syst. 10(3), 227–252 (1998)
3. Heymans, S., Nieuwenborgh, D.V., Vermeir, D.: Semantic web reasoning with conceptual logic programs. In: Antoniou, G., Boley, H. (eds.) RuleML 2004. LNCS, vol. 3323, pp. 113–127. Springer, Heidelberg (2004)
4. Adamus, R., Habela, P., Kaczmarski, K., Lentner, M., Stencel, K., Subieta, K.: Stack-Based Architecture and Stack-Based Query Language. In: Proceedings of the First International Conference on Object Databases, ICOODB 2008, pp. 77–95 (2008)
5. Chang, E., Das, S., Eadon, G., Srinivasan, J.: An efficient SQL-based RDF quering scheme. In: Proceedings of the 31st international conference on Very large databases, pp. 1117–1127 (2005)
6. Goczyła, K., Waloszek, W., Zawadzka, T., Zawadzki, M.: Design World Closures of Knowledge-based System Engineering. In: Software engeneering: evolution and emerging technologies, pp. 271–282. IOS Press, Amsterdam (2005)

7. Wardziak, T.: Orłowska M. E., Semantically Sensitive Execution of Relational Queries. In: Schlender, B., Frielinghaus, W. (eds.) GI-Fachtagung 1974. LNCS, vol. 7, pp. 436–446. Springer, Heidelberg (1974)
8. Macioł, A.: An application of rule-based tool in attributive logic for business rules modeling. Expert Systems with Applications, Expert Syst. Appl. 34, 1825–1836 (2008)
9. Ligeza, A.: Logical Foundations for Rule-Based Systems. Springer, Heidelberg (2006)
10. Zmijewski, M.: Methodological issues related to the estimation of financial distress prediction models. Journal of Accounting Research 22(1), 59–82 (1984)

The Knowledge Generation about an Enterprise in the KBS-AE (Knowledge-Based System - Acts of Explanation)

Jan Andreasik

Zamość University of Management and Administration
Akademicka Str. 4
22-400 Zamość, Poland

Abstract. In the paper, we present the main elements of the KBS-AE (Knowledge-Based System - Acts of Explanation) oriented toward the knowledge generation about enterprise condition. This knowledge is represented by the relation $R : \langle STRATEGY \rangle \rightarrow \langle ENTERPRISE\ CONDITION \rangle$. The strategy is expressed in terms of the competence system. The concept apparatus is expressed by enterprise ontology worked out by the author on the basis of Roman Ingarden's theory of individual object. Procedures of information extraction and creation of acts of explanation are presented by means of the UML diagrams. Acts of explanation generate the knowledge about an enterprise concerning interpretation of competence potential assessment and competence gap assessment and also result trajectories.

1 Introduction

The basic task of a system with the knowledge base is to gather data, information and relationship resources defined on these resources in order to generate the new knowledge according to appropriate action. Such a transformation from information resources to the newly created knowledge is realized according to the 4C framework [1]:

- **Comparison**: how does information about this situation compare to other situations we have known?
- **Consequences**: what implications does the information have for decisions and actions?
- **Connections**: how does this bit of knowledge relate to others?
- **Conversation**: what do other people think about this information?

The knowledge about enterprise condition is the knowledge required to make a strategic decision concerning investments, transactions, contracts. Searching for the relation (R) between strategy elements and enterprise state parameters is essence of knowledge discovering about enterprise condition:

$$R : \langle STRATEGY \rangle \rightarrow \langle ENTERPRISE\ CONDITION \rangle.$$

N.T. Nguyen et al. (Eds.): New Challenges in Compu. Collective Intelligence, SCI 244, pp. 85–94.
springerlink.com

The Altman's model is the oldest model of knowledge discovering about an enterprise [2]. This model determines a relationship between the financial strategy of an enterprise and two distinguished states: the bankruptcy state and the good condition state. Until now, there are created such models with the use of statistical methods, data mining methods, artificial intelligence techniques based on fuzzy set theory, rough set theory, neural networks, multicriteria decision making methods [3,4]. The financial strategy is determined by the set of financial indexes calculated on the basis of the year reports: a profit and loss account and a balance. The conception of intellectual capital introduced by L. Edvinsson and M.S. Malone [5] is the another approach to determining the relation R. N. Bontis presented in [6] a model of the relation R describing an influence of the strategy oriented to intellectual capital on enterprise results. N. Bontis created a set of indexes expressing characteristic of employee capital, client capital, structural capital. The following items are numbered among enterprise results:

- P1 - Industry leadership.
- P2 - Future outlook.
- P3 - Profit.
- P4 - Profit growth.
- P5 - Sales growth.
- P6 - After-tax return on assets.
- P7 - After-tax return on sales.
- P8 - Success rate in new product launch.
- P9 - Overall business performance.

K. Kim, T. Knotts, S. Jones presented in [7] an influence of comprehensive assessment of the enterprise strategy taking into consideration assessment of a firm in the range of marketing management, strategic management, operating management, financial management, and assessment of market activity on a length of the enterprise life (over 8 years). The relation R can be identified by means of the SEM (Structural Equation Model) method. S. Sohn, H. Kim, T. Moon presented in [8] the enterprise strategy by means of a set of indexes including knowledge and experience of manager, operation ability of manager, level of technology, marketability of technology, technology profitability. Enterprise condition is defined by means of indexes: growth ratio of total assets, growth ratio of stockholder's equity, growth ratio of sales, net income to total assets, net income to stockholder's equity, net income to sales, total assets turnover, stockholder's equity turnover, debt ratio.

Currently, the widest considered conception of creating the knowledge about an enterprise is conception worked out by R. Kaplan and D. Norton - Balanced Scorecard (BSC) [9]. The BSC method defines the enterprise image from four perspectives: learning and growth perspective, internal business process perspective, customer perspective, and financial perspective. A key characteristic of the BSC method is a hierarchical system of indexes determining strategic tasks included in four mentioned perspectives. This system forms a fouradic cause-effect relationship. This means that projects and ventures determined in the remaining perspectives have an influence on financial performances. The whole system can be expressed by a value chain scheme. H. Huang presented construction of a system based on a knowledge base with a use of the BSC method [10].

Mentioned above approaches to creating systems based on a knowledge base oriented to analysis of enterprise condition show the way from one-sided financial analysis to strategic analysis including the main categories of strategic planning along with intellectual capital analysis. There is difficulty with determining the relation R in the case of more complex representation of information about an enterprise. Therefore, the CBR (Case-Based Reasoning) methodology [11] is proposed for creating the KBS systems. H. Li and J. Sun presented in [12] an application of the CBR methodology for creating enterprise condition assessment systems. Enterprise condition assessment requires construction of appropriate concept apparatus, thanks to which one can present expert assessments as well as introduce proper data concerning financial performances in individual accounting years. Expert assessments should concern both assessment of enterprise potential and assessment of risk of an enterprise activity. Introduction of such concept apparatus requires working out appropriate enterprise ontology. In different application areas, the KBS systems based on ontologies are created, for example, the system presented by W.L. Chang in [13].

In this paper, the author presents selected elements of the original KBS system based on ontology constructed by the author. A general ontology system is presented in Section 2. In Section 3, a subsystem of information extraction about assessment of enterprise competence is described. In Section 4, characteristics of the so-called explanation acts are shown. These acts make up reasoning procedures in a system according to the CBR methodology.

2 Enterprise Ontology

A process of modeling the knowledge about enterprise condition requires construction of such ontology, which will allow the KBS system to carry out automatic reasoning concerning condition of an enterprise. The author considers this process at three levels of conceptualization:

- Level I: Formal ontology.
- Level II: Domain ontology.
- Level III: Operating models.

The highest level of abstraction in modeling an enterprise requires accepting concept apparatus from the range of object theory. One of the most complex object theories was presented by Polish philosopher Roman Ingarden [14]. He defined three types of objects: individual object, intentional object and object determined in time. Ontologies worked out until now concentrate on depiction of an enterprise as a set of three types of objects: processes, events, and objects lasting in time. The author adopts from R. Ingarden concept apparatus defining an individual object. The structure of concepts of definition of an individual object according to R. Ingarden's object theory is presented in Figure 1.

The author presents his own conception of enterprise ontology based on definition of R. Ingarden's individual object. Figure 2 shows the structure of concepts in this ontology. Exact definitions of individual concepts are presented by the author in [15].

Domain ontology makes up such concept system, which is adequate to the language of a given domain. So far, the following ontologies are most popular: the EO ontology

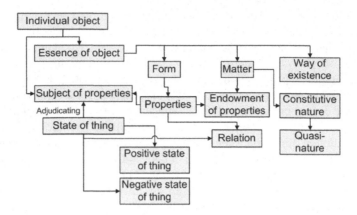

Fig. 1. A concept system of definition of object according to R. Ingarden's theory of object [14]

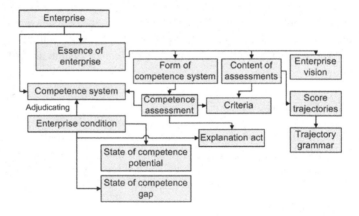

Fig. 2. A concept structure of enterprise ontology modeled on R. Ingarden's formal ontology

worked out by M. Uschold, M. King, S. Morale, Y. Zorgios [16], the TOVE ontology [17], the REA ontology [18] and the enterprise ontology worked out by J. Dietz. In [19], J. Dietz presented classification of approaches distinguishing:

- Domain approach.
- System approach.
- Planning approach.
- Transactional approach.
- Identification approach.
- Diagnostic approach.

Ontologies worked out until now include a concept system concerning processes and events occurring in an enterprise. However, other concept apparatus oriented to assessment of a state is required for determining enterprise condition. Therefore, in author's conception, an enterprise is represented by a competence system appropriately defined. Exact definitions of concepts are presented by the author in [15]. The key concept in the presented ontology is a subject of properties. The subject of properties is defined in order to determine enterprise condition on the basis of comparative analysis with enterprise vision presented in the strategy or with a leader of the market. In author's conception, the subject of properties is the enterprise competence system. This system is defined from two points of view: competence potential and competence gap. According to R. Ingarden, both positive and negative states should be distinguished in the object. A process of gathering information about enterprise condition according to assessment of competence system is presented in Section 3.

3   Subsystem of Information Extraction about Enterprise Condition

The intelligent system with a knowledge base (KBS-AE) designed by the author consists of three fundamental parts:

1. Subsystem for supporting assessment of enterprise competence - AC (Assessment of Competences).

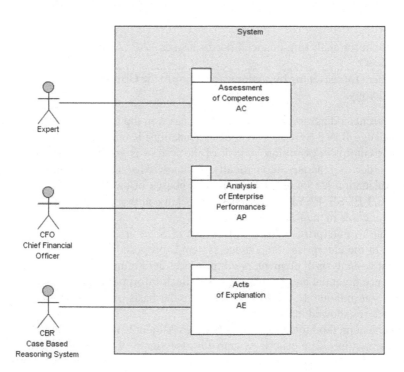

Fig. 3. A diagram of packages of use cases in UML

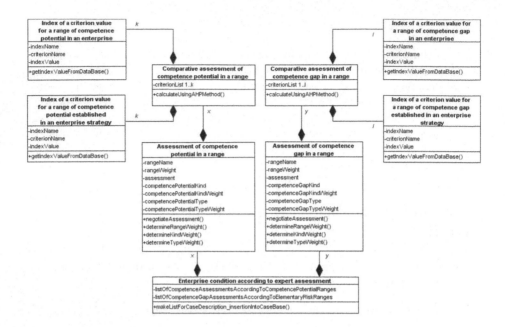

Fig. 4. A diagram of classes in the "Competence assessment" package

2. Subsystem for analyzing financial performances - AP (Analysis of Enterprise Performances).
3. Subsystem for reasoning by analogy according to the CBR (Case-Based Reasoning) methodology.

Figure 3 presents a diagram of packages of use cases in the UML language. The KBS-AE (Knowledge Based System - Acts of Explanation) is oriented to a process of explanation of enterprise condition. A task of the system is gathering information from experts in order to generate automatically the knowledge about enterprise condition. Acts of explanation are included in individual phases of the CBR cycle, i.e., phases: RETRIEVE, REUSE, REVISE, RETAIN. In [20], the author presented a procedure for indexation of cases in the RETAIN phase.

According to enterprise ontology presented in Section 2, information concerning assessment of the enterprise competence system is proposed by an expert on the basis of comparative analysis of elementary competences determined in the individual ranges of competence potential and competence gap. Such information is obtained by expert by comparison individual indexes calculated on the basis of data from the enterprise base and indexes showed in the strategy.

A convenient method for determining final assessment is the AHP method [21]. The author presented in [22] a criteria system for assessment of enterprise competence in each range of competence potential and competence gap. Taxonomies of competence potential and competence gap are presented in [23].

Figure 4 shows a diagram of classes in the "Competence assessment" package. The UML language is frequently and frequently used to present ontology in the KBS systems [24,25].

4   The Knowledge Generation in the KBS-AE System

In the KBS-AE system designed by the author, the knowledge about enterprise condition can be obtained thanks to procedures of assessment aggregation and clustering/classification. This knowledge answers the following questions:

- To which class or cluster does an enterprise belong in the competence potential - competence gap system.
- What characteristic does a given class or cluster have?

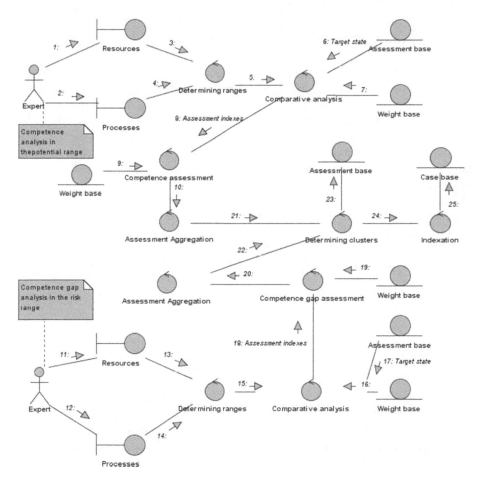

Fig. 5. A communication diagram in UML illustrating a process of creating a case base for the CBR system

- What does a similarity degree to the class or cluster representative like?
- What does the pattern of results of a given enterprise like?
- What does the predicted pattern of results of a given enterprise like?
- How the competence strategy of an enterprise should be changed in order to change the pattern of results?

A communication diagram in UML illustrating a process of creating a case base for the CBR system is presented in Figure 5. Each case in the base is appropriately classified. An index of the case is the determinant of a class or cluster which is equivalent to determining position of a given enterprise in the competence potential - competence gap space. Construction of suitable computational procedures enables us to answer the above questions. These procedures are named acts of explanation. Acts of explanation make up a fundamental property of the CBR systems [26]. In [27], the author presented an algorithm for an explanation act of an enterprise position.

5   Conclusions

In this paper, selected fragments of the designed KBS-AE system have been presented. The task of this system is explanation of enterprise condition. The knowledge about an enterprise is generated by the system on the basis of cases inserted into the case base. Each case makes up a set of assessments of competence potential of an enterprise and the so-called trajectories of results. On the basis of the CBR methodology, a set of procedures (the so-called acts of explanation) has been worked out. These procedures enable us to answer key questions concerning explanation of enterprise condition in the concrete case.

Figure 6 shows the structure of the KBS-AE system. Especially, the following phases are distinguished: gathering data by the multi-agent system, gathering information from experts for generation of the knowledge in explanation acts.

Within the confines of the EQUAL Project No. F0086 [28], the author worked out repository of cases in research of 220 enterprises of the SME sector from south-east Poland.

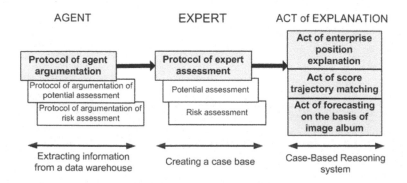

Fig. 6. A diagram of the KBS-AE system for generating the knowledge about an enterprise

References

1. Davenport, T.H., Prusak, L.: Working Knowledge. Harvard Business School Press (2000)
2. Altman, E.I.: Financial rations, discriminant analysis and the prediction of corporate bankruptcy. The Journal of Finance 4, 589–609 (1968)
3. Altman, E.I.: Corporate financial distress and bankruptcy. John Wiley & Sons, Inc., Chichester (1993)
4. Zopounidis, C., Dimitras, A.I.: Multicriteria decision aid methods for the prediction of business failure. Kluwer Academic Publishers, Dordrecht (1998)
5. Edvinsson, L., Malone, M.S.: Intellectual Capital. HarperCollins Publishers, Inc. (1997)
6. Bontis, N.: Intellectual capital. An exploratory study that develops measures and models. In: Choo, C.W., Bontis, N. (eds.) The strategic management of intellectual capital and organizational knowledge, pp. 643–655. Oxford University Press, Oxford (2002)
7. Kim, K.S., Knotts, T.L., Jones, S.C.: Characterizing viability of small manufacturing enterprises (SME) in the market. Expert Systems with Applications 34, 128–134 (2008)
8. Sohn, S.Y., Kim, H.S., Moon, T.H.: Predicting the financial performance index of technology fund for SME using structural equation model. Expert Systems with Applications 32, 890–898 (2007)
9. Kaplan, R.S., Norton, D.P.: Translating strategy into action. The Balanced Scorecard. Harvard Business School Press (1996)
10. Huang, H.C.: Designing a knowledge-based system for strategic planning: A balanced scorecard perspective. Expert Systems with Applications 36, 209–218 (2009)
11. Watson, I.: Applying knowledge management. Techniques for building corporate memories. Morgan Kaufmann Publishers, San Francisco (2003)
12. Li, H., Sun, J.: Ranking-order case-based reasoning for financial distress prediction. Knowledge-Based Systems 21, 868–878 (2008)
13. Chang, W.L.: OnCob: An ontology-based knowledge system for supporting position and classification of co-branding strategy. Knowledge-Based Systems 21, 498–506 (2008)
14. Ingarden, R.: Spór o istnienie świata. Tom II, Ontologia formalna, cz.1. Forma i Istota (in Polish). PWN, Warszawa (1987)
15. Andreasik, J.: Enterprise ontology according to Roman Ingarden formal ontology. In: Proc. of the ICCMI 2009 (in press)
16. Uschold, M., King, M., Moralee, S., Zorgios, Y.: The Enterprise Ontology. The Knowledge Engineering Review 13, 47–76 (1998)
17. Gruninger, M., Atefi, K., Fox, M.S.: Ontologies to Support Process Integration in Enterprise Engineering. Computational & Mathematical Organization Theory 6, 381–394 (2000)
18. Dunn, C., Cherrington, J.O., Hollander, A.S.: Enterprise Information Systems. A Pattern-Based Approach. Mc Graw-Hill (2003)
19. Dietz, J.L.G.: Enterprise ontology. Springer, Heidelberg (2006)
20. Andreasik, J.: A Case-Based Reasoning System for Predicting the Economic Situation on Enterprises. In: Kurzynski, M., et al. (eds.) Tacit Knowledge Capture Process (Externalization), Computer Recognition Systems 2. ASC, vol. 45, pp. 718–730. Springer, Heidelberg (2007)
21. Saaty, T.L.: The analytic hierarchy process. Mc Graw-Hill, New York (1980)
22. Andreasik, J.: Intelligent System for Predicting Economic Situation of SME. In: Józefczyk, J., Thomas, W., Turowska, M. (eds.) Proc. of the 14th International Congress of Cybernetics and Systems of WOSC, Wroclaw, Poland (2008)
23. Andreasik, J.: Enterprise Ontology. Diagnostic Approach. In: Proc. of the HSI 2008, Kraków, Poland, pp. 497–503 (2008)

24. Gasevic, D., Djuric, D., Devedzic, V.: Model driven architecture and ontology development. Springer, Heidelberg (2006)
25. Rhem, A.J.: UML for developing knowledge management systems. Auerbach Publications (2006)
26. Schank, R.C., Kass, A., Riesbeck, C.K.: Inside Case-Based Explanation. Lawrence Erlbaum Associates Publishers, Mahwah (1994)
27. Andreasik, J.: Decision support system for assessment of enterprise competence. In: Kurzynski, M., Wozniak, M. (eds.) Computer Recognition Systems 3. AISC, vol. 57, pp. 559–567. Springer, Heidelberg (2009)
28. e-barometr manual, http://www.wszia.edu.pl/eng/files/e-barometr-manual.pdf

Information Retrieval in the Geodetic and Cartographic Documentation Centers

Tomasz Kubik[1] and Adam Iwaniak[2]

[1] Institute of Computer Engineering, Control and Robotics, Wrocław University of Technology,
Wybrzeże Wyspiańskiego 27, Wrocław 50-370, Poland
`tomasz.kubik@pwr.wroc.pl`
[2] Institute of Geodesy and Geoinformatics, Wrocław University of Environmental and Life
Sciences, Grunwaldzka 53, Wrocław 50-357, Poland
`adam.iwaniak@up.wroc.pl`

Abstract. The paper addresses the problem of geo-referenced documents management and discovery, concerning the use of metadata and ontology. It shows, that the metadata profile developed for the INSPIRE does not meet the requirements set out for the documents retrieving in the documentation centers as well as the majority of the public administration offices. It highlights the need of building the tools for indirect and hidden references and metadata attributes exploration of the geodetic documents like field sketches and others. It indicates the possibility of using ontology, which extending the metadata profile would allow to link different types of documents that correlate indirectly in the space and time.

Keywords: GIS, SDI, ontology.

1 Introduction

The increasing attention for the deployment of web sites providing access to a geographic content is observable since late 1990s. The efforts undertaken evolved from the implementation of closed, proprietary based solution into building open infrastructures conforming SOA paradigm. These activities speeded-up with the publication of open standards, like ISO and OGC standards. Recently these movements have gained a legal strength. They have been given a legal framework in form of the INSPIRE (INfrastructure for SPatial InfoRmation in Europe) directive of the European Parliament and of the Council, published in the official Journal on the 25th April 2007 and entered into force on the 15th May 2007. The main objective of INSPIRE is data interoperability and data sharing. The directive defines needs, and the member countries are obliged to adapt the national law and the administrative procedures to meet them. The technical documentation setting out basic requirements for data and services assuring the conformity and usefulness of the spatial data infrastructures becomes available in the form of Implementation Rules (IRs).

The importance of geospatial information increases in a daily life, and becomes crucial for the business and administrative processes. The ability to provide geospatial content within corporate information systems, accessing it remotely and in an automatic manner, becomes the base of day-to-day utility operations. This is especially true

N.T. Nguyen et al. (Eds.): New Challenges in Compu. Collective Intelligence, SCI 244, pp. 95–106.
springerlink.com   © Springer-Verlag Berlin Heidelberg 2009

in the public administration, which is responsible for the registration, maintenance and distribution of legal documents containing spatial references. There are different public authorities and governmental agencies that manage number of different types of such documents. The geodetic and cartographic documentation centers (surveying documentation centers) are responsible for topographic and cadastral information in Poland. The organizational structure of the centers has three levels: central, regional and local. The centers are the nodes of NSDI aimed at performing the duties resulting from the Law on geodesy and cartography. There are several tasks assigned to them, including realization of state policy in the field of geodesy and cartography, registration of legal and actual status of properties, managing geodetic and cartographic resources (records of land and buildings, evidence of public utilities, address points, spatial plans, cadastral sketches and surveying results, topographic descriptions, maps, photogrammetrical materials, etc.). All these tasks incorporate producing and processing geodetic documents. The extensive search within these documents is very important, especially for the proper execution of geodetic work by authorized surveying companies.

Usually the geodetic work carries on at the local level in three stages: registration and preparation for the work, surveying in the field, reporting and archiving of results compilation. The first stage includes preparation of technical specification and analysis of the existing documents, especially those from the work already performed (surveying networks, topographic surveys, etc.), related to the spatio-temporal extent and the subject of the current work. The realization and outcomes of the next two stages strongly depend on the results of the first stage. The lack of valid information might cause repetition of the work done once, or, in worse case, might lead to falsified, incorrect and erroneous results. Therefore, the local documentation center should provide on the request all related information without omitting any valuable documents.

To fulfill this expectation the documentation center must keep the documents well organized. In most cases, they use analog documentation storages and supporting IT systems for the digitized resources management. Nowadays the IT systems offer Internet access for the humans (web pages) and computers (services interfaces), and make a technological bridge with other nodes of NSDI. However, not all internal functions of these systems are open for the remote access. More over, the search algorithms and methods implemented do not assure full and exhaustive document discovery. Usually the systems offer possibility of defining spatial location of the documents content, allow their text based search and indexing, provide the tools for the document circulation management. But the use of standardized metadata for the documents description (according to the INSPIRE or ISO 19115 rules) is almost not evident.

The paper addresses the problem of document management and discovery in the geodetic and cartographic documentation centers – the nodes of NSDI. It concerns the use of metadata and ontology, and highlights the need of building the tools for indirect and hidden references and metadata attributes exploration of the geodetic documents like field sketches and others. The paper is organized as follows. It starts with a brief review of related works. Then it gives some insights into SDI and the role of metadata and catalogue services. Next, it describes functions of geodetic and cartographic documentation centers associated with to the geodetic work execution. It underlines the issue of the documentation management and discovery by using spatio-temporal extent information. The paper ends up with concluding remarks.

2 Related Works

Many authors discussed the role and the use of ontologies in the SDI. Depending on context, intention of use, and level of abstraction, the ontologies serve for different purposes in different ways. They serve as a tool for logical description and conceptual model creation, a base for construction of taxonomy and thesauri, a background of an intelligent discovery and search, a part of semantic web services and multi-agent systems. Beside subjective, particular propositions there are also standardizing initiatives undertaken in these fields related to the geospatial domain, as ISO/TC 211's proposition of a new project 19150 - Ontology. The aim of this project is to collect and compile information, and investigate how ontology and Semantic Web approaches can benefit ISO/TC 211 objectives, i.e. geographic information interoperability.

The authors of [1] focused on the role of ontologies in the construction of models of spatial data. They addressed practical issues of implementation with some notes on the use of different XML based modeling languages (BPML, GML, CityGML, LandGML) and software editors. The role of ontologies in the context of the geospatial domain was also discussed in [2]. The authors referred to the five dominant universes for geospatial applications. They proposed two distinctive ontologies: application-level and computer-level ontology, constituting a system for interpreting and solving geospatial queries together with an ontological mediator. Similar subject was discussed in [3]. The authors focused on contextualization of geospatial database semantics for user-system interaction and presented philosophical and knowledge engineering based approaches for ontology development.

The authors of [4] proposed to incorporate geospatial semantics for three major types of geospatial relations: topological relations, cardinal direction, and proximity relations. More over, they referred to different approaches to the development of a representative ontology of time and to different languages (OpenCyc, Context Broker Architecture, Region Connection Calculus, RDF Geo vocabulary). The possibility of using OWL and GML languages was mentioned in [5]. The authors proposed some spatial relationship to solve their particular problem.

The authors of [6] discussed the role of geoportals in the NSDI, remaking that portals rely only on metadata and do not support formal semantics. A problem of knowledge discovery in the geoportals was a subject of [7]. The authors demonstrated how to exploit the RDF language in order to describe the geoportal's information with ontology-based metadata. The problems related to the geospatial semantic web were discussed in [8,9]. Various projects connected with geospatial information retrieval were presented in [10].

The authors of [11] proposed the architecture of a centralized ontology service that could be integrated within the OGC Web Service Architecture. The service enables uniform management of lexical ontologies and gives ontology-based support to SDI components. The authors remarked that the ontologies may be used to profile the metadata needs of a specific geospatial resource and its relationships with metadata of other related geospatial resources, or to provide interoperability across metadata schemas.

The proposition of asymmetric and context aware similarity measurement theory for information retrieval and organization within SDIs was provided in [12]. The architecture of a system for geographic information retrieval was presented in [13]. The system makes use of an index structure, that combines an inverted index, a spatial index, and an ontology-based structure.

Several works treat on solving the problem of geospatial web services discovering supported with ontologies. The author of [14] showed the method for geospatial service discovery based on mapping between ontological description of requirements and service capabilities. In [15] relevant semantic properties of services and processes are stored in semantic repositories. The algorithm proposed allows value based service selection.

The cadastral information exchange using Web services was concerned in [16]. The prototype presented there exemplified the concept for the idea of query translation based on semantic relations between different information models. The problems of management of geographical changes in the context of a corporate cadastre application was discussed in [17]. The model proposed based on lineage metadata.

3   Spatial Data Infrastructure

The construction of spatial data infrastructure (SDI) started in the United States in the beginning of the 90-years. The primary aim of this activity was to provide an easy access to the data sets collected in different institutions, to reduce the cost of data collection, to increase the benefits of using and promote the reuse of existing information. The idea of SDI spread over the world what resulted in the implementation of several national spatial data infrastructures (NSDI). The dynamic growth of the SDIs, observed at regional, continental and global level, enforced the creation of open international standards to fulfill the requirements of their interoperability. These standards play a crucial role in the SDI integration and building cross-country solutions.

> SDI – "Technologies, policies, and people necessary to promote sharing of geospatial data throughout all levels of government, the private and non-profit sectors, and the academic community." http://www.fgdc.gov/nsdi/ nsdi.html/

European countries have begun construction of NSDI in different period. Their experience in this field led to the establishment of the INSPIRE Directive on 14 March 2007. The directive is a legislative and imposes an obligation on government bodies responsible for the acquisition, collection and sharing of information, take concrete actions related to the development and implementation of a number of metadata standards. To make the most of NSDI offerings the implementation of each NSDI had to incorporate mechanisms, which allow users to discover, review, and retrieve existing geospatial information resources. The spatial-aware catalogue services are the solutions that made this possible for both humans and software.

> "Catalogue services are singled out as a special type of geo-processing service: they provide the functionality to allow the organization, discovery and access

to geo-spatial data and services, and as such, form the core of any SDI. A catalogue consists of a collection of indexed, searchable catalogue entries each providing a description of some resource. Entries usually take the form of a subset of the complete metadata element set of the resource they describe." [18]

The catalogue services support the ability to publish, search and retrieve collections of descriptive information (metadata) for data, services, and related information objects, together with the references to the location of original resources for binding purposes. The metadata can be queried and presented for evaluation and further processing by both humans and software [19].

"Metadata is the term used to describe the summary information or characteristics of a set of data. This very general definition includes an almost limitless spectrum of possibilities ranging from human-generated textual description of a resource to machine-generated data that may be useful to software applications. <...> In the area of geospatial information or information with a geographic component this normally means the What, Who, Where, Why, When and How of the data. <...>" http://gsdidocs.org/ GSDIWiki/index.php?title=Chapter_3

The research on metadata for the geospatial information resources resulted with a well-defined standards. Thus at the end of 1990s the cartographic agencies of developed countries made efforts to establish national profiles of metadata for data sets. The standardization committee of the International Cartographic Association developed and published the exact description of the national metadata profile. In 1994 the first, and in 1997 the second version of the profile metadata FGDC (The Federal Geographic Data Committee) was developed that has become standard not only in the U.S., but South Africa and Australia. Another standard developed by the DCMI (Dublin Core Metadata Initiative), provide a set of standard core elements specifically for discovery metadata, i.e., for cross-domain information resource description, and includes a 'coverage' element for defining spatial location and temporal period. The most comprehensive set of geographic information related standards comes from ISO 19100 series. A suite of metadata standards is rich enough for describing most of geospatial data and associated sensors and platforms. The suite of metadata standards consists of: the main standards ISO 19115, ISO 19115-2, ISO 19119 and ISO/TS 19130; and the quality standards ISO 19113, ISO 19114, and ISO 19138; and others.

ISO 19115 provides an extensive and generic definition of metadata for geo-spatial data sets described in UML and defined in the data dictionary, but without encoding rules (the principles of metadata implementation in XML were included in ISO 19139). The definitions of ISO 19115 consist of declarations of optional and conditional metadata elements as well as definitions of a set of generalized code lists (over 400 metadata elements collected in 14 packages). Several other ISO standards were used as definition providers of information model parts (as temporal and spatial extend information defined in ISO 19107, ISO 19108, ISO 19111, ISO 19112). The metadata elements defined might serve as collection of information about identification, extent, quality, spatial and temporal schema, spatial reference, and distribution of digital geographic data. The metadata elements for geographic information service description are defined in ISO 19119. The role of service metadata is to provide the description allowing services combination (orchestration).

The history of the catalogue services implementation correlates with the history of the advances of the information technologies and is closely associated with the development of the Internet. In the initial period, the solutions built made use of library systems. The primary communication protocol was Z39.50 and the common metadata profile was Dublin Core.

The catalogue service is one of the five service types being defined as the core of the INSPIRE Network Services Architecture. The INSPIRE Network Service Architecture is designed as Service-Oriented Architecture with service consuming and service providing components communicating via an INSPIRE (enterprise) service bus. The catalog service is referred to as Discovery Service while the other service types are referred to as: View, Download, Transform and Invoke. The detailed taxonomy of the services recognized by INSPIRE is much richer (the regulation 1205/2008 EC of the European Parliament and of the Council lists 70 different services) and follows the taxonomy defined in ISO 19119 standard. Some of the services are individual instances and some can be a part of service chain for both synchronous and asynchronous requests execution through a (web) service orchestration engine.

4   Nodes of NSDI

The main components of the NSDI are SDI nodes with own entity. The SDI nodes deliver services on the Service Bus to portals and applications. For the organizational, legal, and competence reasons the majority of SDI nodes in Poland are the geodetic and cartography documentation centers. The geodetic and cartographic documentation centers maintain records of geodetic and cartographic information, (records of land and buildings, evidence of public utilities, address points, spatial plans, cadastral sketches and surveying results, topographic descriptions, maps, photogrammetrical materials, etc.) and are responsible for their management. Apart from basic services for the public administration and citizens, the centers must also provide services to the majority of their clients: the companies performing geodetic and cartographic works.

The execution of geodetic and cartographic work should follow the rules of geodetic and cartographic law. Thus, each work must be submitted to and registered in the documentation center. The center, in turn, should provide the materials and documentation necessary for the work execution. The work execution finalizes with the geodetic documents being verified and incorporated then into resources of the documentation center.

Registration of the geodetic work includes indication of its range and spatial (and temporal) extent. The range of the documentation prepared by the center depends on this indication. The documentation usually consist of copies of the map, list of geodetic points and border points, and field sketches of the earlier measurements made in the range specified. Only complete documentation can guarantee the proper geodetic work execution. Any missing information might cause repetition of the work done once, or, in worse case, might lead to falsified, incorrect and erroneous results. Thus, the documents search within documentation archives should be the most comprehensive it could be.

5   Spatio-temporal Related Documents Retrieval

When analyzing the official documents archives one can observe that there are several interdependencies between document contents. For example, the building permit refers to the investment localization by indicating the number of cadastral parcels, referring to its spatial extent indirectly. Sometimes the transitive relations of this kind are nested, as in the case of referring to the particular property using property owner ID. The spatio-temporal extent of such property has to be inferred then from the other documents that use the same owner ID.

In general geographic information contains spatial references which relate the features represented in the data to positions in the real world. Somewhat more formally, there are three types of spatial references:

- direct reference – defined by the spatio-temporal extent in the known coordinate reference system (CRS). The example is a map representing objects as buildings, roads, geodetic points, parcels, etc. The location of the objects can be read directly from the map. Another example is a list of geodetic points kept within geodetic center resources. The list contains geodetic points numbers and X,Y coordinates.
- indirect reference – defined by the geographic identifier that can be mapped onto location. The example is a field sketch with hand drawn information (buildings with addresses, parcels with numbers). The coordinates cannot be measured directly, but they can be retrieved from a land and property register, or a system of the address information, or a gazetteer. The references of this type may vary in time causing some troubles. They can get a new scope and alive status (when changing the street names, or dividing or merging the parcels).
- hidden reference – defined by the associations and facts that might be analyzed in order to deduce correct location assignment. The example is real estate information providing owner name and address. In this case the other registries has to be searched for the parcel's number, and then for the location. This reference depends on the current and past information collected in the archives. The difficulties might arise in the case when the owner possess more then one parcel or had changed his/her name.

From the spatio-temporal referencing point of view there are two kinds of individuals:

- reference object – an object with a well defined spatio-temporal location. Frequently this is an address point, cadastral parcel, street, administrative boundary or physiographic object registered in a gazetteer.
- intermediary object – an object without direct spatio-temporal location, but staying in a relationship with a reference object. In general, the relationship can be transitive, leading to the reference object trough the sequence of intermediary objects. An intermediary object example is a person with a given ID, who possesses a property (of known location). The person and the property relates to each other by posses and is_possessed relation.

To get comprehensive information from the geodetic and cartographic documentation center several documents require processing. To automate this task and offer an access to the geodetic documents through the web services, it is necessary to digitize the archives. This process must incorporate also the generation of computer readable description for all documents collected. This is the place for the use of metadata and ontology.

For several reasons the documents should be described by ISO 19115 compliant metadata, with the spatial and temporal extents semantics, model, and encoding defined in the accompanying standards. The INSPIRE Implementing Rules for metadata stated that at least one temporal reference chosen from one of these four categories: temporal extent, data of publication, date of last revision or the date of creation is required. ISO 19115 requires that at least one of the following is provided: date of publication, the date of last revision or the date of creation.

For the spatio-temporal information retrieval ISO 19115 standard defines metadata elements as in the figure 1 (the ISO UML models are available at http:// www.isotc211.org/hmmg/HTML/root.html). The extent is a data type class that is an aggregation of vertical extent, geographic extent (which can be a bounding polygon, or bounding box, or geographic description for indirect referencing) and spatial temporal extent. This model meets the requirements for the direct and indirect spatio-temporal referencing (the spatial scheme includes enumeration with items that match operators' names of Allen's interval algebra, see figure 2). While the direct referencing supported by the model is exhaustive, the indirect referencing is weak. The model uses geographic identifier that refers to the physical phenomena rather then describes associations among different geodetic documents. It plays a role of a key in gazetteers (see figure 3).

In the ISO 19115 there are number of other metadata elements defined, like MD_ContentInformation and MD_Identification, that would be used to model the indirect associations. However, these metadata elements suit general geographic domain and therefore they do not meet the requirements for the indirect referencing for the geodetic documents retrieval. Therefore, the catalogue services based on standard metadata set only are not capable to solve this problem. There is a need to create a dedicated profile for this purpose - the profile incorporating ontological concepts from the cadastral domain.

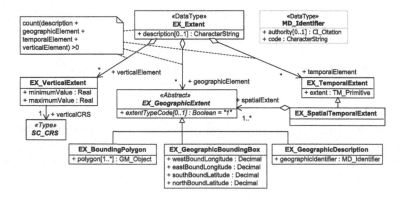

Fig. 1. UML model of extent information classes (according to ISO 19115)

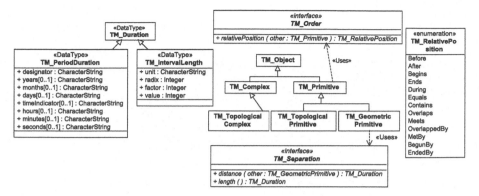

Fig. 2. UML model of temporal classes (according to ISO 19108)

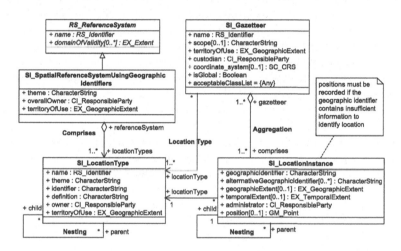

Fig. 3. UML model of classes for spatial referencing using geographic identifiers (according to ISO 19112)

One way to start with ontology design would be analysis of published acts and regulations. Journal of Laws of the Republic of Poland, 1999, No 49, item 49 divides resource materials of the local documentation center into two parts: those obtained as a result of geodetic and cartographic work (with 10 groups defined) and other materials (with 9 groups defined). The resource materials are collected in the assortment groups (there are 25 assortment group defined), sorted according to the basic units of territorial division of the country, taking into account their assignment to the functional groups. Functional group reflects the characteristics, the purpose, and the way of use of resource materials. There are 3 functional groups defined: basic resource (materials serving as the basis for subsequent studies collected in the resource) usable resource (materials for the direct use, available publicly), intermediate resource (auxiliary materials that are not qualified as a base nor usable resource). All resource materials are recorded in

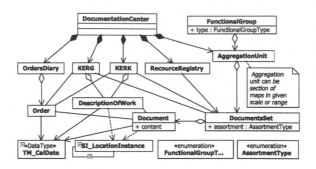

Fig. 4. Simplified UML model of the documentation center

the listing documents as: cadastral books, maps and sketches, card-indexes, registers and lists. For that purpose, each material has a registration number assigned. A separate registry is used to keep track of the materials given out. Hence, the materials collected in the resource and made available on a media have identifiers well defined.

The technical documents issued by the Head Office of Geodesy and Cartography provide more details on the law interpretation and give practical instruction. According to O3-O4 Technical Instruction, 2001, the local documentation center maintains: diary of orders, registration book of geodetic work (KERG), registration book of cartographic work (KERK), and resource registry (see figure 4).

There are systems, which allow searching within documentation center repositories trough the web services - geoportals. The clients can discover and view some of the documents under assumed security policy. The search is done by identifiers (corresponding to the data recorded in the mentioned registries), text (corresponding to documents description), and location (corresponding to the location of the objects the documents describe). Searching by location is supported usually with the aid of map, on which user can select the area of interest (like in the case of searching for topographic description of geodetic points). However, not all of the documents are available on-line. It is not only because of policy applied, but also because of missing methodology for spatio-temporal relevant documents discovery.

6   Concluding Remarks

The problem of the spatio-temporal based documents retrieval is not limited to the geodetic and cartographic domain. It has much wider impact on several offices of public administration that collect or use this kind of information. Analysis of several studies shows that 80% of information maintained by the offices includes references to the location or position in space. The spatial references usually are indirect or hidden, and take a form of address, parcel numbers, the border or the holder of the property. Considering the practical issues of spatio-temporal referencing the question arises: What is the advantage of the use of catalogues services and metadata schemes over indexing data locally? The answer seams to be simple. Thanks to the standardization, the catalogue services assure interoperability and provide clues to solve distributed query problems (one query may retrieve data from various documentation centers). This is particularly important for areas located in the border districts. Using ontology is most suitable for indirect or hidden reference discovering. Ontology describes the relationships between objects. These relationships may be comprehensive, numerous and redundant. Taking the advantage of that, the ontology can help to solve the problem of inconsistency in the set of object identifiers being analyzed, caused by the changes of the objects they point to (removal, renaming, reassigning, merging or dividing). It can help in checking whether two objects with different identifiers are different really or they are identical. The paper defines the research area and its context, practical implementation of which is strongly expected.

References

1. Albrecht, J., Derman, B., Ramasubramanian, L.: Geo-ontology tools: The missing link. Transactions in GIS 12(4), 409–424 (2008)
2. Peachavanish, R., Karimi, H.A.: Ontological engineering for interpreting geospatial queries. Transactions in GIS 11(1), 115–130 (2007)
3. Cai, G.: Contextualization of geospatial database semantics for human?gis interaction. GeoInformatica 11(2), 217–237 (2007)
4. Arpinar, I.B., Sheth, A., Ramakrishnan, C., Usery, E.L., Azami, M., Kwan, M.P.: Geospatial ontology development and semantic analytics. Transactions in GIS 10(4), 551–575 (2006)
5. Abdelmoty, A.I., Smart, P.D., Jones, C.B., Fu, G., Finch, D.: A critical evaluation of ontology languages for geographic information retrieval on the internet. Journal of Visual Languages and Computing 16(4), 331–358 (2005)
6. Wiegand, N., García, C.: A task-based ontology approach to automate geospatial data retrieval. Transactions in GIS 11(3), 355–376 (2007)
7. Athanasis, N., Kalabokidis, K., Vaitis, M., Soulakellis, N.: Towards a semantics-based approach in the development of geographic portals. Computers and Geosciences 35(2), 301–308 (2009)
8. Weis, M., Müller, S., Liedtke, C.E., Pahl, M.: A framework for gis and imagery data fusion in support of cartographic updating. Information Fusion 6(4), 311–317 (2005)
9. Brodaric, B.: Geo-pragmatics for the geospatial semantic web. Transactions in GIS 11(3), 453–477 (2007)
10. Silva, M.J., Martins, B., Chaves, M., Afonso, A.P., Cardoso, N.: Adding geographic scopes to web resources. Computers, Environment and Urban Systems 30(4), 378–399 (2006)
11. Lacasta, J., Nogueras-Iso, J., Béjar, R., Muro-Medrano, P., Zarazaga-Soria, F.: A web ontology service to facilitate interoperability within a spatial data infrastructure: Applicability to discovery. Data & Knowledge Engineering 63(3), 947–971 (2007)
12. Janowicz, K.: Similarity-based information retrieval and its role within spatial data infrastructures. In: Scharl, A., Tochtermann, K. (eds.) The Geospatial Web - How Geo-Browsers, Social Software and the Web 2.0 are Shaping the Network Society, pp. 235–245. Springer, Heidelberg (2007)
13. Luaces, M.R., Paramá, J.R., Pedreira, O., Seco, D.: An ontology-based index to retrieve documents with geographic information. In: Ludäscher, B., Mamoulis, N. (eds.) SSDBM 2008. LNCS, vol. 5069, pp. 384–400. Springer, Heidelberg (2008)
14. Lutz, M.: Ontology-based descriptions for semantic discovery and composition of geoprocessing services. GeoInformatica 11(1), 1–36 (2007)

15. Palmonari, M., Viscusi, G., Batini, C.: A semantic repository approach to improve the government to business relationship. Data & Knowledge Engineering 65(3), 485–511 (2008)
16. Hess, C., de Vries, M.: From models to data: A prototype query translator for the cadastral domain. Computers, Environment and Urban Systems 30(5), 529–542 (2006)
17. Spéry, L., Claramunt, C., Libourel, T.: A spatio-temporal model for the manipulation of lineage metadata. Geoinformatica 5(1), 51–70 (2001)
18. Tuama, E., Best, C., Hamre, T.: A web-based distributed architecture for coastal zone management. In: Fifth International Symposium on GIS and Computer Cartography for Coastal Zone Management, Genova, Italy, October 16-18 (2003)
19. OpenGIS Catalogue Services Specification, Version 2.0.2, Corrigendum 2 Release, OGC 07-006r1 (2007)

Visualization Framework of Information Map in Blog Using Ontology

Nurul Akhmal Mohd. Zulkefli, Inay Ha, and Geun-Sik Jo

Intelligent E-Commerce Systems Laboratory,
Department of Computer and Information Engineering, Inha University,
253 Yonghyun-dong, Incheon, Korea 402-751
{nurul_teens85,inay}@eslab.inha.ac.kr, gsjo@inha.ac.kr

Abstract. In this paper, we introduce our proposed framework for visualization of the blog's content using ontology method to let blogger know the existing relationship between the blogs. The proposed framework is to explain the details on how the important information from the blog such as category, friend and content will be extracted, how the ontology will be implemented in this system and how visualization framework will be done. From this proposed visualization, we highly expect to provide an interactive tool for information retrieval where blogger can have multiple ways to catch the information from the favorite blogs.

Keywords: Visualization, Blog, Ontology, Mapping.

1 Introduction

Nowadays, Blog becomes a popular virtual diary among internet bloggers. Blog or "Web Log" had started to grow around 90's and until now there are thousands million of blogs exist in the World Wide Web. According to a survey report [1], in the US, about 12 million people maintain blogs, and about 57 million Internet bloggers read blog. In this research, we focused on one blogger which has relationships with some friends or favorite friends. Normally, blogger will open the blog more than one time per day to check whether blogger's favorite friends have updated or posted any new entries. Besides, blogger also might want to know whether blogger's post entry has a respond or not. From our experience, blogger's friends usually come from blogger's life background, for example, John's favorite friends are Anna and Bee and both of them are John's ex-schoolmate. This environment shows that one of the reasons why John joins the blog is because he is interested to know his friend's story and social life activities. It is quite interesting if there is a system that can help John to easily organize and get the information from his friends besides getting to improve the relationship between him and his friends. This situation motivates us to come out with an effective and interactive visualization for blogger.

There are two contributions in this paper: first, a proposed framework for blog visualization and second, the personal ontology that will be implemented in visualization blog. Detail contributions are listed as below:

N.T. Nguyen et al. (Eds.): New Challenges in Compu. Collective Intelligence, SCI 244, pp. 107–118.

- We propose visualization framework for bloggers to improve the usability of information retrieval in blog. By providing an interactive graphic and animation in visualization, blogger can easily get the important and needed information in his blog and his favorite blog. In our research, we specify that, only a new entry and non-read entry will be displayed in each visualization node so that blogger can easily find out the needed information especially the newest entry. In another hand, this concept will help blogger to save their time in reviewing some new entries for each blogs without going through blog by blog.
- Second contribution is the implementation of ontology in the visualization. As a blogger, it is important to classify the content to each category to make all entries flexible and systematic. We propose our own personal ontology based on blogger and blog behavior which it refers to the categories defined by blogger. Using the main personal ontology, another blogger can make his personal ontology by referring to the main personal ontology. From main personal ontology, we can identify relationship that occurs between each personnel's ontology. Once the relationship is known, blogger can figure out what relationship exists between each entry. Identifying the relationship is important and beneficial for blogger especially if they use the blog for sharing knowledge and experiences. This motivation is very useful for us in order to improve the information retrieval in blog.

The subsequent sections are organized as follow: The next section contains a brief review of related works. In Section 3, we describe about overview of visualization framework and ontology. A proposed system is presented in Section 4. Finally, conclusions are presented and future work is discussed in Section 5.

2 Related Works

2.1 An Overview of Blog

Blog is well known as a virtual diary for bloggers. There are nearly 16 million active blogs on the Internet with more launched every day. Although much of what's discussed in the blogosphere is of little consequence, increasingly, blogs are emerging as powerful organizing mechanisms, giving momentum to ideas that shape public opinion and influence behavior. Blog becomes a popular place where everyone can share their own experience, knowledge and anything about life. For example, Malaysian bloggers have recently used blogs as a medium to publish their own opinion towards political environment in Malaysia.

Blogspot.com, Wordpress.com, LiveJournal and Multiply are some of the biggest blog communities today. Novel bloggers can easily use the blog to publish the entry using the simple basic interface. Some blogs have a complete tool for blogger usability. For example, Blogspot was designed to be a very simple blog for blogger. It contains a blogger profile, blog archive, blogger favorite link, and entry at the right side. Bloggers can edit their templates either using the basic templates prepared by Blogspot or free downloading from template websites.

In this blog, blogger can define labels for each entry. For example in Figure 1, on the left side, blogger defines his own categories using his own word. He also makes an archive collection where the archive can be viewed through monthly category. Blogger also can upload video, music and picture to the post entry. For novel and expert bloggers, they can use some html codes to edit their entry for example <b></b>, <italic></italic> or embed some video using code <embed></embed>. This is quite interesting for bloggers where they are totally free to use and edit their own personal blog.

Fig. 1. Example of blog

Nowadays, blogs are just not limited to personal diary, they are already spread in business environment where many bloggers use blogs as a free platform to sell the goods. For example, "Kedai Gunting" or "Barber Shop[1]" is using Blogspot to sell the hiljabs and shawls while "One Stop Babyz n Mom[2]" sells some products for pregnant mother and baby. There are also blogs for preview some movies and music where blogger can catch the information of the newest movies and music.

However, to visit each favorite blog will take a lot of time and although blogosphere is used, it is still difficult to find the interesting topic. Some bloggers might be interested in certain topic and some may prefer to read the entries from their friend's blog only. Although blogger has introduced Blog List where blogger can schedule post publishing on blogger, there is still a problem where blogger only can read the latest news from blogger's friends. So, how to allow blogger to read either latest entry or non-read entry without spending too much time on searching for it? This question can be easily answered by our proposed visualization that will be explained in the next section.

---

[1] http://guntingstore.blogspot.com/
[2] http://onestopbabyznmom.blogspot.com/

2.2 Visualization for Social Network and Blog

Visualization has become an interactive tool to display many things such as data, network and blogger profile. Nowadays, there are a few visualization network applications such as TouchGraph's Facebook Browser and Vizster. The TouchGraph's Facebook Browser lets bloggers visualize their friends and photo. While the Vizster is built as a visualization system that end-bloggers of social networking services could use to facilitate discovery and increase awareness of their online community [2]. Both of these tools are used to display networking between blogger and his friends in one big community such as Facebook and Friendster. In blog case, there are some visualization tools developed by researchers. For example, Takama and his colleagues [3] attempted to develop a visualization map that contributed to interactive blog search. The visualization map was facilitated by linking keywords of blogs. While Karger and Quan [4] presented a visualization system that displayed messages from multiple blogs together as a reply graph where a diagram described relationships between a message and all comments related to the message and the result showed that bloggers were able to understand how the relevant issues are constructed and related. However, in previous researches, fewer researchers focus on blogger's behavior where blogger can visualize his or her blog based on favorite blog and favorite friends. Besides using the interactive visualization, we can help blogger to identify the relationship that occurs between blogger contents and blogger favorite contents.

2.3 Ontology in Visualization

Classifying the different categories with similar contents is one of the major problems for this proposed system. Normally, bloggers will categorize each entry with their own tag such as "Life", "Movie", "Sport", "Friend" and "Work Life". It probably happens that different bloggers have different tags but similar contents. For example two blogs below (Figure 2) had posted "Happy New Year" entry and each of them is categorized with different labels where the first blog labeled as New Year and second blog labeled as Aktiviti (activity).

There are many similar contents labeled by different categories and these existing contents should be classified together. In our case study, the blogger friend's content is classified into same label with blogger so that he can view the content by using his category. To classify the contents in same blogger's category, ontology method needs to be implemented. Several existing ontology have been proposed and among all ontology-related research issues, comparing ontology [5, 6, 7] and ontology mapping [8, 9] are probably the most studied one. Various techniques have been proposed for the latter problem. For example, Quick Ontology Mapping is proposed to trade off between effectiveness and efficiency of the mapping generation algorithms [9]. Schmidtke et al. [10] and Kalfoglou et al. [8] discussed the state of art in ontology mapping and ontology integration and Pei-Ling Hsu et al [11] proposed a framework to automatically map blogger-defined categories in blog. Although many techniques have been proposed, none of them addressed the issue of classifying the similar contents with different labels between blogs. Most of the previous systems used to compare the similarity and relationship between nodes by labels or blogger-defined

Fig. 2. Example for different label but same entry

category. While in our research, we focus on how to map the contents in blogger defined category via implementing the ontology.

3 Visualization of Information Map Techniques

The goal for our proposed system is to build a visualization system that can enhance the information retrieval between blogger and friends' blog. We want blogger to easily access the newest updated entry and non-read entry from the visualization system. In other words, blogger can save a lot of time to read the newest and non-read entry from all blogger favorite blogs. Blogger also can figure out the interested topic using category defined by them.

Our visualization tool is made for blogger where they can use this visualization as a main tool to surf the blogs. Enhance to visualization tools, we add some functions for blogger to view the blog's content whether by category or member, view friend's profile through the same window, and view the entry using pop-up box. Blogger also can choose to view the picture, video and music based on category or member.

From Figure 3, Blogger has one or many categories and category has one or many entries. Bloggers defined the category based on their interest, such as "Life", "Music", "Games" and "Programming". There is also case when blogger define one entry to many categories. Example, as shown in Figure 3 blogger labeled Entry 5 with Category 1 and Category 3.

To help improving the matching content between blog, ontology is implemented to classify the category. The categories act as the main node in visualization to help blogger identify the relation of each entry in using one category. Besides, ontology is

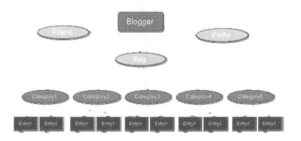

Fig. 3. Blog structure

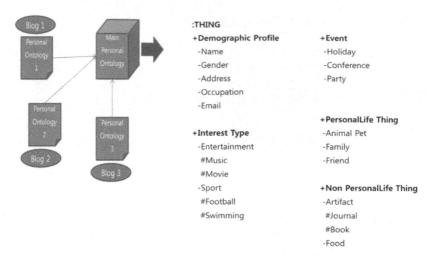

:THING

+Demographic Profile	+Event
-Name	-Holiday
-Gender	-Conference
-Address	-Party
-Occupation	
-Email	
	+PersonalLife Thing
+Interest Type	-Animal Pet
-Entertainment	-Family
#Music	-Friend
#Movie	
-Sport	+Non PersonalLife Thing
#Football	-Artifact
#Swimming	#Journal
	#Book
	-Food

Fig. 4. Personal Ontology structure

used to match the appropriate category where case of one entry has many categories occurred. In this problem, we developed a Main Personal Ontology as a main reference for making a blogger's personal ontology. Using Main Personal Ontology, we can find the similarity between each entry by mapping the related concept and attribute in personal ontology. Overall structure for building Personal Ontology that will be implement in the system is shown in Figure 4.

Some blog platforms such as Blogspot have gadgets where bloggers can simply adds the gadgets to their blogs. One of the gadgets is the "Blog List[3]" where it is used to show off what we read with a blogroll for favorite blogs. Blog List shows the newest entry based on favorite blogs (followers of blogger's blog). Figure 5 shows the example of Blog List where only the newest entry is appears in this blog.

Figure 5 shows an example of application reader used to display the newest entry by blogger's favorite friends. However, in our research, we focus on unread entry by blogger where any unread entry will be display in visualization and blogger still can read the newest entry as long as the entries did not read by blogger.

Fig. 5. Application used for read new entry by blogger defines friends

---

[3] http://help.blogger.com/bin/answer.py?hl=en&answer=97628

4 Framework

4.1 Overview of Visualization Framework

This section explains how the proposed system is working. For information, we used Malaysian Bloggers and some of the words in this paper are Malay Word and the meaning of each word can refer to the reference at the last section. We will start by blogger registration. Blogger registers the blog using the propose system. Once the registration is complete, system will analyze all contents in blogger's blog.

The overall system architecture is as Figure 6 below. Generally, it has eight processes starting from extracting the categories from blog and finally making visualization for blogger.

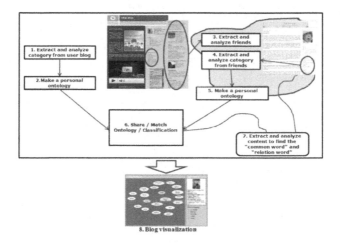

Fig. 6. Visualization Blog architecture design

Step 1:
After registration, system will extract the data from blogger's categories and these categories will be displayed as a node as a visualization map (Figure 7).

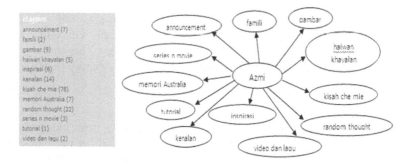

Fig. 7. Categories (Chapters) and Visualization

System will proceed to build blogger personal ontology. This personal ontology as shown in Figure 8 is used as main reference ontology to match friend's category with blogger category.

```
:THING
+Demographic Profile          +Event
 -Name                         -Holiday
 -Gender                       -Conference
 -Address                      -Party
 -Occupation
 -Email
                              +PersonalLife Thing
+Interest Type                 -Animal Pet
 -Entertainment                -Family
  #Music                       -Friend
  #Movie
 -Sport                       +Non PersonalLife Thing
  #Football                    -Artifact
  #Swimming                     #Journal
                                #Book
                               -Food
```

Fig. 8. Building a main Personal Ontology

Step 2:
Extract category from blogger's friend.
In this step, system will scan blogger's friend. Normally, blogger already defined his friends at the right or left side of blog, see Figure 9.

BLOG'S NAME	
<TITLE> <CONTENT> <TAG> <COMMENT>	<USER PROFILE> <CATEGORY> <ARCHIEVE> **<LIST OF FRIENDS>**

Fig. 9. List of friend

Here, system will filter blogger's friends and the process will continue by analyzing the friend's category. For each blogger, system will build personal ontology, meaning each blogger has his own personal ontology. This ontology is built in the same way as blogger personal ontology where we use category to define a class and concept.

Step 3:
Next, system will find the newest entry and non-read entry from blogger's friend. Once system found this group of entries, it will map based on the category that already classified in the first step. To map the entry to the exact category, matched

category needs to implement. As we know, the categories between blogger and friend are probably not same and to solve this problem, we will use matching method. There are three cases to match the category. Below are the cases that we identified in our research:

i. Belong to same category but use different name declaration.

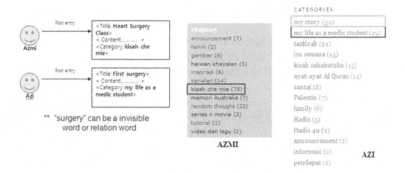

Fig. 10. Different label but belongs to same category

It means that favorite friend uses different name to label the category but actually the category is same as the blogger category. From Figure 10 blogger Azmi has "kisah che mie" category while blogger Azi has "my story" and "my life as a medic student" categories. These categories have the same meaning, example 'stories about their life'. For example, Azmi sent entry under category "kisah che mie" and Azi sent entry under category "my life as a medic student". In this case, we know that both categories are telling about their life so we assume that these entries should be in the same class. We use Azmi's personal ontology as a Refer Ontology as we explained in previous section; each entry from friends will be mapped to Azmi's category.

ii. One entry has many categories

Fig. 11. Example of one entry with more than one category

Figure 11 represents the case of one entry is labeled to more than one category. This case happens when blogger label the entry using many categories while in our propose system, the aim is to map one entry for one category only so that the blogger can consistently know that the entry is about the category it belongs to.

iii. Same labels but different categories.

Fig. 12. Two entries with same title and same category but have different meaning

Figure 12 shows the case where some word has more than one meaning and it is under unpredictable case. In this example, Azi posted one entry title AIA and categorized under AIA but in this case, Azi referred AIA as a Research field. While Azmi post the entry title AIA and also categorized under AIA but his entry is referred to Football field. To solve this problem, we will extract the content to find the match word and similar word based on meaning. For example, Azi explained about Research field, so naturally, she would use some words related to the research area such as "paper", "conference", "professor" and "thesis". On another side, Azmi is telling about Football game, so he would probably use words such as "game", "Ronaldo" for player, "Brazil" for country and "1-0" for football result. So, with these terms or words, we can match the correct category for each entry using the match ontology.

This is an important step in finding and matching the entry to the bloger's category. As the previous explanation, some entries have some problems and these problems should be solve before the matching category process done. In order to evaluate the enhancement of information retrieval between blogger and friend's blog, the same method to measure the semantic similarity in previous paper [11] will be implement in this system.

Step 4: Making visualization
This is the last step where system will make visualization for blog. The framework for visualization is shown in Figure 13.

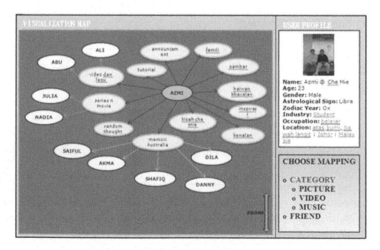

Fig. 13. Framework of visualization blog

4.2 Technical Specification

In making a visualization of blog, The Prefuse Visualization toolkit is used. Prefuse is a set of software tools for creating rich interactive data visualizations (prefuse.org). Prefuse is chosen as the visualization toolkit because it is successfully written using Java programming language, Java 2D graphics library and also it is easily integrated into Java Swing applications or web applets. Prefuse also can support a rich set of features for data modeling, visualization, and interaction.

To improve the visualization, we implement Fisheye View. It is developed based on the distortion approach. Tory and Moller 12] claimed that Fish-eye lenses magnify the center of the field of view, with a continuous fall-off in magnification toward the edges and the degree-of-interest value determine the level of detail to be displayed for each item and are assigned through blogger interaction.

5 Conclusion and Future Work

In this paper, we proposed a framework for visualization blog using ontology. With main goal to provide an interactive graphic for bloggers we hope that this system can help bloggers to get the information by interactive way using better interface interaction. In our visualization, blogger can reduce time to find the favorite content in blogs because our visualization is appeared the contents based on the blogger's define category. Besides, based on blogger behavior that only interested to know the latest entry content from favorite blog, we implement the concept where only the latest and non-read entry will appear in the visualization.

Another advantage for whether for novel or expert bloggers is identifics the relationship existing between entries, bloggers and friends. Implementation of ontology concept can help this system works more successfully in defining and detecting the relationship in blog architecture.

In our future research we will define our new approach in ontology where we can use this approach to map the same data in same node like we have explained in section 4. Besides with this system can help blogger in organizing time context management where blogger can save their time using one interactive interface to gain the information from their favorite blogs. In other hand this system can provide the overall context view of the data information from favorite blogs.

Some features such as communication and interactive tools will be added in this system so that blogger can directly post, add, and edit some media using this tool. Our next plan also will focus on e-business blog where we know many people use blog to conduct their online business.

References

1. Lenhart, A., Fox, S.: Bloggers: a portrait of the internet's new storytellers (2006), http://www.pewinternet.org/PPF/r/186/report_display.asp
2. Heer, J., Boyd, D.: Vizster: Visualizing Online Social Networks. IEEE Symposium on Information Visualization, InfoVis (2005)

3. Takama, Y., Kajinami, T., Matsumura, A.: Blog search with keyword map-based relevance feedback. In: Fuzzy systems and knowledge discovery, pp. 1208–1215. Springer, Heidelberg (2005)
4. Karger, D.R., Quan, D.: What would it mean to blog on the semantic web? In: McIlraith, S.A., Plexousakis, D., van Harmelen, F. (eds.) ISWC 2004. LNCS, vol. 3298, pp. 214–228. Springer, Heidelberg (2004)
5. Wang, J.Z., Ali, F.: An efficient ontology comparison tool for semantic web applications. In: The 2005 IEEE / WIC / ACM International Conference onWeb Intelligence (WI 2005), pp. 372–378 (2005)
6. Maedche, A., Staab, S.: Measuring similarity between ontologies. In: Gómez-Pérez, A., Benjamins, V.R. (eds.) EKAW 2002. LNCS (LNAI), vol. 2473, p. 251. Springer, Heidelberg (2002)
7. Wang, Z., Ali, F., Appaneravanda, R.: A web service for efficient ontology comparison. In: Proceedings of the IEEE International Conference on Web Services (ICWS 2005), pp. 843–844 (2005)
8. Kalfoglou, Y., Schorlemmer, M.: Ontology mapping: The state of the art. The Knowledge Engineering Review (1), 1–31 (2003)
9. Ehrig, M., Staab, S.: Efficiency of ontology mapping approaches. In: InternationalWorkshop on Semantic Intelligent Middleware for the Web and the Grid at ECAI 2004 (2004)
10. Schmidtke, H.R., Sofia, P.H., Gomez-Perez, A., Martins, J.P.: Some issues on ontology integration. In: Proceedings of the Workshop on Ontologies and Problem Solving Methods during IJCAI 1999 (1999)
11. Hsu, P.L., Liu, P.-C., Chen, Y.-S.: Using Ontology to Map Categories in Blog. In: Proceeding of the International Workshop on Integrating AI and Data Mining, AIDM 2006 (2006)
12. Fisheye View,
    http://www.infovis-wiki.net/index.php?title=Fisheye_View
13. Duong, T.H., Nguyen, N.T., Jo, G.S.: A Method for Integration across Text Corpus and WordNet-based Ontologies. In: IEEE/ACM/WI/IAT 2008 Workshops Proceedings, pp. 1–4. IEEE Computer Society Press, Los Alamitos (2008)
14. Duong, T.H., Nguyen, N.T., Jo, G.S.: A Method for Integrating Multiple Ontologies. Cybernetics and Systems 40, 123–145 (2009)

Semantic Data Integration in the Domain of Medicine*

Beata Jankowska

Poznań University of Technology, Institute of Control and Information Engineering,
Pl. M.Sklodowskiej-Curie 5,
60-965 Poznan, Poland
beata.jankowska@put.poznan.pl

Abstract. Finding an efficient solution to a problem of medical data integration is an important challenge for computer science and technology. Medical data integration is necessary both to retrieve missing data of an individual patient (horizontal integration) and to merge similar data from varied medical repositories (vertical integration) in order to increase data reliability. In the paper, we propose to solve the problem of medical data integration by means of an algebraic approach. In this approach also heterogeneous data can be considered. If only we know data taxonomies, can interpret data schemas and design schema mappings, then semantic and syntax differences are not an obstacle to integration.

Keywords: knowledge bases, domain ontology and taxonomy, schema mapping, data integration, algebraic approach.

1 Introduction

The problem of data integration constitutes one of the most important challenges for contemporary computer science and technology. If we have an access to a number of databases that keep data from the same domain, it emerges a possibility of combining these data, in order to: verify their correctness, retrieve missing components, or increase their reliability.

The verification of data correctness consists mainly in detecting potential contradictions between data coming from various sources. Retrieving missing components is the most frequent job of data integration. Assuming that each data can be expressed by means of elementary components whose values are fully known, partially known, or unknown, we can say that data retrieval leads to some change in the status of these components (at least – one of them): from unknown – to fully or partially known ones, and from partially known – to partially (to a greater extent than previously) or fully known ones. This kind of data integration we often call as "horizontal", due to the increase of data quantity, particularly – in the horizontal direction. The process of increasing data reliability consists in merging the same (or similar) data acquired from different sources. The greater the number of sources, the higher the reliability of data.

---

* The research has been partially supported by the Polish Ministry of Science and Higher Education under grant N516 369536.

N.T. Nguyen et al. (Eds.): New Challenges in Compu. Collective Intelligence, SCI 244, pp. 119–131.
springerlink.com   © Springer-Verlag Berlin Heidelberg 2009

Due to the increase of data quality, this new kind of integration is called as "vertical". The two mentioned kinds of data integration can be performed also for heterogeneous data, written in different formats. If only mappings between pairs of these formats are known, the semantic and syntax differences are not an obstacle to integration.

In the last decade data integration techniques received a great deal of attention and were reported in many papers. Most of them are focused on integrating data horizontally. The proposed solutions are based on relational data model and SQL query language [1,2,3], XML data model and XPath language [4,5], or description logics with OWL language [6]. The specificity of medical data, in particular – aggregate data specifying results of clinical experiments, leads us to think of yet another data model, i.e. feature structures [7,8]. They are suitable for representing aggregate data that consist of components taking elementary or compound values (also – set values). Adopting the open world assumption, they admit ignorance of particular components' values. Feature structures can be compared one with the other – by means of subsumption, and can be processed – by means of unification.

Irrespective of its form, a feature structure can be viewed as a list of key-value pairs, where each key is identified by its unique location in the structure. Such a viewpoint is close to a multisort algebra model, with disjoint sorts corresponding to the keys mentioned above. Since values in the pairs can be seen as set values (it is very justified with reference to aggregate data), then it would be proper to use some union and intersection operations for their processing. Being similar to classical set theory operations, they should be however based on data semantics and consider data 'counts'. In order to express full data semantics one has to use a domain ontology [9,10], but for the purpose of data integration – a domain taxonomy provides sufficient expressiveness. A special set's attribute will be needed for counting data occurrences.

In the paper we point at some algebra (lattice) as a tool that is strong enough to precisely define the rules of heterogeneous medical data integration.

2   Two Cases of Medical Data Integration

Medicine is one the domains where data integration is a task of great importance. In order to demonstrate the necessity of using the two mentioned kinds of data integration in medicine [11], let us consider the two following hypothetical cases.

First, let us assume that an elderly, 72-year man comes to the eye doctor's office. He complains of a serious visual field disorder. After having him examined, the doctor diagnoses glaucoma in its advanced stages [12]. In such a situation, the doctor suggests to the patient that he should undergo an aggressive pharmacotherapy. However, it is excluded in a case of cardiac insufficiency. Unfortunately, the patient cannot say if he suffers from this disease. He remembers nothing but the fact that he was hospitalized in a cardiology ward some years before. Having heard it from the patient, the eye specialist makes an attempt to find out missing data in the hospital repository. If he succeed, he will order their integration with the data stored in his own patients' files (under the constraint that the both data are consistent at their components of *name*, *surname* and *PESEL*, meaning a unique personal number in General Electronic Register of Polish Citizens.)

Next, let us consider another situation, taking place in an emergency room of state children's hospital. A doctor, being on hospital duty, admits a 8-year-old boy with moderate asthma exacerbation [13]. If he had no doubts, he would order a routine treatment, consisting in administering intravenous glucocorticosteriods (GCS) to such the patient. However, he knows from the medical interview that the boy suffers not only from moderate bronchial asthma but also from diabetes. The doctor remembers a case of 12-year-old asthmatic girl, with bronchial asthma accompanied by diabetes, who was admitted to hospital with a sudden asthma exacerbation. Immediately after taking intravenous GCS she went into coma. The doctor did not verify the hypothesis of the relation between the diseases and the response to the administered drug. That is why, he is looking for similar medical cases in all known repositories. This time, the constraint of data consistency concerns not *Personal_data* components (at most - patients' A*ge_range*, limited to 1- to 18-year-olds) but the diseases diagnosed in a patient (Moderate_asthma_exacerb and Diabetes). In response to his request the doctor obtains two medical case descriptions. They raise no doubts about using GCS in such a case. Having awareness of low statistical value of the two descriptions, the doctor weakens constraints imposed on medical data records and searches for the pediatric patients with Asthma_exacerb (more general than Moderate_asthma_exacerb) and Diabetes. Now, he succeeds in obtaining an extensive response, comprising 21 different medical cases (including the two previous cases), with 8 raising doubts among them. These 8 records includes Coma or Diabetic_coma as an element of their *Side_effect* components.

3   Foundations of Heterogeneous Data Integration

All discoveries in biology, genetics and fundamental medical sciences can improve as well diagnostic as therapeutic methods and influence the efficiency of medical care. However, a great advancement in medicine is also due to medical information exchange.

It is very desirable that medical knowledge is broad and fast to access. The two demands can be satisfied with the aid of efficient procedures of medical data integration. The integrated data represent knowledge that was acquired from many different sources and encapsulated in a small, useful form. They can be obtained "on demand", as temporary data, or systematically, as permanent data, ready for multiple use. The integrated data can be applied for designing the knowledge base of an expert system that helps doctors make diagnostic and therapeutic decisions [14].

Checking the possibility of data integration and also performing the whole process of integration must be based on data syntax and semantics. In order to define data semantics, designers conceptualize the domain knowledge in a form of domain ontologies or taxonomies. The ontologies can substantially differ one from the other, as well in sets of concepts and categories as in relationships between concepts and categories, hence it is yet necessary to define precise mappings between them (or, possibly, between each such ontology and some reference ontology, recognised as a pattern). Further on, for simplicity, we will consider one general set of concepts, common for all the domain ontologies. On the other hand, we will admit differences between categories and relationships proposed in various ontologies.

3.1 Medical Data Sources and Formats of Data Representation

From among all medical data sources, we are interested in those only that store data in some electronic, structuralized form. Speaking "structuralized", we mean a tuple form, that is characteristic of relational databases, or a hierarchical form, expressed by means of XML records [15]. As a matter of fact, if using an appropriate set of attributes, each of them being an equivalent of some path in XML data record, it is possible to determine one-to-one correspondence between XML records and tuples. In the current section, to make evident differences in the structures (schemas) of heterogeneous data, we will use XML data records. Beginning from the section 3.2, we will use the form of tuples that are not only readable but also convenient to store, process and access.

Let us assume that the eye specialist (chap. 2) keeps his electronic patients' files in a form of data records adapted to specific ophthalmology needs. Figure 1 shows a fragment of an XML record with data describing the elderly man with glaucoma. The hospital cardiology ward, that the doctor approaches for missing data of the man, stores patients' files in a form of other data records, different in schema from the previous ones. The one interesting for the eye specialist is illustrated in the Figure 2.

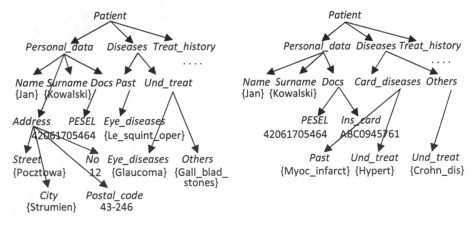

Fig. 1. Individual data - a record from eye patients' file

Fig. 2. Individual data - a record from cardiology ward file

As well schemas as contents of our records are simplified, but sufficient to illustrate all the problems met with when integrating heterogeneous data. Firstly, the records are linked to different ontologies (we could them call as 'eye ontology' and 'heart ontology', respectively). The both ontologies use different ontological categories (e.g. an eye category represented by the path *Patient -> Diseases -> Und_treat -> Eye_diseases* is absent in the heart ontology). Secondly, categories represented by similar paths *Patient -> Diseases -> Und_treat -> Others* and *Patient -> Diseases -> Others -> Und_treat* in the two records have two different ontological meanings.

In order to integrate the target cardiology record with the source eye record it is necessary to know a schema mapping specifying which way data structured under the target schema should be transformed into the source schema. Namely, the schema mapping assigns a target ontological category to each source ontological category in such a way that the target category is given directly (from the set of elementary target categories), or as a result of some operation (performed on elementary target categories), or as an empty category (meaning that the source category has not its counterpart in the target ontology).

Next, constraints imposed on the integration process ought to be examined. To this end, for each source category that is recognized as a key one, the consistency of two sets of individuals (ontological concepts, numbers, strings, etc.) is being tested: the first one, assigned to the category in the source record, and the second one, assigned to its counterpart in the target record.

Let us assume the schema mapping between schemas of our records to be a function taking for the source category *Patient -> Diseases -> Und_treat -> Others* - a target value *union(Patient -> Diseases -> Card_diseases -> Und_treat, Patient -> Diseases -> Others -> Und_treat)*, of obvious semantics. It is allowed to integrate the both records because of their consistency at key categories of *Patient -> Personal_data -> Name, Patient -> Personal_data -> Surname*, and *Patient -> Personal_data -> Docs -> PESEL*. That is why, the integration will be performed and its result at the examined category *Patient -> Diseases -> Und_treat -> Others* will be as follows: {Gall_blad_stones, Hypert, Crohn_ Dis}.

Similar data integration process will be performed in case of the 8-year-old boy admitted with moderate asthma exacerbation. However, this time it will comprise all available patients' files. The most valuable among the files will be registers of clinical trials (CTRs) [16]. CTRs contain aggregate results of experiments, carried out on groups of participants, for verifying various medical hypotheses. Integration of data from a few such registers satisfying consistency relations can be sufficient for obtaining a reliable answer. Each CTR is stored in a form of XML record comprising guidelines (personal and clinical features of participants, treatment rules) and results (values of clinically essential outcomes) of some experiment. Obviously, the schema of such aggregate record is somewhat different from the schema of that individual one.

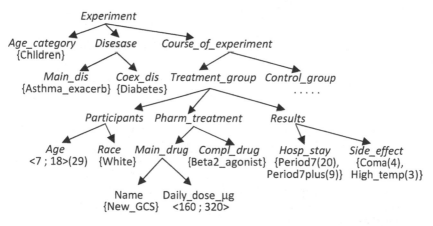

Fig. 3. Aggregate record – the evidence of an experiment with New_GCS

In the Figure 3 we can see a fragment of an exemplary aggregate record obtained for a group of children, with asthma exacerbation and coexisting diabetes, that were administered a modern GCS, called New_GCS. Among the outcomes tested were the length of *Hosp_stay* and *Side_effect* of the experiment. Concepts assigned to the both categories (*Period7*, *Period7plus*, *Coma*, and *High_temp*, respectively) have been enriched by numerical attributes, counting the concepts occurrences. Thanks to these attributes (being equivalents of aggregating 'count' operators from query languages), it is possible to use one compact aggregate record instead of many individual ones. Let us notice that our aggregate record contains data that can be immediately used by the doctor on hospital duty. From among 29 children that were administered New_GCS, as many as four children showed a side-effect of Coma. Thus, by an access to one aggregate record, the doctor acquired just as rich knowledge as can be obtained by means of integration of many individual records. However, if he is not sure enough about its quality [16], he can order integration of this record with further ones from various repositories.

3.2 A Course and Results of Data Integration Process

For clarity, let us restrict our considerations to aggregate records only. Furthermore, from now on, let us present the records in the syntax form of tuples (instead of previous XML ones).

Let us return to the case of our 8-year-old boy with moderate asthma exacerbation. In response to the question about a record with components of: *Main_dis*≥ Asthma_exacerb, *Coex_dis*≥Diabetes, *Age_category*=Children, and *Results_side_effect* ='*'(a claim for the record to contain *Results_side_effect* component), the doctor can obtain a tuple of the form illustrated in the Figure 4.

(*Main_dis*={Asthma_exacerb}; *Coex_dis*={Diabetes}; *Age_category*={Children};
Main_drug_name={New_GCS};*Main_drug_daily_dose*=<160;320>;*Compl_drug*={Beta2_agonist};
Participants_age=<7; 18>(29); *Participants_race*={White};
Results_hosp_stay={Period7(20), Period7plus(9)};
Results_side_effect={Coma(4), High_temp(3)})

Fig. 4. A result of quick searching of medical repositories

The results made him think so the doctor orders a full search of available repositories. It is possible that in case of administering another drug instead of New_GCS, coma will not occur at all. In order to check his hypothesis, the doctor asks for integration of all the records fulfilling the previous constraints and also two new ones, of the form: *Main_drug_name* = 'const' (a claim for the records to have all the same concepts assigned to *Main_drug_name* components), and Coma(?) in *Results_side_effect* (a claim for the records to contain Coma(?) in *Results_side_effect* components; if data relating to the side-effect of coma is lacking, then a default value of 0 for its 'count' attribute is applied: Coma(0)). Two answers obtained are illustrated in the Figure 5.

Now, the doctor has a broad view of side-effects of administering GCS to children with asthma exacerbation and coexisting coma. Although New_GCS has a great influence on the duration of the treatment period (decrease in the number of long hospitalizations), yet it gives relatively often (on average – in 17 from the global number of 102 cases) complications in the form of serious coma. All forms of itch (that is a frequent side-effect of taking Old_GCS) can be easily controlled.

(Main_dis={Asthma_exacerb}; *Coex_dis*={Diabetes}; *Age_categ*={Chidren};
Main_drug_name={New_GCS};*Main_drug_daily_dose*=<160;320>;*Compl_drug*={Beta2_agonist};
Partcipants_age=<4; 18>(102); *Participants_race*={White, Black};
Results_hosp_stay={Period7(72), Period7plus(30)};
Results_side_effect={Coma(17), High_temp(15)})

(Main_dis={Asthma_exacerb}; *Coex_dis*={Diabetes}; *Age_categ*={Chidren};
Main_drug_name={Old_GCS};*Main_drug_daily_dose*=<160;240>;
Partcipants_age=<3; 16>(73);
Results_side_effect={Itch(31), Coma(0)})

Fig. 5. A result of full searching of medical repositories

How does the process of obtaining the above integrated records get on? From all the records examined, those and only those are chosen for integration that satisfy the constraints defined in the question. The constraints refer to record's components and their number and contents are arbitrarily set by the user. The doctor on hospital duty searched for the data applying to children (the age range from 1 to 18 years): *Age_category* =Children, suffering from asthma exacerbation: *Main_dis*≥Asthma_exacerb, with coexisting diabetes: *Coex_dis*≥Diabetes, who underwent treatment with any but the same drug: *Main_drug_name*='const', and were under observation regarding side effects of the treatment applied: *Results_side_effect*='*'. In such the case, if only coma occurred, it would be stated for sure. So, in order to count the number of coma cases, it is sufficient to formulate an extra constraint of the form: Coma(?) in *Results_side_effect*.

In the above question, it was very important to use the constraint *Main_dis*≥ Asthma_exacerb, instead of the following one: *Main_dis*= Moderate_Asthma_exacerb. Thanks to the "greater-or-equal" sign it is guaranteed that all patients with asthma exacerbation will be taken into account (as well with a mild one, as a moderate or severe one). As a consequence, the integrated records are of great statistical importance (they relate to 102 and 73 children, respectively).

From the other side, their expressive power is lower in comparison to that of the individual record obtained while quick searching. The both ranges <4; 18> and <3; 18> of *Partcipants_age* are more general than the previous one <7; 18>, comprising only school-age children. Next, the information about *Participants_race* is now lacking or less precise (White or Black) in comparison to the previous, individual record (White). Also, the lack of *Compl_drug* component in the second integrated record proves that the reported 73 children underwent treatment with different complementary drugs (combinations of drugs), contrary to the first group of 29 children who underwent identical treatment, with the same complementary drug -Beta2_agonist.

As we shall see, in the process of integration, data components are treated diversely. Some of them, being of primary importance, must fulfil the constraints imposed. The others, being of secondary importance, can take any values. While integrating, total values of secondary components are being calculated by means of union or intersection operations, applied to corresponding partial values, depending on whether they represent disjunction (*Participants_age*, *Results_side_effect*) or conjunction (*Coex_dis*) of individuals.

4 Algebra of Medical Concepts and Categories

Knowledge interpretation, processing and exchange should be based on its conceptual model. That is why, for the last two decades, a number of health organizations have done research on designing medical ontologies.

The widely known UMLS (Unified Medical Language System) [17] is an effect of a long-term project initialized in US National Library of Medicine, in 1986. In next years, many further medical ontologies have been designed. We mean here as well general (e.g. SNOMED CT - Systematized Nomenclature of Medicine – Clinical Terms [18], built in 2002) as specialized ontologies (e.g. Disease Ontology [19] - a dictionary of diseases, designed in 2005 in a form of acyclic graph).

Lately, a great importance has been attached to the Gene Ontology [20], conceptualising genetic knowledge that is closely related to medicine. Although its dictionary includes a great number of concepts (about 25 thousands), yet only 5 types of semantic relations are used to reflect dependencies between each other. These are, among others: *is_a* – a simple subtype relation, where "A *is_a* B" means that concepts' category A is a subclass of concepts' category B, and *part_of* – a part-to-whole relation, where "A *part_of* B" means that if only category A is defined, then it must be a part of category B. From the other side, B does not necessarily contain A. The both mentioned relations are transitive.

4.1 Algebraic Data Specification Based on Medical Taxonomy

As we have seen, medical data differ one from each other in syntax form. However, major differences exist between ontologies that are the basis for defining data semantics. On the other hand, if only we had limited our considerations to data taxonomies, the last differences would have blurred. In most taxonomies, the following three types of semantic relations (and only these) are being considered: *is_a* and *part_of*, as they are defined in the Gene Ontology, and *not_greater*, a relation between two elements of the same category. Obviously, these are the most important relations for comparing, processing and merging data records. So, let us make a simplified assumption that the domain knowledge is modelled by means of taxonomies. Having accepted the simplification, we can get a useful, formal tool for designing the whole data integration process. We mean here an algebra-lattice, with its classical union and intersection operations. Further on, in the successive steps, we will present the algebra and also the way of its use in integration of heterogeneous medical data.

Let us first define the syntax of homogeneous medical data. It can be done by means of a multisort algebra, with elementary sorts representing concrete concepts (such as *Name*, *PESEL* from fig.1, or *Participants_age* from fig.3), compound sorts representing abstract concepts (such as *Personal_data* from fig.1 or *Results_side_effect* from fig. 3), and a special sort, representing two-element set of values $\{\top, \bot\}$ (denoting "the most general" and "the most particular", respectively). The algebra has the following form:

$$A = (\{S_i\}_{i \in I}, \{s_{i\_}is\_a\_s_j\}_{i,j \in I}, \{s_{i\_}is\_part\_s_j\}_{i,j \in I}, \{s_{i\_}is\_not\_greater\_s_i\}_{i \in I}, \qquad (1)$$
$$\{s_{i\_}union\_s_j\}_{i,j \in I}, \{s_{i\_}intersection\_s_j\}_{i,j \in I}) ,$$

where I stands for a set of sort indices, S_i - a sort with index i, $s_{i\_}is\_a\_s_j$-a relation of the type *is_a*, defined for the product of the domains S_i by S_j, $s_{i\_}is\_part\_s_j$ - a relation of the type *part_of*, defined for the product of the domains S_i by S_j, $s_{i\_}is\_not\_greater\_s_i$ - a partial order relation, defined for the product of the domain S_i by itself, $s_{i\_}union\_s_j$-a union function, defined from the product of the domains S_i by S_j to a codomain $S_k \cup \{\bot\}$, $s_{i\_}intersection\_s_j$-an intersection function, defined from the product of the domains S_i by S_j to a codomain $S_k \cup \{\top\}$.

If semantics of full programming language can be defined by means of a mapping between similar algebras (algebras of the same signature) [21], all the more semantics of homogeneous data can be defined in such a way. How to build an algebra defining data semantics? It should have a number of semantic sorts, one for each syntax sort mentioned above. The following rules of correspondence between sorts must be fulfilled:

- elementary sorts match simple valuations of the form: $[z_i <- D_{Si}]$, where D_{Si} is a semantic domain for the elementary sort S_i, it can be a subset of integers, a set of strings, a set of keywords, etc.
- compound sorts match multiple valuations of the form $[z_1 <- D_{S1}, ..., z_n <- D_{Sn}]$,
- elements of the special sort $\{\top, \bot\}$ match logical values, respectively true and false.

The semantic counterparts of the relations $s_{i\_}is\_a\_s_j$, $s_{i\_}is\_part\_s_j$ and $s_{i\_}is\_not\_greater\_s_i$ are subsumption relations, respectively: $r1_{\leq ij}$, $r2_{\leq ij}$ and $r3_{\leq i}$, stating the possibility to match arguments being valuations or logical values. In turn, the semantic counterparts of the functions $s_{i\_}union\_s_j$ and $s_{i\_}intersection\_s_j$ are the functions $\cup_{ij}$ and $\cap_{ij}$ respectively, of classical semantics.

4.2 From the Multisort Algebra to a Lattice

Let us keep the assumption about one set of medical concepts, common for all the domain ontologies (taxonomies). The hierarchical structure of medical data changes with each change in the subjective taxonomy. As a result, although the set of elementary sorts remains unchanged, yet the set of compound sorts usually undergoes a substantial change. However, in the light of the above considerations, the change does not spread to semantic domains. Since semantic counterparts of compound sorts are multiple valuations, being actually sets of simple valuations, then the way of organizing elementary sorts within a compound one is – from the semantic point of view – of no importance. Following this reasoning, we come to the conclusion that the algebra A can be a base for defining semantics of all medical data, regardless of format of their representation. However, before we use the algebra for this purpose, we will convert it into a more convenient form.

Let us focus our attention on dependences between elementary and compound sorts within the algebra A. Assume that the set of elementary sorts has a form of $\{S_{ei}\}_{ei \in EI}$, where $EI \subset I$ stands for a set of elementary sorts' indices, satisfying the condition (2):

$$\forall(S_{ei}, S_{ej} \in \{S_{ei}\}) \ (ei \neq ej => S_{ei} \cap S_{ej} = \varnothing) . \tag{2}$$

Then, each sort S_i can be defined as a product (3):

$$S_i = S_{e(i1)} \times S_{e(i2)} \times \ldots \times S_{e(in)} , \tag{3}$$

where $e(i1), e(i2), \ldots, e(in)$ are indices from the set EI, different one from each other.

Considering that positions of elementary sorts within a compound sort have no meaning, we can represent each sort by means of the same product (4):

$$SORT = SA_{e1} \times SA_{e2} \times \ldots \times SA_{es} , \tag{4}$$

where s stands for the number of all elementary sorts $S_{e1}, S_{e2}, \ldots, S_{es}, e1, e2, \ldots, es$ make an ordered sequence of indices: $e1 < e2 < \ldots < es$, and SA_{ei} ($1 \leq i \leq s$) $= S_{ei} \cup \{\top\}$. The most general value $\top$ occurring in the position ei means that the elementary sort S_{ei} has no influence on the form of the considered sort S_i.

Now, in a similar way, the relations $s_{i\_is\_a\_s_j}$, $s_{i\_is\_part\_s_j}$ and $s_{i\_is\_not\_greater\_s_i}$ can be reduced to the only relation $is\_subsumed$: SORT $\times$ SORT $\rightarrow \{\top, \bot\}$, such that:

$\forall(sort_i = (sa_{ie1}, sa_{ie2}, \ldots, sa_{ies}), sort_j = (sa_{je1}, sa_{je2}, \ldots, sa_{jes}) \in SORT)$

$$sort_i \ is\_subsumed \ sort_j = \begin{cases} \top, \ if \ \forall(1 \leq k \leq s)((sa_{iek} = \top) \vee (sa_{jek} = \bot) \vee \\ \qquad \vee (sa_{iek} \ s_{ej\_is\_not\_greater\_s_{ej}} \ sa_{jek})) \\ \bot, \ in \ opp.case . \end{cases} \tag{5}$$

At last, we can reduce all the union and intersection functions $s_{i\_union\_s_j}$ and $s_{i\_intersection\_s_j}$ to the generalized forms of $union$ and $intersection$, respectively, that are defined as follows (6):

$\forall(sort_i = (sa_{ie1}, sa_{ie2}, \ldots, sa_{ies}), sort_j = (sa_{je1}, sa_{je2}, \ldots, sa_{jes})) \in SORT)$ \hfill (6)

$sort_i \ union \ sort_j = sup(sort_i, sort_j) =$
$\qquad (sup_{Se1}(sa_{ie1}, sa_{je1}), sup_{Se2}(sa_{ie2}, sa_{je2}), \ldots, sup_{Ses}(sa_{ies}, sa_{jes}))$
$sort_i \ intersection \ sort_j = inf(sort_i, sort_j) =$
$\qquad (inf_{Se1}(sa_{ie1}, sa_{je1}), inf_{Se2}(sa_{ie2}, sa_{je2}), \ldots, inf_{Ses}(sa_{ies}, sa_{jes})) ,$

where sup_{Sek} and inf_{Sek}, $k = 1, \ldots, s$ are the functions of supremum and infimum, defined on the elementary sort S_{ek}, in accordance with the partial order relation $s_{k\_is\_not\_greater\_s_k}$.

As a result, after having done the above transformations, we obtain a generalized algebra g-A (7), that is actually a product of similar algebras A_k, $k = 1, \ldots, s$, of the form $A_k = (S_{ek}, s_{ek\_not\_greater\_s_{ek}}, s_{ek\_union\_s_{ek}}, s_{ek\_intersection\_s_{ek}})$:

$$g\text{-A} = (SORT, \textit{is\_subsumed, union, intersection}) = \prod(1 \le k \le s) \, A_k. \qquad (7)$$

As a matter of fact, the relation *is_subsumed* is given indirectly by means of the functions *union* and *intersection*. This way, it is redundant in the above signature g-A. After having it removed, we obtain an algebra of destination l-A, that is a lattice:

$$\text{l-A} = (SORT, \textit{union, intersection}). \qquad (8)$$

4.3 Examples of Using the Algebra to Do Data Retrieving and Integration

Finally, let us illustrate our considerations with a few examples of using the algebra **l-A** for medical data integration. To this end, let us refer to the case of our ophthalmological patient and his heterogeneous data given in the figures 1 and 2. As it was considered in 3.1, we will use Web Service technology to perform integration of ophthalmological data (schema in the fig. 1) with cardiological data (schema in the fig.2). The goal of integration is to learn of the diseases of 'Jan Kowalski', PESEL: 42061705464.

It is necessary that we know a schema mapping $map_{1\text{-}2}$: $SORT_1 \rightarrow terms(SORT_2) \cup$ $\{\varepsilon\}$, where $SORT_1$ is a set of all concepts' categories from the schema 1, $terms(SORT_2)$ - a set of all terms built from concepts' categories from the schema 2, by means of operators $union(\cup)$ and $intersection$ $(\cap)$, and ε is an empty value, meaning that a category from the schema 1 has no equivalent in the schema 2. For elementary categories shown in the fig.1, the mapping $map_{1\text{-}2}$ is as follows:

(1) $map_{1\text{-}2}(Personal\_data\_Name_1) = Personal\_data\_Name_2$
(2) $map_{1\,2}(Personal\_data\_Surname_1) = Personal\_data\_Surname_2$
(3) $map_{1\text{-}2}(Personal\_data\_PESEL_1) = Personal\_data\_PESEL_2$
(4) $map_{1\text{-}2}(Personal\_data\_Address\_Street_1) = \varepsilon$
(5) $map_{1\text{-}2}(Personal\_data\_Address\_No_1) = \varepsilon$
(6) $map_{1\text{-}2}(Personal\_data\_Address\_City_1) = \varepsilon$
(7) $map_{1\text{-}2}(Personal\_data\_Address\_Postal\_code_1) = \varepsilon$
(8) $map_{1\text{-}2}(Diseases\_Past\_Eye\_diseases_1) = \varepsilon$
(9) $map_{1\text{-}2}(Diseases\_Past\_Others_1) = \varepsilon$
(10) $map_{1\text{-}2}(Diseases\_Und\_treat\_Eye\_diseases_1) = \varepsilon$
(11) $map_{1\,2}(Diseases\_Und\_treat\_Others_1) =$
$\qquad Diseases\_Card\_diseases\_Und\_treat_2 \cup Diseases\_Others\_Und\_treat_2$

Fig. 6. Schema mapping between the schemas from figures 1 and 2

In turn, it must be examined whether constraints imposed on the process of integration are satisfied. We require that:

$$Personal\_data\_Name_1(=) \wedge Personal\_data\_Surname_1(=) \wedge Personal\_data\_PESEL_1(=), \qquad (9)$$

ordering the components *Name*, *Surname* and *PESEL* to be identical in source and target data being integrated. Next, for each elementary category of the source schema, we should appoint an operator of integration:

$$(3): \cap; \quad (1), (2), (4), (5), (6), (7), (8), (9), (10), (11): \cup. \qquad (10)$$

A query about cardiac insufficiency in 'Jan Kowalski', *PESEL*: 42061705464 will be as follows:

$$\{\text{Card\_insuff}\} \leq Diseases_1 + (?) \,, \tag{11}$$

where $Diseases_1+$ stands for a conjunction of individuals being the result of integration in the compound category *Diseases*. An answer obtained will be 'yes', because $Diseases_1+$ has a form {Le_squint_oper, Myoc_infarc, Glaucoma, Gall_blad_stones, Hypert, Crohn_dis} and the following relation holds: Card_insuff ≤ Myoc_infarc.

We can also formulate a more general query about connections between occurrences of Glaucoma and other diseases. Constraints imposed on the process of integration will be now as follows:

$$Diseases\{\text{Glaucoma}\}_1 \, (\leq) \,, \tag{12}$$

ordering each integrated category *Diseases* to contain the concept subsuming Glaucoma. Operators for all the elementary categories should be assigned in the following way:

$$(1), (2), (3), (4), (5), (6), (7): \cap \,; \quad (8), (9), (10), (11): \cup \,. \tag{13}$$

After having done integration, an answer can be found in $Diseases_1+ (?)$.

5 Conclusions and Future Research

The proposed method of medical data integration shall apply both in case of individual data, coming from individual patients' files, and in case of aggregate data, coming from CTRs. Based on applying the algebraic union and intersection operations to the categories from medical taxonomies, the method proves to be useful as well in discovering the knowledge that answers specific questions as in formulating general medical hypotheses. By means of the method, also heterogeneous data can be integrated. If only we have a medical Semantic Web (we know data, data schemas and all schema mappings), then semantic and syntax differences are not an obstacle to integration.

The hypotheses obtained can be transformed to the form of production rules, used for designing knowledge bases of medical RBSs. To this end, it is not only necessary to formulate rules' premises and conclusions but also estimate the level of their uncertainty. That is why, in future research, we will focus our attention on calculating the levels by means of known statistical methods.

The method should go through a thorough testing phase. At present, data of patients from the Clinic of Gynecology and Obstetrics of Poznan University of Medical Sciences are being gathered and preprocessed. We hope that the method will turn out to be effective and the data obtained will be a good base for defining medical production rules in the specialty of Obstetrics.

Obviously, the proposed method of data integration can be applied in all the domains where aggregate data are being stored and analyzed.

References

1. Miller, R.J., et al.: The Clio Project: Managing Heterogeneity. SIGMOD Record 30(1), 78–83 (2001)
2. Haas, L.M., Hernández, M.A., Ho, H., Popa, L., Roth, M.: Clio grows up: from research prototype to industrial tool. In: Proceedings of the ACM SIGMOD Int. Conf. on Management of Data, Baltimore, Maryland, USA, pp. 805–810 (2005)
3. Chen, H.: Rewriting Queries Using View for RDF/RDFS-Based Relational Data Integration. In: Chakraborty, G. (ed.) ICDCIT 2005. LNCS, vol. 3816, pp. 243–254. Springer, Heidelberg (2005)
4. Arenas, M., Libkin, L.: XML Data Exchange: Consistency and Query Answering. In: Proc. of the 24th Symposium on Principles of Database Systems, Baltimore, Maryland, USA, pp. 13–24 (2005)
5. Pankowski, T.: XML data integration in SixP2P a theoretical framework. In: ACM Data Management in P2P Systems, pp. 11–18 (2008)
6. Calvanese, D., De Giacomo, G., Lenzerini, M.: Description Logics for Information Integration. In: Kakas, A.C., Sadri, F. (eds.) Computational Logic: Logic Programming and Beyond. LNCS (LNAI), vol. 2408, pp. 41–60. Springer, Heidelberg (2002)
7. Ait-Kaci, H., Nasr, R.: LOGIN: A logic programming language with built-in inheritance. Journal of Logic Programming 3(3), 185–215 (1986)
8. Carpenter, B.: The Logic of Typed Feature Structures. Cambridge University Press, Cambridge (1992)
9. Anjum, A., et al.: The Requirements for Ontologies in Medical Data Integration: A Case Study. In: Proc. 11th Int. Database Engineering and Applications Symposium (IDEAS 2007), pp. 308–314. IEEE Computer Society, Canada (2007)
10. Poggi, A., et al.: Linking data to ontologies. In: Spaccapietra, S. (ed.) Journal on Data Semantics X. LNCS, vol. 4900, pp. 133–173. Springer, Heidelberg (2008)
11. Jankowska, B.: Medical Data Integration an Algebraic Approach. Journal of Medical Informatics and Technologies 12, 241–247 (2008)
12. Niżankowska, M.H. (ed.): Glaucoma (in Polish). Basic and Clinical Science Course (BCSC 10). Elsevier Urban & Partner (2006)
13. Global Strategy for Asthma Management and Prevention -Global Initiative for Asthma GINA Report (2008)
14. Jankowska, B., Szymkowiak, M.: How to Acquire and Structuralize Knowledge for Medical Rule-Based Systems? In: Knowledge-Driven Computing. SCI, vol. 102, pp. 99–116. Springer, Heidelberg (2008)
15. Jankowska, B.: Specificity and Methods of Medical Data Integration. Polish Journal of Environmental Studies 17(2A), 24–28 (2008)
16. Mrukowicz, J.: Foundations of Evidence Based Medicine (EBM), or the Art of Making Right Decisions in Nursing of Patients (in Polish). In: Medycyna Praktyczna Ginekologia i Położnictwo, June 2004, pp. 7–21 (2004)
17. Unified Medical Language System (1999-2009),
    http://www.nlm.nih.gov/research/umls/
18. Systematized Nomenclature of Medicine-Clinical Terms,
    http://www.ihtsdo.org/snomed-ct/
19. Disease Ontology, http://diseaseontology.sourceforge.net/
20. The Gene Ontology (1998-2008), http://www.geneontology.org/
21. Dembiński, P., Małuszyński, J.: Mathematical Methods of Defining Programming Languages (in Polish). WNT, Warszawa (1981)

Part III
Social Networks

Analysis of Social Network's Structural Properties in Huge Community Portal

Bernadetta Mianowska, Marcin Maleszka, and Krzysztof Juszczyszyn

Institute of Computer Science
Wroclaw University of Technology,
Wyb. Wyspianskiego 27, 50-370 Wroclaw, Poland
bmianowska@gmail.com, maleszka@gmail.com, krzysztof@pwr.wroc.pl

Abstract. The structure Internet-based social network of Nasza-klasa.pl – a huge (7.5 million users) social community portal was analyzed with respect to popular social network structural properties. We have found out that the network structure is typical for affiliation network and proposed its hypergraph representation. The temporal changes of the network structure were also analyzed.

1 Introduction

Usually social networks are viewed as graphs with nodes representing people and edges representing their relations. These edges are almost always one- or two-way connections of interest in other person, appreciation of his work, common interests of others work or more or less defined friendship. Sometimes also so-called weak links are marked, representing more distant forms of relation. In this paper we work on a social network consisting entirely of weak links.

Nasza-klasa.pl [13] is a Polish community portal resembling Facebook [14] and allowing grouping of people into school classes (schools – elementary schools, colleges, universities etc. – as groups of common interests are possible as well) and connecting them by a link of 'friendship' that must be confirmed by both sides. At this time nasza-klasa.pl contains more than 15 million user profiles and more than half of those are profiles of real people (due to large numbers of fake profiles those are now marked). This means that the database on nasza-klasa.pl servers is the second largest database of personal data in Poland (first being social insurance). At this point it is virtually impossible to crawl full user data (which means friend connections), as the social network grows faster than any outside tool can view and save it. Thus we have tried a different approach to acquiring and understanding data from this site. In this paper we concentrated on weak links, which may be downloaded faster.

The data we gathered were lists of all users from a particular classes in a particular schools (where fictional groups where downloaded this represents groups of interest). We downloaded about 0,5-1% of the created schools. This is a small part of the overall network structure, which is hardly possible to crawl due to its size and dynamics. From the point of view of local network topology it constitutes a complete sample allowing to analyze the structures typical for this network. As stated above, collecting data from

N.T. Nguyen et al. (Eds.): New Challenges in Compu. Collective Intelligence, SCI 244, pp. 135–146.
springerlink.com　　　　　　　　　　　　　　　　　　© Springer-Verlag Berlin Heidelberg 2009

the entire portal was impossible due to the scale and dynamics of the system. The basic way to create social network from this data is to connect each user in the same class with a weak link – a bidirectional edge in the graph. This approach is justified because the users have to openly declare themselves as members of classes. Connections between those user clusters (classes) may then be done by tracking users, who belong to two classes (i.e. in primary school and in high school). Usually teachers also join different classes thus establishing connections a school. An average high school with over 2500 profiles and over 100.000 links is presented below (full graph on Fig.1, connected center on Fig.2. Note, that most visible nodes on Fig.1 and 2 are in fact groups of nodes which form a fully connected clique and thus are drawn together).

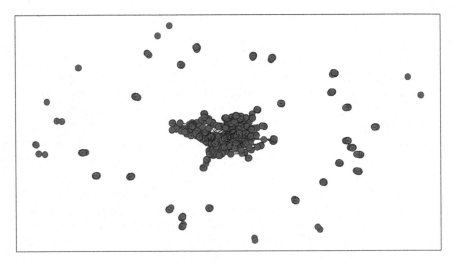

Fig. 1. Random single school as represented by the NetworkX Python package

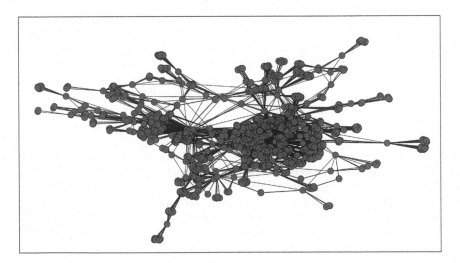

Fig. 2. A zoom on the connected part of the single school represented in Fig.1

Note that small schools in low internet availability areas or newly created school groups may have less than 100 users and large universities up to 50.000 of users. The entire social structure stored in nasza-klasa.pl portal differs from classic Internet-based social sites – there is an implicit structure (schools and classes) which influences the actual connection patterns rendering traditional analysis methods misleading and inadequate, as we will show in the following sections.

The structure of the portal may be also viewed in a different manner. If graph nodes were to represent school classes and edges – users that belong to both of them, the result network would be eligible to standard research. Moreover, in this approach we face also an implicit temporal dimension, obviously membership in an elementary school class and a college class does not happen in the same time. In this case the analysis allows discovering personal "education paths" of the users which may be of use (for example) for universities which want to know where their students come from.

As this network is highly unreadable both visually and imposes difficulties for standard measures like centrality indices, we propose to use hypergraphs to work on it and other, similar networks.

This paper is organized as follows: section 2 presents a standard analysis for nasza-klasa.pl weak link network, section 3 presents a different approach to network representation with people being represented by edges instead of nodes, as an affiliation network and as a hypergraph, and finally section 4 concludes the paper with final remarks.

2   Standard Network Analysis of Nasza-klasa.pl

The standard analysis of nasza-klasa.pl social network was performed in two time points, June 2008 and September 2008. About 0,5-1% of the weak link network data was downloaded. But in this section we will focus on a single school – one of over 200.000 schools registered on this site – in order to present the structural properties of social networks emerging on nasza-klasa.pl.

The single school (the other, randomly selected for analysis, returned similar results) had almost 2.700 registered participants in June 2008 and over 2.800 in September 2008 (a big part of those new users consisted on pupils just after the summer recruitment to the school, but there were also some people who finished it decades ago). With over 100.000 edges between nodes, about 90% of the nodes belonged to Largest Connected Component (LCC) of the social network graph. Table 1 presents the dynamics of the network growth within three holiday months. We may see significant growth in the number of nodes and edges, which is also the common feature of all the schools on nasza-klasa.pl. Along with the 7.5 millions of personal accounts this rate of changes implies the above-mentioned difficulties with crawling the entire site.

Those graphs were then used to calculate some standard properties of the networks and evaluate them in light of social network analysis.

An important number for each social network is its centrality. While one should have no doubt that the analyzed network was a social one, the value of centrality for it points in entirely different direction. Three approaches to calculate centrality were used: Brandes betweenness centrality, Newman betweenness centrality and classical

Table 1. The size of single school graph from nasza-klasa.pl

	Nodes	Edges	Edges per node
June 2008	2695	107621	39,9
September 2008	2815	116348	41,3
Change (%)	4,5%	8,1%	3,5%
Change (abs)	120	8727	1,4

betweenness centrality. The first one is defined as the fraction of number of shortest paths that pass through each node [3]. The second one is defined as load centrality for nodes - the fraction of number of shortest paths that go through each node counted according to the algorithm in [4]. The classical centrality was calculated using the algorithm presented in [5] and is simply the fraction of number of shortest paths that pass through each node. Each algorithm used returned the same results up to 10th decimal place. The average node betweenness results were 0,00067 for June data and 0,00057 for September. This and the following results were compared with the values typical for social networks [1] and the results obtained for spontaneously formed e-mail based social network of the employees of Wroclaw University of Technology (from here on denoted as "WUT network") described in [2]. The e-mail based WUT social network contains 3514 nodes and 141670 edges (so the complexity is similar) but there are no groups and classes of users, and the links are formed just by mutual communication between the individuals. This network will serve us as an example to show that the social structure of nasza-klasa.pl differs significantly from typical networks emerging from Internet communication.

The average node betweenness for the WUT network is 0,00063 which is almost exactly the same as for nasza-klasa.pl. However, the distribution of the individual node betweenness centralities is very different. The existence of densely connected cliques (in fact: fully connected subgraphs) in nasza-klasa.pl network implies that the nodes belonging to large cliques have betweenness centrality close to zero, while a few (0.5% of all) bridging nodes, adjacent to links between cliques, have abnormally high centrality. Despite similar average values the highest measured betweenness centrality was 0,0534 for the WUT network and 0,414 for nasza-klasa.pl network.

Additionally, closeness centrality and degree centrality for the networks were calculated. The first of those is defined as one divided by average distance to all nodes. The latter is the fraction of nodes connected to. Closeness centrality equaled 0,39 for June and 0,41 for September and degree centrality was 0,00887 and 0,00889 respectively. Graph density, defined as in (1) below was as low as 0,00887 and 0,00889 as well.

$$density(G) = \frac{size(G)}{order(G) * \frac{(order(G) - 1)}{2}} \qquad (1)$$

Where:

$size(G)$ – denotes the number of nodes in graph G
$order(G)$ – is the highest degree of a node in graph G

Apart from centrality measures which address network connectivity in global sense, two important structural measures of local graph connectivity were computed: clustering coefficients *CC1* and *CC2*. The clustering coefficients were defined in the standard way, according to eq. 2 and 3:

$$CC1 = \frac{2|E(G1(n))|}{\deg(n)(\deg(n)-1)} \qquad (2)$$

$$CC2 = \frac{|E(G1(n))|}{|E(G2(n))|} \qquad (3)$$

where:

deg(n) – denotes degree of node *n*,
$|E(G1(n))|$ – is the number of edges among nodes in 1–neighbourhood of node *n*,
$|E(G2(n))|$ – is the number of edges among nodes in 1 and 2–neighbourhood of node *v*.

We also assume that for a node *n* such that $deg(n) \leq 1$ all clustering coefficients are 0. The intuitional meaning of the CC1 and CC2 is that they represent how many edges exist within 1- (or 2- respectively) edge radius from the node *n* compared to the number of possible edges. CC1 or CC2 equalling 1 mean that the nodes in 1 (or 2) edge distance from *n* form a full graph.

The results obtained for nasza-klasa.pl graph are given in Table 2.

Table 2. The average clustering coefficients for nasza-klasa.pl

	CC1	CC2
June 2008	0.9492	0.2152
September 2008	0.9514	0.1846

In comparison the average values of CC1 and CC2 for the WUT network were CC1 = 0.19567 and CC2 = 0.01217. We may see that the local structure of nasza klasa.pl tends to look like the set of connected cliques, which feature is preserved even within 2-edge radius from given node (with CC2 greater by the order of magnitude when compared to the WUT network).

We see that the overall network connectivity (expressed by the betweenness centrality) is similar to spontaneously emerging social communication networks (excluding a few nodes for which we measured extremely high centrality). The other measures, however, especially these which take into account the local connection patterns return not-typical values. Clustering close to 1 suggests dense network and the existence of cliques which are in fact full graphs.

This suggests that the techniques which build on local network structures may fail or return exotic results.

We confirmed this conclusion by the preliminary analysis of nasza-klasa.pl network with a FANMOD tool [11], dedicated for measuring the distribution of small (3-7 nodes in size) subgraphs (*network motifs*) by network sampling method. In most networks this allows the creation so-called *motif significance profiles* which reflect the typical local connectivity patterns, already measure for various types of social networks [2][10]. However, in the case of nasza-klasa.pl network only full subgraphs (motifs) were represented (with abnormally high frequencies) in the significance profile and the results were not interpretable.

Summing up, the local topology of connections within nasza-klasa.pl also differs significantly from classic internet-based social networks. The structure of connections in our network is distorted by introducing classes and schools which interfere with normal (free-way) patterns of creating relationships in social networks.

We concluded that this type of network should be analyzed as an affiliation network, despite its free mode of creation of relations which resembles many other Internet-based social networks.

3   Nasza-klasa.pl as an Affiliation Network

Affiliation networks are often defined as collections of subsets of entities [1]. There is an inherent duality in this definition because each group (or: event) is defined as a set of actors (users, agents) and each actor defines also a set of groups he belongs to. Previously described graph of nasza-klasa.pl was a social network created by standard means by connecting nodes-people by edges-relations of acquaintance. In this section we describe another approach that may be also analyzed by means of standard graph theory.

We will analyze our network as an affiliation network. A node will now be understood as a whole school class - the group of people that earlier constituted a complete subgraph. Correspondingly, an edge between such nodes will be a relation of acquaintance between any single member of one group (node) and any single member of the second group (node). The edge is in fact a single person or a set of persons who are members of both school classes.

This interpretation strays highly from the classical view of social networks that arose from social sciences, but in this situation it is more precise in showing the actual social connections. Whole school classes, being a group of mutually acquainted people, may be treated as single entities (peer groups).

This approach implies the use of the hypergraph representation of our network. A hypergraph is a graph in which an edge can connect more than two nodes. In other words – an edge is a subset of nodes [6][7]. Hypergraphs have been found useful for analyzing social structures where we experience complex scenarios and relations involving more than two actors. In these cases we must extend a standard definition of social relation which bases on interaction between a pair of actors. Affiliation networks are good example of the structures in which the hypergraphs were applied [8][9].

The data set used in the following analysis was identical to these used in section 2. Thus, a single school with almost 2.700 registered participants in June 2008 and over 2.800 in September 2008 was used. Fig. 3 presents the entire network hypergraph, while Fig.4 – its central connected part. The nodes were organized into 262 school classes (hypernodes) in June and 260 in September (few fake classes were found

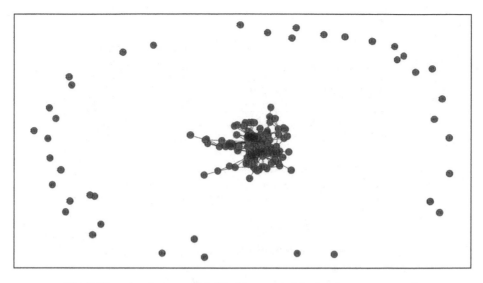

Fig. 3. The school presented in Fig.1 interpreted in the alternate approach

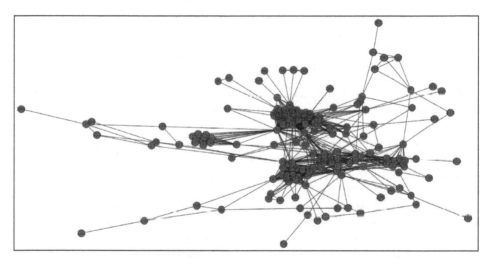

Fig. 4. A zoom on the connected part of the single school presented in Fig.3

and deleted from the school and a few new were created in their stead), of which 227 and 226, respectively, were connected within the network structure. Between those classes (now understood as hypernodes) there were over 12.000 connections and number of those grew by 12% within three months (note that number of edges in the network presented in section 2 grew by only 8% in the same time). The number of edges per node, while being higher to start with, grew over three times faster in the same time. For full data on this figures refer to table 3.

Table 3. Affiliation network created with social network data

	Hypernodes	Edges	Edges per node
June 2008	262	12232	46,7
September 2008	260	13703	52,7
Change (%)	-0,8%	12%	12,9%
Change (abs)	-2	1471	6

When we assume that the classes (now being hypernodes in a hypergraph – affiliation network) are connected by the hyperedges (sets of users belonging to both connected hypernodes) the structure of the network changes and so does the dynamics of growth (Table 4).

Table 4. Affiliation network created with social network data

	Hypernodes	Hyperedges	Edges per node
June 2008	262	2231	8,51
September 2008	260	2750	10,57
Change (%)	-0,8%	23%	24,2%
Change (abs)	-2	519	2,06

Notice that the growth rate, measured in the number of nodes and edges doubles when we switch to the hypergraph representation, showing that despite relatively stable number of nodes, the associations between them emerge more intensively then new user-to-user relations. This suggests that the hypergraph approach is better for analyzing the dynamics of this network.

As previously, the graphs were used to calculate some standard properties of the networks and evaluate them in light of social network analysis.

The analysis of three distinct betweenness centrality measures led, as in section 2, to unsatisfactory results. All three were equal to 0,00475 for the June 2008 dataset and 0,00394 for the September 2008 dataset. Again, these values decreased with time and growing number of edges, which is even more surprising in this network as the number of nodes did not increase.

Additionally, closeness centrality and degree centrality for the networks were calculated. The first one equaled 0,28 for June and 0,33 for September and degree centrality was 0,067 and 0,086 respectively. Graph density, defined as in (1) was as low as 0,07796 and 0,09931 as well.

The values of the average clustering coefficients CC1 and CC2 (as defined in section 2) for the graphs was not as high as in the standard network and equaled 0,59 and 0,63, respectively. During considered three months we experienced the growth of CC1, reflecting the growing number of connections within one-edge radius from given node. From the other hand, the CC2 for September 2008 is smaller which

Table 5. The average clustering coefficients for the affiliation network of nasza-klasa.pl

	CC1	CC2
June 2008	**0.5964**	**0.2144**
September 2008	**0.6323**	**0.1732**

is normal in the case of growing networks (maintaining the same edge density in the 2-edge radius requires more much edges then the same for CC1).

The values of local clustering coefficients are this time much more typical and similar to other social networks investigated (compare with the results for WUT network in section 2).

Following this track of thought, a larger experiment was conducted, using the whole September data set, with schools (clusters of users) represented as nodes and users registered in two schools standing for an edge joining these nodes.

In this representation, the graph consisted of 1021 nodes (distinct schools) with edges derived from 2238915 users registered to multiple schools. The Largest Connected Component (LCC) of the created network consisted of 880 nodes. Partial representation of the created network is presented in Fig.5. This representation may be used for informal sociological interpretation, for example it can be seen that students from multiple high schools attend later to the single university, creating tree-like patterns (e.q. node 8 in the bottom of Fig.5). This kind of information may be useful for university administration, allowing to measure the preferences of college graduates. However, in this paper, only structural properties of this network will be discussed.

The centrality measures – Brandes betweenness centrality, Newman betweenness centrality and classical betweenness centrality (as described in section 2) – yielded

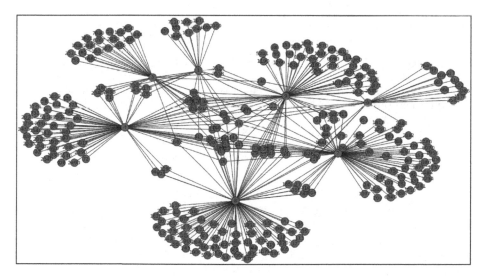

Fig. 5. Partial representation of the schools network in NetworkX Python package

identical results of 0,00163 for this data set (0,00189 when considering just the LCC). The degree centrality was calculated as 0,0366 (0,0315 in LCC) and the closeness centrality as 0,347 (0,299 in LCC). With 49 distinct cliques identified in the graph, the average clustering was 0,57, the average degree of a node 32,16 and the density just 0,0366.

Table 6. Affiliation network created with social network data

	Hypernodes	Edges	Edges per node
Single school	**260**	**13703**	**52,7**
Whole data set	**1021**	**2238915**	**2192,9**

A single school picked for measurements in the September data set consisted of 260 student classes, each of about 30 people, who additionally know some people from outside the class (e.q. teachers, guest or temporary students), equaling at an average of 52,7 edges connected to each node.

On the other hand, in the whole data set consisting of 1021 schools, the number of edges per node leads to a conclusion of an average of between 1000 and 2000 students in each school (depending on a single or double connection, that is the student being registered in two or three different schools). With users age between 5 and 80 this leads to about 15-30 people per tier.

Table 7. Affiliation network created with social network data – other parameters

Parameters	school	school net	ratio
Betweenness	0,00057294	0,00163	2,8
Degree centrality	0,00889109	0,0315	3,5
Closeness centrality	0,39023798	0,299	0,76
Graph clique number	71	49	0,7
Average clustering	0,9499808	0,5691896	0,6
Number connected components	38	36	0,9
Degree	23	32	1,4
Density	0,008869	0,03659375	4,1

Comparing the network by the parameters, one may note the same general order of the values. The closeness centrality is about the same, but the number o cliques is about one third lower in the larger network (this is understandable as multiple schools are a much more distributed network – additionally this is confirmed by almost twice lower average clustering). On the other hand the betweenness values and the degree centrality are three times larger in the whole data set with degree (number of nodes connected to) being just 50% larger. With a large number of edges per node, the high density of the whole school network is also understandably higher.

All these results suggest that in the case of implicit structures (like schools and classes) underlying social network basic structural analysis with measures dedicated to the discovery of the role of single node may be misleading.

4 Conclusions

Nasza-klasa.pl is a huge social community portal, which growth rate for the last 12 months was 5-10% per month (regarding both the number of network nodes and connections). These dynamics paired with the estimated number of 7.5 million users makes the analysis of the entire portal social network structure very difficult and complex task.

However, even mining of its partial structure (which may be understood as random network sampling) reveals some of its properties which may be important for the future analysis of this and other community portals of this kind.

The most important conclusions coming from our experiments are:

- The structure of the portal, which is a mix of free-association social network and an affiliation network, results in dense structure of local connections between pairs of users. The reason is that (along with the possibility of establishing free relations) the users are organized in classes which naturally support clique formation.
- This feature implies the limited usability of structural social network measures which are based on local node connectivity.
- The affiliation network model, based on hypergraph approach, better reflects the structure and dynamics of the network.

For future experiments we plan to apply existing hypergraph partitioning methods and modify them in order to use under condition that only a partial knowledge of the network structure is available.

The other promising area of future research is the educational history of single users which may be mined from the portal community structure. Users belong to classes which are associated with elementary schools, colleges, universities and so on. This gives the possibility of mining of the temporal data describing the history of single user and – along with the size of the portal – allows conducting the statistical research on educational preferences and their changes (the age of the users varies from 5 up to 80 years).

References

1. Wasserman, S., Faust, K.: Social network analysis: Methods and applications. Cambridge University Press, New York (1994)
2. Juszczyszyn, K., Kazienko, P., Musiał, K.: Local Topology of Social Network Based on Motif Analysis. In: Lovrek, I., Howlett, R.J., Jain, L.C. (eds.) KES 2008, Part II. LNCS (LNAI), vol. 5178, pp. 97–105. Springer, Heidelberg (2008)
3. Brandes, U.: A Faster Algorithm for Betweenness Centrality. Journal of Mathematical Sociology 25(2), 163–177 (2001)

4. Newman, M.E.J.: Scientific collaboration networks: II. Shortest paths, weighted networks, and centrality. Phys. Rev. E 64, 016132 (2001)
5. Brandes, U.: A Faster Algorithm for Betweenness Centrality. Journal of Mathematical Sociology 25(2), 163–177 (2001)
6. Berge, C.: Graphs and Hypergraphs. North-Holland, Amsterdam (1973)
7. Berge, C.: Hypergraphs: Combinatoics of Finite Sets. North-Holland, Amsterdam (1989)
8. McPherson, J.M.: Hypernetwork Sampling: Duality andDifferentiation Among Voluntary Organizations. Social Networks 3, 225–249 (1982)
9. Estrada, E., Rodriguez-Velasquez, J.A.: Subgraph Centrality and Clustering in Complex Hyper-networks. Physica A 364, 581–594 (2006)
10. Milo, R., Itzkovitz, S., Kashtan, N., Levitt, R., Shen-Orr, S., Ayzenshtat, I., Sheffer, M., Alon, U.: Superfamilies of evolved and designed networks. Science 303(5663), 1538–1542 (2004)
11. Wernicke, S., Rasche, F.: FANMOD: a tool for fast network motif detection. Bioinformatics 22(9), 1152–1153 (2006)
12. Hagberg, A.A., Schult, D.A., Swart, P.J.: Exploring network structure, dynamics, and function using NetworkX. In: Varoquaux, G., Vaught, T., Millman, J. (eds.) Proceedings of the 7th Python in Science Conference (SciPy2008), Pasadena, CA USA, August 2008, pp. 11–15 (2008), http://math.lanl.gov/~hagberg/Papers/hagberg-2008-exploring.pdf
13. Community portal Nasza klasa, http://www.nasza-klasa.pl
14. Community portal Facebook, http://www.facebook.com/

Efficient Construction of (d+1,3d)-Ruling Set in Wireless Ad Hoc Networks*

Krzysztof Krzywdziński

Faculty of Mathematics and Computer Science, Adam Mickiewicz University,
60–769 Poznań, Poland
kkrzywd@amu.edu.pl

Abstract. Let $G = (V, E)$ be a graph and $V' \subseteq V$. We call V' a (d, b)-ruling set if for all $v_i, v_j \in V'$ the distance $dist(v_i, v_j) \geq d$ and V' is b-dominating set (i.e. each vertex of G can be reached form some vertex from V' by a path of length at most b).

Here we consider the problem of finding (d, b)-ruling set in a Unit Disc Graph. Unit Disc Graphs (UDG) are the most natural class of graphs to model wireless ad hoc networks (for example cell phones networks, wireless sensor networks etc.). The set of vertices of UDG is a set of points on the plane and two vertices, say v and w, are connected by an edge if and only if they are at the distance at most one in Euclidean norm ($\|v, w\| \leq 1$). We also assume that every vertex knows its position on the plane and, for such setting, present a deterministic algorithm which finds $(d + 1, 3d)$-ruling set in UDG, and works in $O(poly(d))$ time (where d is a parameter).

As a result we are able improve time complexity (from $O(\log |V(G)|)$ to $O(poly(d))$) of several known algorithms for UDG, such as algorithm determining k-dominating set, maximum matching, minimum connected dominating set, minimum spanning tree and regular clustering.

1 Introduction

In the field of distributed algorithms on graphs it is important to divide any arbitrary graph into "large" connected subgraphs (clusters) with "small" diameter. Such a division enables to solve problems locally and then apply those local results to get a global solution. In such approach we usually first define a set L of, say, l, vertices of G, called "leaders" and next define "clusters" (subgraphs of G) $K = \{K_1, K_2, \ldots, K_l\}$, such that a vertex $v \in V(G)$ belongs to i-th cluster ($v \in K_i$) if and only if i-th "leader" is the nearest to v considering vertices from the set L. Given a problem and a clustering of a graph G determined by a set of "leaders", using simple distributed BFS procedure, we compute a local solution of the problem in each cluster and next use them to work toward a global solution. Ruling sets serve as a tool to obtain a clustering with desired properties.

* This work was supported by grant N206 017 32/2452 for years 2007-2010.

N.T. Nguyen et al. (Eds.): New Challenges in Compu. Collective Intelligence, SCI 244, pp. 147–156.
springerlink.com   © Springer-Verlag Berlin Heidelberg 2009

The first approach to the problem was presented in the paper of Awerbuch, Goldberg, Luby and Plotkin [1], where a distributed algorithm for finding $(d, d \log |V(G)|)$-ruling set in any graph G, with d as an input, is given. Their algorithm works for every graph G in $O(d \log |V(G)|)$ synchronous rounds (for further results in this direction for general graphs see [2] and [3])

In this work we concentrate on a very important class of graphs, namely on Unit Disc Graph (UDG). Unit Disc Graphs (UDG) are the most natural class of graphs to model wireless ad hoc networks (for example cell phones networks, wireless sensor networks etc.). Recall, that the set of vertices of UDG is a set of points on the plane and two vertices of UDG, v and w, are connected by an edge in UDG if and only if they are at the distance at most one in Euclidean norm ($\|v, w\| \leq 1$). Here we also assume that every node is equipped with the Global Positioning System (GPS), or knows its position on the plane by other sources. If in addition we assume that each node has different coordinates, then we can define unique ID determined by its coordinates (x, y) on the plane. Moreover, to simplify arguments let us assume that local clocks of vertices of UDG can be synchronized i.e., assume that we perform computations in rounds (synchronous model). However, it should be mentioned that all our results hold for asynchronous model as well. Precisely we will use model LOCAL defined in [4], in which each vertex knows its position. We will call the discussed model LOCAL+COORD model of computations.

In this paper we give a deterministic algorithm which finds $(d + 1, 3d)$-ruling set in UDG, and works in $O(poly(d))$ time, where d is a chosen parameter. This algorithm has many nice applications, lading to a substantial improvement of time complexity of algorithms dealing with basic properties of Unit Disc Graphs, such as dominating sets and maximum matchings (for details, see 5). Moreover our construction of $(d + 1, 3d)$-ruling sets is instrumental in solving an interesting problem of regular clustering, i.e., finding a partition of a graph into bounded connected components (see [5]).

The key idea behind the construction of the ruling set algorithm, stems from a new approach to the analysis of local properties of Unit Disc Graphs, called here *"three lattices method"*. We should also mention that this method, can be applied in many other cases concerning properties of UDG. For example, it enables us (see [6]) to get $(1 + O(1/d))$ approximation of Minimal Spanning Tree (where weight of an edge is equal to the distance of its vertices) in $O(poly(d))$ time, a substantial improvement over previously known algorithms.

Our paper is organized as follows. In Section 2 we present main idea of our algorithm - the "three lattice method". In Section 3 we prove few simple lemmas, which will be useful in the analysis of algorithms. In Section 4 we introduce algorithms RULINGSETINSQUARE and DELETINGLOCALCOLLISIONS, which formalize ideas of "finding local solutions" and "deleting collisions". In this section we also show correctness of our main algorithm RULINGSET (Theorem 1 stated in Section 2). In Subsection 5.1 we show a simple application of our result for finding a $O(d)$ approximation of the optimal d-dominating set in $O(poly(d))$ time. In Subsection 5.2 we show that our construction of $(d + 1, 3d)$-ruling set improve the time complexity of the algorithms for maximal matching and minimum connected dominating set presented in work [7].

2 Main Result

In this section we present a deterministic distributed algorithm which finds $(d + 1, 3d)$-ruling set in UDG (where every vertex knows its position on the plane), and works in $O(poly(d))$ synchronous rounds. Recall that a ruling set is defined as follows:

Definition 1. *Let d and b be some natural numbers. A (d, b)-ruling set in G is a subset U of $V(G)$ with two properties:*

(a) For any two distinct vertices u, u' from U, $dist(u, u') \geq d$ (where $dist(u, u')$ is the length of the shortest path between vertices u and u').
(b) For any vertex $v \in V \setminus U$ there is a vertex $u \in U$ such that the $dist(v, u) \leq b$.

The main idea of our algorithm is to use three square lattices to compute locally optimal solutions inside the appropriate squares formed by those lattices. We call this approach the "three lattice method".

In this approach we divide the plane into three separate lattices which define three classes of squares. In the first step we compute optimal solutions inside squares of the first class. In the second step we correct those solutions inside squares of the second class and, next we perform corrections inside squares of the third class.

Consider a lattice $L_a^{(0,0)}$ (with origin in $(0, 0)$ and which consists of parallel horizontal and vertical lines at distance a) and two other lattices $L_a^{(a/3,a/3)}$ and $L_a^{(2a/3,2a/3)}$ (obtained from $L_a^{(0,0)}$ by moving it by vectors equal respectively $(a/3, a/3)$ and $(2a/3, 2a/3)$ (see Figure 1). The interiors of squares determined by lattice $L_a^{(\nu_1,\nu_2)}$ are denoted by $S_a^{(\nu_1,\nu_2)}$.

The algorithm RULINGSETINSQUARE computes the optimal solutions in the first class of squares. It finds $(d + 1, d)$-ruling set in every $\mathcal{S} \in \mathcal{S}_a^{(0,0)}$ in time $O(d^3)$. To "delete collisions" (to remove all vertices which are at distance less or equal d) we use DELETINGLOCALCOLLISIONS. This algorithm also works in time $O(d^3)$.

We will show that after moving the lattice for the first time and using DELETINGLOCALCOLLISIONS we get $(d + 1, 2d)$-ruling set in each of the squares $\mathcal{S} \in \mathcal{S}_a^{(a/3,a/3)}$, as well as that after having moving the lattice for the second time and using DELETINGLOCALCOLLISIONS we obtain $(d + 1, 3d)$-ruling set in a whole graph.

Let $\mathcal{S}$ be an arbitrary square of side a. By $G[\mathcal{S}]$ we denote a subgraph of G induced by vertices $\{v \in V(G) : v \in \mathcal{S}\}$, and assume that $d > 0$. Our main algorithm works in the following way:

RULINGSET
Input: A Unit Disc Graph G, constant d and a
Output: $(d + 1, 3d)$-ruling set of the graph G.

(1) In parallel, find a $(d + 1, d)$-ruling set in all connected components of $G[\mathcal{S}]$, for every $\mathcal{S} \in \mathcal{S}_a^{(0,0)}$ using RULINGSETINSQUARE. Denote a union of output sets as R.
(2) Apply algorithm DELETINGLOCALCOLLISIONS to connected components of a graph $R[\mathcal{S}]$, for every $\mathcal{S} \in \mathcal{S}_a^{(a/3,a/3)}$. Denote a union of output sets as R'.

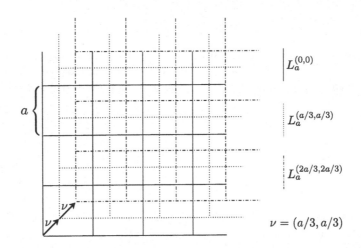

Fig. 1. Idea of three lattices method

(3) Apply algorithm DELETINGLOCALCOLLISIONS to all connected components of $R'[\mathcal{S}]$, for every $\mathcal{S} \in \mathcal{S}_a^{(2a/3,2a/3)}$. Denote a union of output sets as R''.
(4) Return R''.

The main task of our paper is to prove the following result.

Theorem 1. *Let G be a Unit Disc Graph and let $a = 3d + 3$. Algorithm RULINGSET finds $(d + 1, 3d)$-ruling sets of G in $O(d^3)$ synchronous rounds in LOCAL+COORD model of computations.*

3   Useful Lemmas

We shall prove first a sequence of simple but useful observations.

Lemma 1. *For any three points $q_0 \in L_a^{(0,0)}$, $q_1 \in L_a^{(a/3,a/3)}$ and $q_2 \in L_a^{(2a/3,2a/3)}$ we have $\|q_0, q_1\| + \|q_0, q_2\| \geq a/3$.*

Proof. Let $\left\| q_0, L_a^{(\nu_1,\nu_2)} \right\| = \min_{q' \in L_a^{(\nu_1,\nu_2)}} \|q_0, q'\|$. We prove our lemma for an arbitrary point $q_0 = (x, y)$ lying on the segment with ends in $(0,0)$ and $(0,a)$. In this case:

$$\left\| q_0, L_a^{(a/3,a/3)} \right\| = \begin{cases} a/3 - y & \text{if } y \in [0, a/3] \\ y - a/3 & \text{if } y \in [a/3, 2a/3] \\ a/3 & \text{if } y \in [2a/3, a] \end{cases}$$

$$\left\| q_0, L_a^{(2a/3,2a/3)} \right\| = \begin{cases} a/3 & \text{if } y \in [0, a/3] \\ 2a/3 - y & \text{if } y \in [a/3, 2a/3] \\ y - 2a/3 & \text{if } y \in [2a/3, a]. \end{cases}$$

Therefore

$$\left\| q_0, L_a^{a/3, a/3)} \right\| + \left\| q_0, L_a^{(2a/3, 2a/3)} \right\| = \begin{cases} 2a/3 - y & \text{if } y \in [0, a/3] \\ a/3 & \text{if } y \in [a/3, 2a/3] \\ y - a/3 & \text{if } y \in [2a/3, a] . \end{cases}$$

Thus in the considered segment this sum is always at least $a/3$. For other cases the proof is analogous.

Next lemma derives a bound on a size of a Maximal Independent Set of $G[\mathcal{S}]$, denoted by $\text{MIS}(G[\mathcal{S}])$ and its diameter, denoted by $diam(G[\mathcal{S}])$.

Lemma 2. *If $\mathcal{S}$ is a square of side a, then $\text{MIS}(G[\mathcal{S}]) \leq \frac{4(a+1)^2}{\pi}$. Moreover if $G[\mathcal{S}]$ is connected then $diam(G[\mathcal{S}]) \leq \frac{8(a+1)^2}{\pi}$.*

Proof. Denote by $C_r(v)$ a circle of radius r with the center in a vertex v. If we take two vertices $g_i, g_j \in \text{MIS}(G[\mathcal{S}])$ then $\|g_i, g_j\| > 1$, so any two different circles $C_{0.5}(g_i)$ and $C_{0.5}(g_j)$ are disjoint. Obviously every $C_{0.5}(g_i)$ lie in a square of side $a + 1$. But such square can contain at most $\frac{(a+1)^2}{\pi/4}$ disjoint circles of radius 0.5 therefore $\text{MIS}(G[\mathcal{S}])$ is at most $\frac{4(a+1)^2}{\pi}$. Let $v, w \in G[\mathcal{S}]$ be two vertices such that the shortest path $p_{v,w}$ between v and w has length equal to the diameter of $G[\mathcal{S}]$. It is obvious that if we choose every second vertex on that path, it will form an independent set. Therefore $diam(G) \leq 2|\text{MIS}(G[\mathcal{S}])| \leq \frac{8(a+1)^2}{\pi}$

Let $N_k(u)$ denote a subgraph of G induced by vertices at distance at most k from vertex u and let $\tilde{N}_{r_i} \equiv N_{\lfloor (d-1)/2 \rfloor}(r_i)$.

Lemma 3. *Let G be connected graph $d \geq 2$ and R be a $(d + 1, b)$-ruling set in G. If $r_i \neq r_j$ then there is no edge between $\tilde{N}_{r_i}$ and $\tilde{N}_{r_j}$, for any distinct vertices $r_i, r_j \in R$. Moreover if $|R| > 1$ then for every $r_i \in R$ $diam(\tilde{N}_{r_i}) \geq \lfloor (d-1)/2 \rfloor$ and $MIS(\tilde{N}_{r_i}) \geq \lfloor \frac{1}{2} \lfloor (d-1)/2 \rfloor \rfloor + 1$.*

Proof. Suppose that there is an edge between $\tilde{N}_{r_i}$ and $\tilde{N}_{r_j}$, then there exists a path of length less than $2 \lfloor (d-1)/2 \rfloor + 1 \leq d$ between r_i and r_j. This is a contradiction to the assumption that R is a $(d + 1, b)$-ruling set. Since G is connected and $|R| > 1$ there must be a path between r_i and r_j . Suppose that p_{r_i, r_j} is the shortest path between r_i and r_j. Obviously first $\lfloor (d-1)/2 \rfloor$ edges on p_{r_i, r_j} belong to $\tilde{N}_{r_i}$. Therefore $diam(\tilde{N}_{r_i}) \geq \lfloor (d-1)/2 \rfloor$. Similarly, if a Unit Disc Graph G has a diameter k then $|\text{MIS}(G)| \geq \lfloor k/2 \rfloor + 1$, so $MIS(\tilde{N}_{r_i}) \geq \lfloor \frac{1}{2} \lfloor (d-1)/2 \rfloor \rfloor + 1$

4   Proof of Main Theorem

First we introduce algorithm for finding $(d + 1, d)$-ruling set in $G[\mathcal{S}]$.

RULINGSETINSQUARE
Input: Connected Unit Disc Graph G, square $\mathcal{S}$ of side a and a constant d.
Output: $(d + 1, d)$-ruling set of graph $G[\mathcal{S}]$.

(1) Let $M := \emptyset$ be an output set and $Z := \emptyset$ denote the set of vertices which are d-dominated by some $s \in M$.

(2) Repeat $t = \left\lceil \frac{16(a+1)^2}{\pi d} \right\rceil$ times:

    (a) Choose a vertex w from $V(G[\mathcal{S}]) \setminus (M \cup Z)$ with the smallest ID and $M := M \cup w$.

    (b) Add to set Z all vertices $u \in V(G[\mathcal{S}])$ such that $dist(w, u) \le d$.

(3) Return M

Lemma 4. *The algorithm* RULINGSETINSQUARE *finds a* $(d+1, d)$-*ruling set in* $O(a^4/d)$ *synchronous rounds.*

Proof. It is easy to observe that elements of set M have to be of distance at least $d+1$, so the first condition of definition of a ruling set is trivially satisfied.

Let us show that the maximum size of $(d+1, d)$-ruling set in $G[\mathcal{S}]$ is bounded by $t = \left\lceil \frac{16(a+1)^2}{\pi d} \right\rceil$. Let R be a $(d+1, d)$-ruling set. Let $\tilde{N}_{r_i}$ denotes vertices at distance at most $\lfloor (d-1)/2 \rfloor$ from $r_i \in R$ in graph $G[\mathcal{S}]$. We know from Lemma 3 that there is no edge between $\tilde{N}_{r_i}$ and $\tilde{N}_{r_j}$ where $r_i, r_j \in R$, $r_i \ne r_j$. Therefore $MIS(\tilde{N}_{r_i})$ and $MIS(\tilde{N}_{r_j})$ are disjoint sets and there are no edges from $MIS(\tilde{N}_{r_i})$ to $MIS(\tilde{N}_{r_j})$. From Lemma 3 it also follows that $MIS(\tilde{N}_{r_i}) \ge \lfloor \frac{1}{2} \lfloor (d-1)/2 \rfloor \rfloor + 1$. On the other hand Lemma 2 shows that $MIS(G[\mathcal{S}]) \le \frac{4(a+1)^2}{\pi}$, so

$$|R| \le \frac{4(a+1)^2}{\pi} \left(\left\lfloor \frac{1}{2} \lfloor (d-1)/2 \rfloor \right\rfloor + 1 \right)^{-1} \le \frac{16(a+1)^2}{\pi d}$$

To prove the second postulate from Definition 1, suppose that, after algorithm RULINGSETINSQUARE ends, there exists a vertex u which is not d-dominated. Obviously $u \notin Z$ because Z is the set of vertices d-dominated by vertices from M. So we can enlarge a $(d+1, d)$-ruling set adding u to M. But it contradicts the choice of t as greater or equal then maximum size of $(d+1, d)$-ruling set of the graph $G[\mathcal{S}]$.

Finally observe that parts (a) and (b) of step (2) of the algorithm RULINGSETINSQUARE can be implemented via simple distributed BSF procedure which takes $O(diam(G[\mathcal{S}]))$ synchronous rounds. By Lemma 2, $diam(G[\mathcal{S}]) = O(a^2)$, therefore the algorithm needs $O(a^4/d)$ rounds to get a solution.

As we mentioned before, in the second and third step of the algorithm RULINGSET, we "delete collisions" (remove from R all the vertices which are at distance less or equal d). We will achieve this goal applying the following procedure.

DELETINGLOCALCOLLISIONS

Input: Connected graph G, $R \subseteq V(G)$ and constant parameter d.

Output: Subset $M \subseteq R$

(1) Let $M := \emptyset$ and $Z := R$ denote vertices which may be added to M.

(2) While $Z \ne \emptyset$ do:

    (a) Choose vertex w in $Z \setminus M$ with the smallest ID and $M := M \cup w$, $Z := Z \setminus w$.

    (b) Remove from Z all vertices $u \in Z$ such that $dist(w, u) \le d$.

(3) Return M.

Lemma 5. *If R is $(0, b)$-ruling set (b-dominating set) then output set M of algorithm* DELETINGLOCALCOLLISIONS *is $(d+1, d+b)$-ruling set. Moreover algorithm* DELETINGLOCALCOLLISIONS *finds output set M in at most $(O(|V(R)|a^2))$ synchronous rounds.*

Proof. First we prove that for all vertices $v, w \in M$ we have $dist(v, w) > d$. Suppose that there exist $v, w \in M$ such that $dist(v, w) \leq d$. Without loss of generality we may assume that v was added to the set M before the vertex w. Nevertheless at the step (b) of the algorithm DELETINGLOCALCOLLISIONS all vertices $u \in G$ such that $dist(u, v) \leq d$ are removed from Z and since $dist(v, w) \leq d$ then $w \notin M$, a contradiction. Now we check that if $r \in R$ and $dist(r, M) = min_{s \in M}dist(r, s)$ then $\forall_{r \in R}dist(r, M) \leq d$. Obviously, for every $r \in R$ there exists $s \in M$ such that $dist(r, s) \leq d$ (if such s does not exist then DELETINGLOCALCOLLISIONS algorithm would add vertex r to M in part (a) od step (2) of the algorithm).

Complexity follows from observation that an algorithm goes through part (a) of step (2), it removes one vertex from Z. Therefore we have to do at most $|R|$ iterations. In each of those iterations, we use BFS to find the vertex of the smallest ID in a subgraph of G (Lemma 2 says that it takes $O(a^2)$ synchronous rounds). Moreover each time we perform part (b) of step (2), the BFS procedure for finding all vertices $u \in Z$ such that $dist(w, u) \leq d$ takes obviously less time than the diameter of $G[\mathcal{S}]$.

Recall that in the algorithm RULINGSET, first we find a $(d + 1, d)$-ruling set locally in squares $\mathcal{S}_a^{(0,0)}$. Next we move the lattice twice by the vectors equal $(a/3, a/3)$ and $(2a/3, 2a/3)$ respectively as shown in Figure 1. After each translation of the lattice we delete all collisions in squares using DELETINGLOCALCOLLISIONS.

We will show that after the first translation of the lattice, we obtain a $(d + 1, 2d)$-ruling set in each of the squares and after moving it for the second time we obtain $(d + 1, 3d)$-ruling set in the whole graph.

Now we are redy to prove our main result.

Proof (Proof of Theorem 1)
To show that R'' is indeed a $(d + 1, 3d)$-ruling set we start with checking if R'' satisfies the first condition of Definition 1. Suppose that there are two points $r_1, r_2 \in R''$ such that $dist(r_1, r_2) \leq d$. Then there exists a path p_{r_1, r_2} of length at most d between r_1 and r_2. This path has to cross all three lattices $L_a^{(0,0)}$ $L_a^{(a/3, a/3)}$ and $L_a^{(2a/3, 2a/3)}$. If p_{r_1, r_2} does not cross the lattice $L_a^{(0,0)}$ then the path p_{r_1, r_2} lies entirely in one of the squares of $\mathcal{S}_a^{(0,0)}$. It means that in step (1) of the algorithm RULINGSET by Lemma 4 we add at most one point r_1 or r_2 to output set R, a contradiction. Similarly, if the path p_{r_1, r_2} does not cross the lattice $L_a^{(a/3, a/3)}$ then the path p_{r_1, r_2} lies entirely in one of the squares of $\mathcal{S}_a^{(a/3, a/3)}$. Then from the Lemma 5 step (2) of the algorithm RULINGSET certainly add to output set R' at most one of the points r_1 or r_2, again a contradiction. From the Lemma 5 it follows, in analogues way, that p_{r_1, r_2} has to cross lattice $L_a^{(2a/3, 2a/3)}$ too.

Without loss of generality, assume that the p_{r_1, r_2} first crosses $L_a^{(a/3, a/3)}$ at point q_1 then it crosses $L_a^{(0,0)}$ at the point q_0 and finally p_{r_1, r_2} crosses $L_a^{(2a/3, 2a/3)}$ at the point q_2. By Lemma 1 we know that $\|q_1, q_0\| + \|q_2, q_0\| \geq a/3$. However $\|v, w\| \leq$

$dist(v, w)$, so $\|q_1, q_0\| + \|q_2, q_0\| \leq dist(r_1, r_2)$ which imply that $a/3 \leq dist(r_1, r_2) < d + 1$, a contradiction with the assumption that $a = 3d + 3$.

The second condition of the definition of the $(d + 1, 3d)$-ruling set simply follows from Lemmas 4 and 5. Denote by $dist(v, R) = min_{r \in R} dist(v, r)$. First observe that after step (1) of the algorithm RULINGSET $\forall_{v \in G} dist(v, R) \leq d$. If there exists $v \in V(G)$, such that $dist(v, R) > d$ then the algorithm RULINGSETINSQUARE always add v to the output (d, d)-ruling set in $\mathcal{S}_a^{(0,0)}$. The second step of the algorithm RULINGSET gives us $\forall_{v \in G} dist(v, R') \leq 2d$. Suppose that there exists $v \in G$, such that $dist(v, R') > 2d$. We know that $\exists_{r \in R} dist(v, r) \leq d$. On the other hand Lemma 5 implies that $\exists_{r' \in R'} dist(r, r') \leq d$ so $dist(v, r') \leq dist(v, r) + dist(r, r') \leq 2d$. This is a contradiction with $dist(v, R') > 2d$. Similar reasoning leads us to the conclusion that after the third step $\forall_{v \in G} dist(v, R'') \leq 3d$.

Now we prove time complexity. From Lemma 4 step (1) of the algorithm RULINGSET takes $O\left(a^4/d\right)$ synchronous rounds. We also know that in each $\mathcal{S} \in \mathcal{S}_a^{(0,0)}$ the output set M of the algorithm RULINGSETINSQUARE is of size $O(a^2/d)$. Each $\mathcal{S} \in \mathcal{S}_a^{(a/3, a/3)}$ contains output sets from four sets $\mathcal{S} \in \mathcal{S}_a^{(0,0)}$ thus by Lemma 5 it follows that steps (2) and (3) of the algorithm RULINGSET take $4|R|\left(O(a^2) + |d|\right) = O\left(a^2/d\right)(O(a^2))$ synchronous rounds. Since $a = 3d + 3$, therefore RULINGSET has time complexity $O\left(d^3\right)$.

5   Applications

5.1   Approximation of k-Dominating Set

In paper [8] Fernandess and Malkhi present a distributed algorithm which approximates k-dominating set of a Unit Disc Graph G with approximation factor $O(k)$ in time $O(V(G))$ in LOCAL+COORD model of computations. In this section, applying our approach to the construction of ruling sets, we substantially improve their result in terms of time complexity.

Theorem 2. *Le G be Unit Disc Graph. There exists deterministic algorithm which approximates every k-dominating set of G with $O(k)$ factor in $O(poly(k))$ time.*

Proof. First we prove that if $d \geq 1$ then $(d + 1, 3d)$-ruling set is $O(d)$ approximation of the smallest $3d$-dominating set.

Obviously, $(d + 1, 3d)$-ruling set is $3d$-dominating set. Let H be the optimal $3d$-dominating set and $h \in H$ be a dominating vertex. Denote by R the output $(d + 1, 3d)$-ruling set from the algorithm RULINGSET.

Let $N_k(h) = \{v \in V(G) : dist(v, h) \leq k\}$. We will show that $|N_{3d}(h) \cup R| = O(d)$, which implies that $|R| = O(d)|H|$.

Denote by $C_r(v)$ the circle with the center at a vertex v and the radius r. Observe that, for every distinct $g_1, g_2 \in \text{MIS}(N_{3d}(h))$, circles $C_{0.5}(g_i)$ and $C_{0.5}(g_j)$ are disjoint and lie inside the circle of a center in h_i and with radius $3d + 1$. Thus

$$|\text{MIS}(N_{3d}(h))| \leq \frac{\pi(3d + 1)^2}{\pi/4} = 4(3d + 1)^2.$$

Let $r \in R$ and define $\tilde{N}_{r_i} = N_{\lfloor (d-1)/2 \rfloor}(r_i)$. If we take distinct $r_i, r_j \in N_{3d}(h) \cup R$ then, by Lemma 3, sets $\tilde{N}_{r_i}$ and $\tilde{N}_{r_j}$ are disjoint and there is no edge between them and $\text{MIS}(\tilde{N}_{r_i}) \geq \lfloor \frac{1}{2} \lfloor (d-1)/2 \rfloor \rfloor + 1$. Hence,

$$|N_{3d}(h) \cup R| \leq \frac{|\text{MIS}(N_{3d}(h))|}{min_{r_i \in R} \left| \text{MIS}(\tilde{N}_{r_i}) \right|} \leq \frac{4(3d+1)^2}{\lfloor \frac{1}{2} \lfloor (d-1)/2 \rfloor \rfloor + 1} = O(d).$$

Similarly, we can prove that $(d+1, 3d)$-ruling set is $O(d)$ approximation of the $3d + 1$ and $3d + 2$-dominating sets. The only case left is the $O(d)$ approximation of the d-dominating set for d equal to 1 or 2. The problem of finding 1-dominating set in constant time has already been solved and in [9], where a distributed algorithm finding $O(1)$ approximation of the optimal dominating set in LOCAL+COORD model is presented. Moreover, using similar arguments to those from the first part of the proof, we can show that in UDG the $O(1)$ approximation of the 1-dominating set is also $O(1)$ approximation of the 2-dominating set. So, we can approximate every k-dominating set with $O(k)$ factor in $O(poly(k))$ time, and the theorem follows.

5.2 Approximation of Maximum Matching and Minimum Connected Dominating Set

In [7] A. Czygrinow and M. Hańćkowiak present approximation schemes for the maximum matching problem and the minimum connected dominating set problem in Unit Disk Graphs. The algorithms are deterministic and run in a $O(poly(\log(V(G))))$ number of synchronous rounds in LOCAL model of computation. An approximation error of their algorithms is $O(\frac{1}{log^c V(G)})$, where c is a positive integer. However, in a slightly more restrictive model LOCAL+COORD of computations, we are able to significantly improve the time complexity of the above–mentioned algorithms.

Theorem 3. *Let c be a positive integer. There exist a distributed algorithms, which find a maximum matching and a minimum connected dominating set in Unit Disc Graph with approximation error $O(\frac{1}{log^c V(G)})$. This algorithm works in $poly(c)$ synchronous rounds.*

Proof. Only two procedures in algorithms from [7] take more than a constant number of synchronous rounds.

The first one, deals with a construction of some auxiliary graph, which in turn is based on the algorithm, presented in [10], which finds a maximal independent set and works in $O(\log \Delta(G) log^*(V(G)))$ rounds. However, adding the assumption that each vertex knows its position on the plane, one can apply, instead of MIS algorithm from [10], the algorithm given in [9], which works in a constant time.

Second procedure in [7], takes $O(poly(\log(V(G))))$ time, and deals with the construction of a clustering, resulting from an application of the Ruling Set algorithm. They use the well known algorithm of Awerbuch, Goldberg, Luby and Plotkin from [1], which finds $(d, d \log |V(G)|)$-ruling set in any graph G, in $O(d \log |V(G)|)$ synchronous rounds , with a constant d given as an input. In our case, when a vertex knows its position on the plane, we can perform clustering using the algorithm RULINGSET, which finds $(d + 1, 3d)$-ruling sets of G in a constant time.

Those two major modifications transform algorithms from [7] into a new procedure which works in $poly(c)$ in LOCAL+COORD model of computations.

References

1. Awerbuch, B., Goldberg, A.V., Luby, M., Plotkin, S.A.: Network decomposition and locality in distributed computation. In: IEEE Symposium on Foundations of Computer Science, pp. 364–369 (1989)
2. Linial, N., Saks, M.: Decomposing graphs into regions of small diameter. In: SODA 1991: Proceedings of the second annual ACM-SIAM symposium on Discrete algorithms, pp. 320–330. SIAM, Philadelphia (1991)
3. Panconesi, A., Srinivasan, A.: On the complexity of distributed network decomposition. J. Algorithms 20(2), 356–374 (1996)
4. Peleg, D.: Distributed computing: a locality-sensitive approach. In: Society for Industrial and Applied Mathematics, Philadelphia, PA, USA (2000)
5. Krzywdziński, K.: Distributed algorithm finding regular clustering in unit disc graphs, http://atos.wmid.amu.edu.pl/kkrzywd/research/REGULAR.pdf
6. Krzywdziński, K.: A local distributed algorithm to approximate mst in unit disc graphs. In: Gębala, M. (ed.) FCT 2009. LNCS, vol. 5699. Springer, Heidelberg (2009)
7. Czygrinow, A., Hańćkowiak, M.: Distributed approximation algorithms in unit-disk graphs. In: Dolev, S. (ed.) DISC 2006. LNCS, vol. 4167, pp. 385–398. Springer, Heidelberg (2006)
8. Fernandess, Y., Malkhi, D.: K-clustering in wireless ad hoc networks. In: POMC 2002 roceedings of the second ACM international workshop on Principles of mobile computing, pp. 31–37. ACM, New York (2002)
9. Czyzowicz, J., Dobrev, S., Fevens, T., González-Aguilar, H., Kranakis, E., Opatrny, J., Urrutia, J.: Local algorithms for dominating and connected dominating sets of unit disk graphs with location aware nodes. In: Laber, E.S., Bornstein, C., Nogueira, L.T., Faria, L. (eds.) LATIN 2008. LNCS, vol. 4957, pp. 158–169. Springer, Heidelberg (2008)
10. Kuhn, F., Moscibroda, T., Wattenhofer, R.: On the locality of bounded growth. In: PODC 2005Proceedings of the twenty-fourth annual ACM symposium on Principles of distributed computing, pp. 60–68. ACM Press, New York (2005)

A Case Study of Building Social Network for Mobile Carriers

Sewook Oh[1], Hojin Lee[1], Yong Gyoo Lee[1], and Kwang-Chun Kang[2]

[1] KT Data Network Development Team, Wireless Laboratory, Seoul, Korea
[2] Nable Communications, Inc R&D Center, Seoul, Korea
{sewook5,horilla,SteveLee}@kt.com, kkckc@nablecomm.com

Abstract. Recently, social networking service(SNS) is migrating to the stage of acquiring new business models from the stage of simple toy of interest. Various Internet portals and service carriers are trying to build SNS as a major service infrastructure which reflects their own strength. In this paper, we will focus on the activity of a Korean mobile carriers in order to consider the mobile carrier specific SNS. To build an SNS, two main procedures should be devised: how to build a relationship among members; how to provide an efficient way of communication to attract people to participate and share. Usually, social relations are built by the user's explicit intention. In this paper, however, we propose the derivation of the relationship based on the information owned by mobile carriers, such as the address book or call record, or else. In addition, we will show that IMS based group, messenger service can be utilized to enrich the communication among the social network members.

Keywords: SNS, IMS, Mobile Carrier Specific Service.

1 Introduction

The number of people using online SNS is increasing rapidly. According to comScore, US-based market research firm, globally 600 million people have visited SNS sites. Major SNS such as Facebook, Friendster, Hi5, MySpace and orkut recorded increase of 8.8 million visitors. Also, use of the services was also found to be very frequent- around 96% of teenage user have visited SNS site at least once a week. Following table shows the growth of visitors and page views of major SNS[1].

Despite the growth of SNS as a major internet service, it has not yet found a significant profit models. However, with firm confidence in the potential of SNS, News Corporation bought MySpace at $ 580 Million. Other IT industry leaders such as IBM and CISCO announced plans to launch SNS. SNS is drawing attention as a effective method of advertisement and especially as a innovative marketing method in consumer-lead market[2][3].

In the past, consumption has been made through the communication between produce and consumer. But now, consumption is made through the consumer's word of mouth and feedback of the products, because people value the frank and candid comments of their friends rather than those of the professionals. Therefore,

N.T. Nguyen et al. (Eds.): New Challenges in Compu. Collective Intelligence, SCI 244, pp. 157–165.
springerlink.com   © Springer-Verlag Berlin Heidelberg 2009

Table 1. Information of Representative International SNS [1]

	Unique Visitor (Thousand)			Page View (Millions)		
	May-06	May-07	YoY Growth	May-06	May-07	YoY Growth Seq.
MYSPACE.COM	61,635	109,535	78%	28,807	50,593	76%
FACEBOOK.COM	14,096	47,208	235%	6,562	29,882	355%
HI5.COM	20,864	28,492	37%	4,776	7,192	51%
FRIENDSTER.COM	14,175	24,684	74%	2,186	8,990	311%
BEBO.COM	6,135	17,246	181%	2,176	11,202	415%

opportunities to strengthen the real-life relationship and sharing of information or contents among friends may provide innovative business opportunities.

As these roles of SNS came into attention, the fact the real-life social relationship is as important as the relationship in the cyberspace. Information such as phone numbers and call records of mobile phone carriers may said to represent real social network. Mobile communication service provider's valuable assets are expediting the provider's entry into the market.

In the following aspects, mobile provider's entry into SNS is expected to be quiet massive.

- Possession of address, phone number and other social relations data which in turn could be utilized to draw potential social relationships.
- Facilitated mobilization of SNS. More than 5 million people are already using mobile Facebook within one year of initial service[3].
- Easily maintain social networks by using instant message, SMS, voice/video call and conference without PC. Especially, Mobile carrier's IMS based communication services such as the Presence can be integrated with web-based SNS to provide strong advantage against portal businesses.

This paper would like to study on the construction of social networks using strengths of mobile service providers, including development of social network index, build up and management of communities using relevant data and IMS based communication methods and system. Chapter 2 will explain the entire system structure of proposed social network service. Chapter 3 will explain the type of source data required to draw social network index and methods to collect the data. On Chapter 4, formation method of social network index from source data will be discussed. Finally, Chapter 5 will describe the build up of the communities using social network index, relevant mapping procedures with IMS and effective IMS-based communication between community members.

2 Social Network Service System

This chapter will describe Social Network Service System for mobile service providers. Proposed system is named SRIE(Social Relation Based Interactive Environment) and composed of three part as seen in Fig. 1.

RCDC(Relation and Context Data Collector) collects source date from internal systems such as CDE(Content Delivery Enabler), LBS(Location Relation Based Interactive Environment), PIMS(Personal Information Management System) and external portals such as Cyworld and NHN. RCDC is interface modules in effectively collecting source data from various systems. SRG(Social Relations Generator) calculates relation index between a person and person based on flexible index calculation policy based on source data. Finally ICM(Intelligent Community Manager) creates community based on friends information, user profiles, messaging history and manages community information.

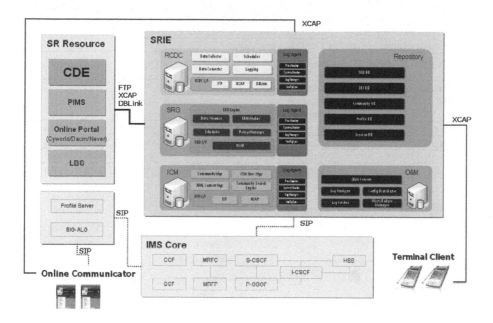

Fig. 1. SRIE System Composition & Concerned System

Each module's role and interface with external systems composing SRIE will be described on Chapter 3, 4 and 5.

3 Collection of Source Data in Creating Social Network Index

This chapter will discuss the source data collection and systems need to draw social network index.

Common SNS site's social network index completely relies on the implicit 'friend' relationship of the users and built relations are used to limit the access rights to mutual information within the site. Since such relationship is limited only to cyberspace, real-life relationship not fully reflected. Therefore, application of general services is limited.

On the other hand, since mobile carrier's address / phonebook and call history information does reflect real-life networks, combination of existing SNS will help forming more useful social networks.

From the observation, RCDC collects various source data from internal sources such as CDE data, SMS/MMS history, PIMS address book, location and external data such as online portal's SNS, blog and cafe joining information to integrate.

Table 2. RCDC Collect Relation Information

Type	Contents	Remark
Telephone Call Record	Number of Telephone Calls	
SMS Transmission Record	Number of SMS Transmission	
MMS Transmission Record	Number of MMS Transmission	
CDE Intimacy Information	CDE Intimacy Value	
Address Book Information	Register with Address Book or not	
Broadcasting Message Record	Number of Broadcasting Message Transmission	
Neighbors	Neighbor or not	Cyworld / Naver / Daum
Gift Record	Number of Giving Gift	Cyworld / Naver / Daum
Cafe Posting Record	Number of Cafe Posting	Cyworld / Naver / Daum
Blog Guestbook Record	Number of Guestbook messages	Cyworld / Naver / Daum

Detailed structure of RCDC is on Fig. 2.

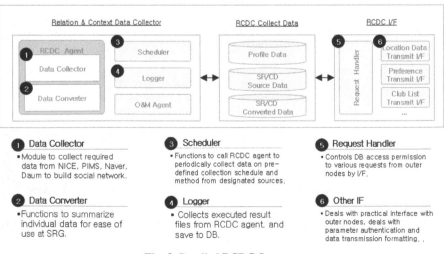

Fig. 2. Detailed RCDC Structure

Since RCDC must read data from various systems, it provides different interfaces such as HTTP, DBLink, FTP and following is the type of data collected.

RCDC's major functions can be classified as below.

1. Source Data Collection
   a. CDE intimacy, SMS/MMS summary collection
   b. CDE preference and broadcasting message summary collection
   c. PIMS address book information collection
   d. Online Portal relations information collection
   e. Online Portal cafe / blog information collection
   f. LBS location information collection
2. Collection Scheduling
   a. Periodical data collection
   b. Real-time data collection
3. Data Management and Transmission
   a. Provide V Community standard information to ICM Modules
   b. Provide user preference information based on community service
   c. Data summarization and SRIE ID management

4 Methods to Collect Social Network Index

Chapter 4 will explain SRG(Social Relation Generator) used to create social network index based on source data collected from RCDC. Fig. 3 depicts process that creates social network in SRG.

First, information acquired from RCDC is normalized by pre-defined criteria by administrative policy. Then, weighted SRI(Social Relation Index) is created. For

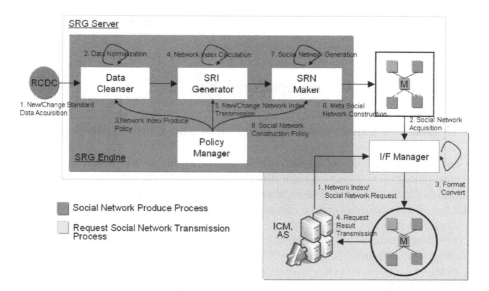

Fig. 3. SRG Process Composition

example, weight on call frequency and weight on online portal relations information can result in a different SRI, with different usage and value. Administrator may set different weight depending on diverse needs. At the end of the process, social network is set according to created network index and provide social network information according to different system's request.

Fig. 4 depicts detailed structure of SRG. SRG produces network index depending on the admin policy or rule-sets based on collected source data.

Generally, buddy list is a non-dynamic information created by user by registering friends and setting his or her own groups. But the buddy list based on social network index turns into a dynamic information by given criteria. For example, if a buddy list is created by call history weighted social network index and transmitted to terminal, user can easily find a friend on his buddy list, confirm presence status and make a call. Businesses may provide more diverse additional services depending on social network index-based buddy list service.

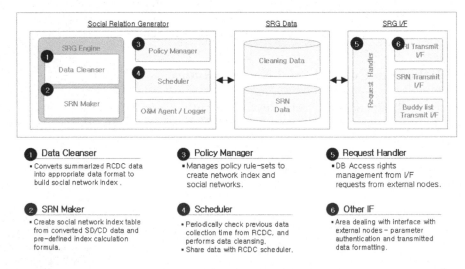

Fig. 4. Detailed SRG Structure

5   IMS-Based Community Management and Communication

Proceeding chapters described various systems creating social network index and social relationship information from different source data. This chapter will introduce communities created from aforementioned information. Furthermore, methods to use IMS-based technology to enable community information management, utilization and member communication are also mentioned.

IMS based community management and communication is provided through ICM (Intelligent Community Manager). ICM extracts community-based user preference information, creates situation-attributed temporary community, broadcast message-based virtual community management based on buddy list created from SRG. Also users may search other users, search communities, notified of new communities, register community contents[4][5][6].

ICM saves community information in XDMS in XML format and use XCAP protocol to inquire and create community information from terminals. In this way, IMS infra may be re-utilized and most effectively manage XML-based community information[7][8].

V-community is one of the most representative services provided by ICM. Consider following situation. A president of bowling club broadcasts a SMS message to notify weekly Friday game. Such repetitious broadcast history implies possible presence of a community. A community may be formed based on the people receiving broadcasted message, the president can immediately utilize the network with little change in name or purpose within the message. Such community information is managed by group information of IMS and can be used to facilitate member communication by SMS, messenger, multilateral video chat and other forms of communication. Fig. 5 depicts process of V-community formation.

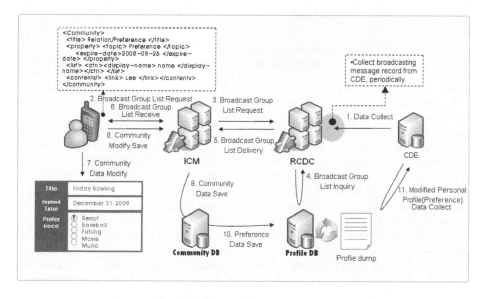

Fig. 5. V-Community Produce Case

As V-community is created and managed as IMS group concept, general IMS client is used to communicate with all members by Group-URI. Since it utilizes pre-structured IMS client, this is more advantageous than the portals which has to create separate member communication measure. Especially, when this is integrated with Presence information, appropriate form of communication can be set at appropriate timing considering member status. This is a huge advantage. As seen from above, when IMS service and community service by social network is integrated, new communication demand will happen and other profit-creating models will emerge.

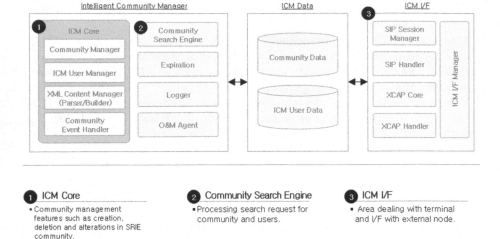

ICM Core
- Community management features such as creation, deletion and alterations in SRIE community.

Community Search Engine
- Processing search request for community and users.

ICM I/F
- Area dealing with terminal and I/F with external node.

Fig. 6. Detailed ICM Structure

Finally, Fig. 6 depicts detailed structure of ICM. ICM is composed of community creation, deletion, member management, search and interface dealing with external nodes.

6 Conclusion

As the social network services are drawing attentions as a new business model, this paper intends to have new perspective of social networking related services from the mobile service provider's point of view. Further methods to differentiate the model from portal-based social network services are also mentioned. When mobile service provider's PIMS information and call history is integrated with denoted relations of existing social relationship, this paper presents new form of social relationship can be presented. Also, based on the social network index derived from such new methods, IMS based XDMS can be utilized to manager buddy list and various community management, that existing IMS infrastructure can be re-utilized. Further, compared with existing portal businesses, this can provide diverse and abundant communication methods such as SMS, Instant Messenger, Presence and video-sharing so that effective communication environment can be structured between community members within the social networks. This will be a significant advantage for mobile service providers in providing social network services.

This work was supported by the IT R&D program of MIC/IITA [2007-S-048-01, Development of Elementary Technologies for Fixed Mobile IP Multimedia Convergence Services on Enhanced 3G Network]

References

1. Peck, R.S., et al.: What Should Yahoo! Do Regarding Social Networks. Bear Sterns (August 2007)
2. EMarketer, Ten Key Online Predictions for 2008 (January 2008), http://www.emarketer.com

3. Yu, J.: The next generation Killer App – SNS KIPA (Korea software Industry Promotion Act) (April 2008)
4. 3GPP TS 22.141, Presence service, State 1, OMA (Open Mobile Alliance), http://www.openmobilealliance.org/
5. 3GPP TS 23.141 Presence Service; Architecture and Functional Description, OMA (Open Mobile Alliance), http://www.openmobilealliance.org/
6. RFC 3261, SIP: Session Initiation Protocol, IETF
7. RFC 4825, The Extensible Markup Language (XML) Configuration Access Protocol (XCAP), IETF
8. Internet-Draft, An Extensible Markup Language (XML) Document Format for Indicating A Change in XML Configuration Access Protocol (XCAP) Resources, IETF

Modelling Dynamics of Social Support Networks
for Mutual Support in Coping with Stress

Azizi Ab Aziz and Jan Treur

Agent Systems Research Group, Department of Artificial Intelligence
Faculty of Sciences, Vrije Universiteit Amsterdam
De Boelelaan 1081, 1081 HV Amsterdam, The Netherlands
{mraaziz, treur}@few.vu.nl
http://www.few.vu.nl/~{mraaziz, treur}

Abstract. This paper presents a computational multi-agent model of support receipt and provision to cope during stressful event within social support networks. The underlying agent model covers support seeking behavior and support provision behaviour. The multi-agent model can be used to understand human interaction and social support within networks, when facing stress. Simulation experiments under different negative events and personality attributes for both support receipt and provision pointed out that the model is able to produce realistic behavior to explain conditions for coping with long term stress by provided mutual support. In addition, by a mathematical analysis, the possible equilibria of the model have been determined.

Keywords: Social Support Networks, Strong and Weak Ties, Stressors, Support Recipient and Provision, Multi-Agent Simulation.

1 Introduction

Persons differ in their vulnerability for stress. To cope with stress, the social ties of the person are an important factor; [2][5]. Such ties are the basis of social networks or communities within which support is given from one person to the other and vice versa. Examples of such social networks are patient communities for persons suffering from a long or forever lasting and stressful disease. Providing and receiving social support within such a network is an intra and interpersonal process, with as a major effect that it improves the quality of life of the members of the social network.

This fundamental form of human functioning is an important aspect of our lives. Research shows that in the event of stress a social support network is able to influence individuals' wellbeing and act as a buffer for the impact of negative events. In recent years, social support with particularly the perception of support seeking and availability (provision), has well documented positive effects on both physical and psychological health. The explication of relationship between support seeking and provision has been studied intensively to explain this relationship. For example, simply knowing that someone is available to support can be comforting and capable to alleviate the effect of negative events [4][8]. More general social support helps its recipients to escalate self-confidence and overcome the risk of stress [5][9].

N.T. Nguyen et al. (Eds.): New Challenges in Compu. Collective Intelligence, SCI 244, pp. 167–179.
springerlink.com
© Springer-Verlag Berlin Heidelberg 2009

However, little attention has been devoted to a computational modelling perspective on social support networks, on how the dynamics of support seeking and providing work at a societal level. In many ways, the availability of social support is still too frequently viewed as a static facet of individual or environment. However, the support seeking and provision process is highly dynamic and it involves substantial changes as demanding conditions occur [2]. From this dynamic process a collective pattern may emerge that costs almost no effort, and is beneficial for all members. While it is difficult to observe such conditions in the real world, a multiagent system model offers a more convenient perspective. This paper is organized as follows. Section 2 describes the theoretical concepts of support receipt and provision. From this perspective, a formal model is designed and developed (Section 3). Later, in Section 4, several simulation traces are presented to illustrate how this model satisfies the expected outcomes. In Section 5, a mathematical analysis is performed in order to identify possible equilibria in the model. Finally, Section 6 concludes the paper.

2   Antecedents of Social Support Receipt and Provision

Research on social support provides useful information from controlled experimental paradigms on several important factors influenced the possibilities of seeking and giving help. During the formation of stress, there is a condition where an individual either will increase the support interaction demands on support providers. It is typically involves many options, such as whether or not a support provider performs particular support, based on what actions to take and in what manner [1]. Furthermore, through a perspective of help seeking behavior, it also related to the answer of which support member is suitable to pledge for help and so forth. In general, support provision is driven by altruistic intentions and is influenced by several factors that related to provide a support. Within social support researchers' community, it has commonly been viewed that social support is related to several characteristics, namely; (1) stress risk factors, (2) receipt factors, (3) relationship factors, (4) provision factors, and (5) motivation in support [1][3][5][9]. For the first point, stress risk factor is related to the recipient ability to recognize the need of support and be willing to accept support assistance. It includes both features of stressors and appraisal of stressors. This factor is influenced by individual's perceptions of stressors, vulnerability (risk in mental illness), and expectations support from the others [7]. Research indicates that the degree of stressors is correlated to amount of support levels. For example, situations considered as stressful by both support recipients and providers are much more probable to trigger support responses than non-stressful events [2][9]. Having this requirement in motion, potential support providers will recognize the need of support assistance and be willing to offer support [1].

Another point that can be made to understand the social support process is a recipient factor. Despites evidence that primarily shows the negative event plays an important role in seeking and providing support, yet severely distress individuals as experienced by major depression patients seems to reduce social support process. It is highly related to the individual's personality. Normally, a neurotic personality tends to attract a negative relationship between social support provider and social engagement [6]. Studies of the personality and support have documented that individuals with high self-esteem (assertive) receive more social support compared to the individuals with neurotic person-ality [1][6]. In relationship factors, characteristics of the relationship (ties) between support recipient and provider are equally to important to activate support selection be-haviours. It includes mutual interest (experiential and situational similarity), and satisfac-tion with a relationship. It is eventually becomes a part of socio-cultural system that has a balance between giving and receiving support. In this connection, it should also be men-tioned that there are two additional antecedents related closely to the relationship factors. These are acceptance of social norms and reciprocity norms [1]. Social norms are highly coupled with the view of individual responsibility, intimate relationship and obligation. An example of this is, it is a common fact that many individuals will feel responsible (personal responsibility) for anyone who is dependent upon them. Because of this, it will increase the likelihood of support offering in a certain relationship (either strong tie or weak tie relationship). Strong tie is a relationship typically between individuals in a close personal network. While, a weak tie is typically occurs among individuals who commu-nicate on relatively frequent basis, but do not consider them as close acquaintances. In reciprocity norms, previous interaction and past supportive exchanges will reflect future willingness of both support recipients and providers [2]. Previous failure and frustration of past efforts may influence to reduce individual's motivation and willingness to provide support. For this reason, if individuals always refuse to receive support, it is more likely to receive less support in future [3].

The fourth factor is related to the support provision attributes. Social support members who are faced with condition to give support will be motivated by several factors. Many research works have maintained that there is a link that support-providers with experience empathy and altruistic attitude will regulate altruistic moti-vation to help the others. In spite of this condition related to the subject of helping people in a weak tie network, it is also useful to understand support's patterns in strong tie network as well. In addition, focus on the other individuals may escalate the potential of providing help through the increasing feeling of empathy, which later develop efficacy. The last factor is the motivation in support. This idea concerns the influence of selecting a support provider from a relationship perspective according to an individual's support need. For example, several studies have shown many indi-viduals with long-term motivation (*future goal orientation*) having difficulty to attain appropriate support from close friends or acquaintances since they feel this group of people has limited skills or knowledge towards the individual's problems [2] [3][7]. However, if the individual's intention to seek for emotional support (*emotional goal orientation*) is higher, then they tend to choose a weak tie support over strong tie [7]. Those antecedents also related to explain several individual and interpersonal charac-teristics that influence an individual's decisions to seek support from particular social network members.

3   A Multi-agent Model for Social Support Networks

To support the implementation of multiagent system interaction, the dynamic model for both receipt and provision is proposed and designed. This model uses social and behavioural attributes as indicated in a previous section.

3.1 Formalizing the Multi-agent Model

In the agent model used as a basis for the multi-agent system, five main components are interacting to each other to simulate support-seeking and giving behaviours of an agent. These agent components are grouped as; individual receipt and provision attributes, support preference generation, relationship erosion process, stress component, and support feedbacks. Fig.1 illustrates the interaction for these components.

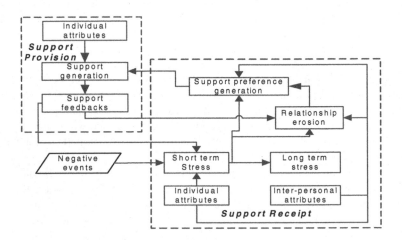

Fig. 1. Overall Structure of the Underlying Agent Model

As illustrated in Fig.1, negative events acts as an external factor stimulus triggers the stress component. Such a stress condition is amplified by individual receipt attributes such as risk of stress (or risk of mental illness) and neurotic personality, which later accumulates in certain periods to develop a long-term stress condition. The short-term stress also plays an important to evoke support preference pertinent to the receipt attributes. Similarly, this triggered information will be channelled to the social erosion component, which acts to diminish individual's ability in seeking help. After the social support-tie preference is selected, then the support generation is regulated. Support provision attributes will determine the level of support feedbacks towards the support recipient. To simplify this interaction process, this model assumes all support feedbacks received provide a positive effect towards the agent's well-being (stress-buffering mechanism). Finally, the channelled social support feedback also will be regulated to reduce the relationship erosion effect within individual. The arrows represent the piece of information that the output of one course of action serves as input for another process. The detailed components of this model are depicted in Fig. 2.

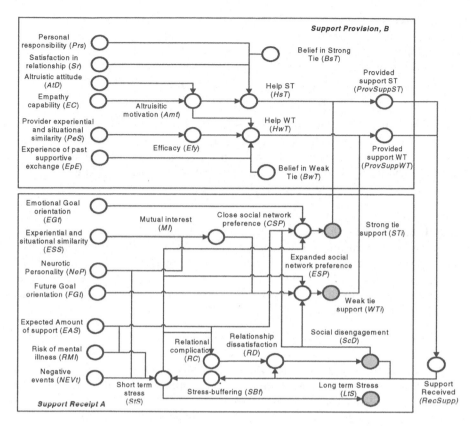

Fig. 2. Detailed Structure and Components of the Agent Model

As can be seen from Fig. 2, several exogenous variables represent individual support receipt and providing attributes. The results from these variables interaction form several relationships, namely instantaneous and temporal relations. To represent these relationships in agent terms, each variable will be coupled with an agent's name (*A* or *B*) and a time variable *t*. When using the agent variable *A*, this refers to the agent's support receipt, and *B* to the agent's support provision. This convention will be used throughout the development of the model in this paper.

3.2 The Agent Component for Support Receipt

This component aims to explain the internal process of support preference during the presence of stress. In general, it combines three main concepts, namely support goal orientation (emotional goal orientation (*EGt*), future goal orientation (*FGt*), expected amount of support *(EAS)*), personality (neurotic (*NeP*), risk of mental illness / vulnerability (*RMI*), experiential and situational similarity (*ESS*)), and external factor (negative events (*NEVt*)). Interactions among these exogenous variables are derived from these formulae.

Mutual Interest: Mutual interest (*MI*) is calculated using the combination of experiential situational similarity (*ESS*) and complement relation of neurotic personality (*NeP*) as opposed to positive personality). That is to say, having a positive personality and a common experience will encourage a better mutual interest engagement.

$$MI_A(t) = ESS_A(t).(1 - NeP_A(t)) \tag{1}$$

Stress Buffering: Stress buffering (*SBf*) is related to the presence of support and the level of social disengagement (*ScD*). Note that, $\eta_{sbf,a}$ regulates the level for both support ties contribution. Note that a high social disengagement level (*ScD*→ 1) will cause stress buffering becomes less effective to curb the formation of stress.

$$SBf_A (t) = RecSupp_A(t).(1\text{-}ScD_A(t)) \tag{2}$$

Short-Term Stress: Short-term stress (*StS*) refers to the combination of negative events, risk in mental illness (vulnerability), and neurotic personality. The contribution of these variables are distributed using regulator parameter $\psi_{sts,\,a}$. If $\psi_{sts,\,a}$ → 1, then the short-term stress will carry only all information from the external environment, rather than individual attributes. In addition, stress-buffering factor eliminates the effect of short-term stress.

$$StS_A(t) = [\,\psi_{stsA} \,.\, NEVt_A(t) + (1 - \psi_{sts\,A}).RMI_A(t).\,NeP_A(t)].(1 - SBf_A(t)) \tag{3}$$

Relational Complication and Relational Dissatisfaction: Relation complication (*RC*) is measured using the contribution rate (determined by γ_{rc}) of the expected support (*EAS*) and short-term stress (*StS*). Related to this, relational dissatisfaction (*RD*) is determined by η_{rd} times relational complication when no support is given.

$$RC_A(t) = \gamma_{rcA}.EAS_A(t).StS_A(t) \tag{4}$$

$$RD_A(t) = \eta_{rdA}.RC_A(t).\,(1\text{-}RecSupp_A(t)) \tag{5}$$

Close and Expanded Support Preferences: Close support preference (*CSP*) depends to the level of emotional goal orientation (*EGt*), short-term stress (*StS*), and social disengagement (*ScD*). In the case of extended support preference (*ESP*), it is calculated using the level of future goal orientation (*FGt*), short-term stress, mutual interest, and social disengagement. In both preferences, the presence of social disengagement decreases the social network preference level. Similar circumstance also occur when *StS* → 0. Parameters β_{csp} and η_{esp} provide a proportional contribution factor in respective social network preference attributes

$$CSP_A(t) = [\beta_{csp,A} \,.EGt_A(t) + (1 - \beta_{csp,A}) \,.(1 - ScD_A(t))] \,.\, StS_A(t) \tag{6}$$

$$ESP_A(t) \ = [\eta_{esp,A} \,.FGt_A(t) + (1 - \eta_{esp,A}) \,.MI_A(t).(1 - ScD_A(t))] \,.\, StS_A(t) \tag{7}$$

Dynamics of Support, Social Disengagement, and Long Term Stress: In addition, there are four temporal relationships are involved, namely strong-tie preference (*Sti*), weak-tie preference (*WTi*), social disengagement (*ScD*), and long-term stress (*LtS*).

The rate of change for all temporal relationships are determined by flexibility parameters, φ_{sti}, ϕ_{wti}, η_{scd}, and β_{lts} respectively.

$$ScD_A(t+\Delta t) = ScD_A(t) + \eta_{scd,A} .(1 - ScD_A(t)) .$$
$$(RD_A(t) - \psi_{scd,A} .ScD_A(t)) . ScD_A(t).\Delta t \qquad (8)$$

$$LtS_A(t+\Delta t) = LtS_A(t) + (\beta_{lts,A} .(1 - LtS_A(t)).$$
$$(StS_A(t) - \xi_{lts,A} .LtS_A(t)) . LtS_A(t).\Delta t \qquad (9)$$

$$STi_A(t+\Delta t) = STi_A(t) + (\varphi_{sti,A} .(1 - STi_A(t)).$$
$$(CSP_A(t) - \varphi_{sti,A}.STi_A(t)) .STi_A(t) .\Delta t \qquad (10)$$

$$WTi_A(t+\Delta t) = WTi_A(t) + (\phi_{wti,A}.(1 - WTi_A(t)).$$
$$(ESP_A(t) - \eta_{wti,A}.WTi_A(t)) . WTi_A(t) .\Delta t \qquad (11)$$

The current value for all of these temporal relations is related to the previous respective attribute. For example, in the case of STi, when CSP is higher than the previous strong-tie preference multiplied with the contribution factor, ψ_{sti}, then the strong-tie preference increases. Otherwise, it decreases depending on its previous level and contribution factor. It should be noted that the change process is measured in a time interval between t and $t+\Delta t$.

3.3   The Agent Component for Support Provision

Another important component to regulate support within social networks is the ability to provide help. In many ways, support provision attributes are often correlated to the amount of support provided to the support recipients. Antecedents of support provision are associated to personal responsibility (PrS), satisfaction in relationship (Sr), altruistic attitudes (AtD), empathy level /capability (EC), provision experiential and situational similarity (PeS), and experience of past supportive exchange (EpE). Combining these factors respectively, instantaneous relationships of altruistic motivation, and efficacy can be derived.

Altruistic Motivation and Efficacy: Altruistic motivation (Amt) is determined by through the combination of individual's attributes in altruistic attitude and empathy capability. In efficacy (Efy), the current contribution to generate efficacy is based on proportional value γ_{efy} towards provision experiential and situational similarity.

$$Amt_B(t) = AtD_B(t).EL_B(t) \qquad (12)$$

$$Efy_B(t) = \gamma_{efyB} .PeS_B(t) \qquad (13)$$

Help Provision of Strong and Weak Tie Support: In help provision, it generates support provision capability to provide help, pertinent to the level of respective attributes and relations. For example, the help provision in strong tie support (HsT) is calculated from the level of altruistic motivation, personal responsibility, and satisfaction in relationship. The contribution from these factors is regulated using regulation parameter μ_{wst}. In addition, belief on strong tie (BsT) controls the help provision towards support recipients. The same concept also applies for help provision in weak tie support (HwT).

$$HsT_B(t) = [(\mu_{wst,B}.Amt_B(t) + (1 - \mu_{wst,B}).Sr_B(t) . PrS_B(t))].BsT(t) \qquad (14)$$

$$HwT_B(t) = [(\mu_{wwt,B}.Efy_B(t) + (1 - \mu_{wwt,B}).AMT_B(t) . PrS_B(t))].BwT(t) \qquad (15)$$

For both cases, these beliefs regulate the level of generated help for later usage in the provided support. Having no belief concerning support causes no support will be provided to the support recipients.

3.4 Social Support Distribution and Aggregation

Within the provided support, there are two main components are implemented to regulate support distribution among agents. The first component is a mechanism to differentiate the strong tie ($ProvSuppST_{B,A}$) or weak tie ($ProvSuppWT_{B,A}$) support provision offered by a support provision agent to multiple support receipt agents. By using this technique, the overall support is distributed over the support receipt agents with the proportional to the level of support that respective agents requested for. Later, the received support ($RecSupp_A$) is aggregated by multiple support provision agents to each support receipt agent accordingly.

$$ProvSuppST_{B,A} = (STi_A / \sum_A STi_A). HsT_B.(1 - \prod_A (1 - STi_A)) \qquad (16)$$

$$ProvSuppWT_{B,A} = (WTi_A / \sum_A WTi_A). HwT_B.(1 - \prod_A (1 - WTi_A)) \qquad (17)$$

$$RecSupp_A = 1 - [(\prod_B (1 - ProvSuppST_{B,A}).(1 - ProvSuppWT_{B,A}))] \qquad (18)$$

4 Results

This section addresses analysis of the multiagent model using several simulation experiments. By variation of the personality attributes for support receipt and provision agents, some typical patterns can be found. Due to the excessive number of possible combinations, this paper shows example runs for four agents under two conditions, namely prolonged and fluctuated stressor events with a different personality profile. Table 1 outlines the values of these profile attributes.

Table 1. Individual Profiles for Each Agent

Support Receipt Agents	Personality Attributes (*EGt, ESS, NeP, FGt, EAS, RMI*)
A1	0.8,0.7,0.8,0.7,0.8,0.8
A2	0.8,0.6,0.2,0.9,0.1,0.3
Support Provision Agents	**Personality Attributes (*PrS, Sr, EL, AtD, PeS, EpE*)**
B1	0.7,0.8,0.8,0.9,0.7,0.9
B2	0.7,0.7,0.3,0.4,0.6,0.7

The duration of the scenario is up to 1000 time points with these simulation settings;

$\Delta t = 0.3$

$\varphi_{sti} = \phi_{wti} = \eta_{scd} = \beta_{lts} = 0.2$

$\psi_{sts} = \mu_{wst} = \beta_{csp} = \eta_{esp} = \mu_{wwt} = 0.5$

$\gamma_{rcA} = \eta_{rd} = \gamma_{efy} = 0.8$

For all cases, if the long term stress is equal or greater than 0.5, it describes the support receipt agent is experiencing stress condition. These experimental results will be discussed in detail below.

Case # 1: Support Provision and Long Term Stress during Prolonged Stressor Events. For this simulation, all support receipt agents have been exposed to an extreme case of stressor events over period of time. It represents individuals that having a difficulty throughout their lifetime. The result of this simulation is shown in Figure 3.

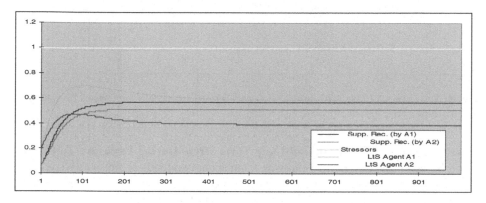

Fig. 3. The Level of Long Term Stress (*LtS*) and Support Received (*Supp. Rec.*) by Agent *A1* and *A2* during Prolonged Stressor

As can be seen from Figure 3, both agents received supports that allow them to reduce their long-term stress throughout time. The amounts of support received by both agents are varied according to their personality attributes. In this case, agent A1 received slightly less support compared to its correspondence long-term stress level. This finding is consistent with [6] who found that an individual with a high neurotic personality received less support from either strong or weak social network tie even during stressful event. Thus, agent *A2* recovers faster compared to agent *A1*.

Case # 2: Support Provision and Long Term Stress during Progression of Stressor Events. In this experiment, both agents are exposed to the progression of stressor event. During this condition, support receipt agent will increase the amount of support needed, and support provision agent will provide certain amount of support with the respect personality attributes. Figure 4 illustrates the progression of stressor, support received, and long term stress for both support receipt agents.

Figure 4 indicates that agent *A2* receives better support compared to *A1* where, the amount support is slightly higher compared to its long-term stress. Throughout time, it decreases the long-term stress, and providing better coping to curb the progression of it. Compared to agent *A1*, agent *A2* is unlikely to develop prolonged stress condition.

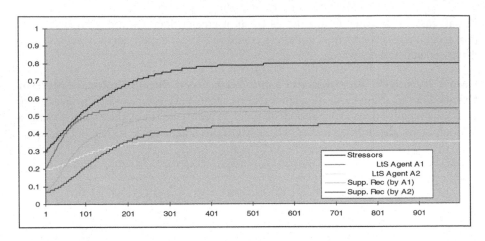

Fig. 4. The Level of Long Term Stress (*LtS*) and Support Received (*Supp. Rec.*) by Agent *A1* and *A2* during Progression Stressor

Case # 3: Support Provision and Long Term Stress During Exposure To Fluctuating Stressor Events. In the following simulation, two kinds of stressors were introduced to agents *A1* and *A2*. The first event contains a very high constant stressor, and is followed by the second event with a very low constant stressor.

As shown in Figure 5, it illustrates the decrease of support level received by both agents. When there is no stressor is experienced by support receipt agents, the lower of support seeking behavior is reduced. It also worth noting that agent *A1* shows slightly declining pattern for the long-term stress, compared to agent *A2* (with considerably decline towards "no stress" condition. This condition explains that individual

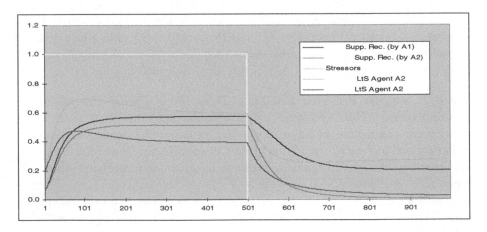

Fig. 5. The Level of Long Term Stress (*LtS*) and Support Received (*Supp. Rec.*) by Agent A1 and A2 during Fluctuated Stressor

with risk in mental illness and neurotic personal is vulnerable towards changes in environment [6]. Having these conditions in motion, more effort in support provision is needed to allow better recovery process to take place [3].

5 Mathematical Analysis

One of the aspects that can be addressed by a mathematical analysis is which types of stable situations are possible. To this end equations for equilibria can be determined from the model equations. This can be done to assume constant values for all variables (also the ones that are used as inputs). Then in all of the equations the reference to time t can be left out, and in addition the differential equations can be simplified by canceling, for example, $ScD_A(t+\Delta t)$ against $ScD_A(t)$. This leads to the following equations.

Agent Component for Support Receipt (by A from some B's)

$$MI_A = ESS_A.(1 - NeP_A) \tag{19}$$

$$SBf_A = RecSupp_A.(1\text{-}ScD_A) \tag{20}$$

$$StS_A = [\,\psi_{sts,A} \cdot NEVt_A + (1 - \psi_{sts,\,A}).RMI_A.\ NeP_A].(1 - SBf_A) \tag{21}$$

$$RC_A = \gamma_{rc,\,A}.EAS_A.StS_A \tag{22}$$

$$RD_A = \eta_{rd,A}.RC_A.\ (1\text{-}RecSupp_A) \tag{23}$$

$$CSP_A = [\beta_{csp,A}\ .EGt_A + (1 - \beta_{csp,A})\ .(1 - ScD_A)]\ .\ StS_A \tag{24}$$

$$ESP_A = [\eta_{esp,A}\ .FGt_A + (1 - \eta_{esp,A})\ .MI_A.(1 - ScD_A)]\ .\ StS_A \tag{25}$$

$$\eta_{scd,A}\ .(1 - ScD_A)\ .(RD_A - \psi_{scd,A}\ .ScD_A)\ .\ ScD_A = 0 \tag{26}$$

$$\beta_{lts,\,A}\ .(1 - LtS_A)\ .\ (StS_A - \xi_{lts,A}\ .LtS_A)\ .\ LtS_A = 0 \tag{27}$$

$$\varphi_{sti,A}\ .(1 - STi_A)\ .\ (CSP_A - \varphi_{sti,A}.STi_A)\ .STi_A = 0 \tag{28}$$

$$\phi_{wti,A}.(1 - WTi_A)\ .(ESP_A - \eta_{wti,A}.WTi_A)\ .\ WTi_A = 0 \tag{29}$$

Agent Component for Support Provision (from B to some A's)

$$Amt_B = AtD_B.EL_B \tag{30}$$

$$Efy_B = \gamma_{efy,B}\ .PeS_B \tag{31}$$

$$HsT_B = [(\mu_{wst,B}.Amt_B + (1 - \mu_{wst,B}).Sr_B\ .\ PrS_B)].BsT \tag{32}$$

$$HwT_B = [(\mu_{wwt,B}.Efy_B + (1 - \mu_{wwt,B}).AMT_B\ .\ PrS_B)].BwT \tag{33}$$

Differentiation of Provided Support from B to A

$$ProvSuppST_{B,A} = (STi_A/\ \textstyle\sum_A STi_A).\ HsT_B.(1 - \textstyle\prod_A (1\text{-}STi_A)) \tag{34}$$

$$ProvSuppWT_{B,A} = (WTi_A/\ \textstyle\sum_A WTi_A).\ HwT_B.(1 - \textstyle\prod_A (1\text{-}WTi_A)) \tag{35}$$

Aggregation of Received Support by A

$$RecSupp_A = 1-[((\Pi_B(1 - ProvSuppST_{B,A}).(1 - ProvSuppWT_{B,A}))] \qquad (36)$$

Assuming the parameters $\eta_{scd,A}$, $\beta_{lts, A}$, η_{scd}, β_{lts} nonzero, from the equations (26) to (29), for any agent A the following cases can be distinguished:

$ScD_A = 1$	or	$RD_A = \psi_{scd,A}.ScD_A$	or	$ScD_A = 0$
$LtS_A = 1$	or	$StS_A = \xi_{lts,A}.LtS_A$	or	$LtS_A = 0$
$STi_A = 1$	or	$CSP_A = \varphi_{sti,A}.STi_A$	or	$STi_A = 0$
$WTi_A = 1$	or	$ESP_A = \eta_{wti,A}.WTi_A$	or	$WTi_A = 0$

For one agent, this amounts to $3^4 = 81$ possible equilibria. Also given the other equations (19) to (25) and (30) to (36) with a large number of input variables, and the number of agents involved, this makes it hard to come up with a complete classification of equilibria. However, for some typical cases the analysis can be pursued further.

Case $ScD_A = 1$
In this case from the equations (20), (24) and (25) it follows:

$$SBf_A = 0, CSP_A = \beta_{csp,A}.EGt_A.StS_A, ESP_A = \eta_{esp,A}.FGt_A.StS_A$$

This can be used to determine values of other variables by (21), (22), (23), for example.

Case $StS_A = LtS_A = 0$
In this case, from the equations (22), (24) and (25) it follows:

$$RC_A = 0, CSP_A = 0, ESP_A = 0$$

from which, for example, by (23) it follows that $RD_A = 0$.

6   Conclusion

In this paper, a computational model is presented that describes the mechanism of support receipt and provision within a social network. The agent model used is composed of two main components: agent receipt and provision. The first component explains how personality attributes affect support-seeking behavior, ties selection, and stress buffering, and the second one explains how personality attributes affect providing support behaviour. The model has been implemented in a multiagent environment, dedicated to perform simulations using scenarios based on different stressful events over time and personality attributes. Simulation results show interesting patterns that illustrate the relation of support seeking behaviours and level of support received, with long-term stress. A mathematical analysis indicates which types of equilibria are indeed a consequence of the model. The model can be used as the basis for a personal software agent that facilitates a person in regulating help within a social network member. In addition, using this model, a personal agent will be able to determine social tie selection, and providing information regarding to the level of support needed with correspondence to personality attributes, for both individuals who are seeking and providing support. Thus, this model could possibly be used as a building block for interventions for individual who are facing stress or as a warning system for social support members.

References

1. Adelman, M.B., Parks, M.R., Albrecht, T.L.: Beyond close relationships: support in weak ties. In: Albrecht, T.L., Adelman, M.B. (eds.) Communicating social support, pp. 126–147 (1987)
2. Albrecht, T.L., Goldsmith, D.: Social support, social networks, and health. In: Thompson, T.L., Dorsey, A.M., Miller, K.I., Parrot, R. (eds.) Handbook of health communication, pp. 263–284. Lawrence Erlbaum Associates, Inc., Mahwah (2003)
3. Bolger, N., Amarel, D.: Effects of social support visibility on adjustment to stress: experimental evidence. Journal of Personality and Social Psychology 92, 458–475 (2007)
4. Both, F., Hoogendoorn, M., Klein, M.C.A., Treur, J.: Design and Analysis of an Ambient Intelligent System Supporting Depression Therapy. In: Proceedings of the Second International Conference on Health Informatics, HEALTHINF 2009, Porto, Portugal, pp. 142–148. INSTICC Press (2009)
5. Groves, L.: Communicating Social Support. Social Work in Health Care 47(3), 338–340 (2008)
6. Gunthert, K.C., Cohen, L.H., Armeli, S.: The role of neuroticism in daily stress and coping. Journal of Personality and Social Psychology 77, 1087–1100 (1999)
7. Lee, J.: Social Support, Quality of Support, and Depression. In: American Sociological Association Annual Meeting, Boston (2008)
8. Neirenberg, A.A., Petersen, T.J., Alpert, J.E.: Prevention of Relapse and Recurrence in Depression: The Role of Long-Term Pharmacotherapy and Psychotherapy. J. Clinical Psychiatry 64(15), 13–17 (2003)
9. Tausig, M., Michello, J.: Seeking Social Support. Basic and Applied Social Psychology 9(1), 1–12 (1988)

Data Portability across Social Networks

Pooyan Balouchian and Atilla Elci

Department of Computer Engineering, and Internet Technology Research Center,
Eastern Mediterranean University, Gazimagusa, North Cyprus
{pooyan.balouchian,atilla.elci}@emu.edu.tr

Abstract. Today, social networks are for much more than just having fun with friends. Millions of dollars are being spent to extract valuable information out of social networks for marketing purposes. Attaching machine readable semantics to social networks, that is absent at present, will lead to a better degree of information extraction. To achieve this goal, user input is required. From users' point of view, providing the same data to different social networks creates a drawback. This is due to the centralized one-of-a-kind nature of social networks. Social networks do not return users' data. In this paper, we exposed proposal and development of a framework toward data reusability and portability across social networks. The proposed framework is built upon "pull" strategy, although "push" strategy is shortly discussed. Test case developed to practically measure the feasibility of the proposed framework will be subject of our discussion. This study was linked to existing research and conclusions were drawn for further development.

Keywords: Semantic Social Network, Data Portability, Data Reuse, Social Network Standardization.

1 Introduction

"Everybody on this planet is separated by only six other people [1]." This sentence expresses the importance of Social Networks (SNs). An SN is a web-based service that allows individuals to construct a public or semi-public profile within a bounded system, articulate a list of other users with whom they share a connection and view and traverse their list of connections and those made by others within the system. SNs have not only attracted the attention of ordinary users, but of late marketing agents discovered its virtues. The reason is the huge amount of 'personal' information available on these services.

Modern SNs allow users to add modules or applications that enhance their profile. Users are interconnected through sending messages, viewing each other's profiles or news feed among others. Different SNs target specific users. Some act as a general-purpose network where users find their friends and share photos and/or videos. Some others are network-oriented where users can create new networks or subscribe to existing ones based on common interests.

N.T. Nguyen et al. (Eds.): New Challenges in Compu. Collective Intelligence, SCI 244, pp. 181–190.
springerlink.com   © Springer-Verlag Berlin Heidelberg 2009

The communication channel in SNs is mainly among users. Are SNs interoperable? What do we need in order to make that feasible? Is semantics involved in current SNs? If at all so, is it machine-readable / understandable? Has there been any serious effort on data portability across SNs? There is no simple yes/no answer to these questions. A great deal of effort has been made in this area of research; nevertheless no serious implementation appears to have formed in the horizon.

In this paper we proposed and developed a user-centric framework toward effecting data reusability and portability across SNs. The rest of this paper is organized as follows: Section 2 discusses the problems with current SNs in terms of data portability and reusability and the reasons involved therein. Section 3 introduces the proposed framework toward data portability across heterogeneous SNs, followed by section 4 discussing the developed framework in the context of a test case implemented to show the feasibility of the framework in practice. Afterwards; in section 5, the attention of the reader is drawn to related works in the area of data portability in the context of social networking. Finally, section 6 concludes the paper based on the findings.

2 Data Reusability and Portability Problems in Existing Social Networks

The first SN was SixDegrees.com, launched in 1997. It was followed by LiveJournal, AsianAvenu, Fotolog, Friendster, LinkedIn, Twitter, Facebook, etc. Recently there has been a huge increase in the number of SNs, indicating a rising demand for such services. Attracting the attention of many users does not mean that SNs are free of criticism. Users are there simply because they have not found any better SN. Besides, different users have different tastes. Therefore, even two close friends may appear in two different SNs. Since various SNs do not have interoperability among themselves, the users have to re-enter their profile and redefine connections from scratch whenever registering for new SN. People provide different data on different sites every time they register to a new service, simply because the process is boring and time-consuming. What would happen if an SN could provide users with their own data to be used in other SNs without having to re-enter all the information from scratch?

No SN is presently taking advantage of data portability and reusability frameworks, although a great deal of effort has been made in this area. They have a passion for power and strive to be unique. They do not like to share their information with other rivals in the market.

Experts point out the security and privacy risks involved in portability of social data on the web. Social data is generally owned and controlled by SNs and not the users. What would happen if the users could take control of their data in a secure way? To answer this question, we were encouraged to develop a user-centric framework, through which SNs would be able to pull users' information from outside their own databases and make use of it, provided that such access is allowed by the user.

3   Developed Social Network Data Portability Framework (SNDPF)

In this section, the architecture of SNDPF is discussed in detail. This framework is tackled from two different viewpoints: "push" and "pull" strategies. Eventually SNDPF has been developed using "pull" strategy. The distinction between "push" and "pull" strategies and the reason why we chose "pull" strategy is further explained. Finally the tools and technologies used to implement SNDPF are introduced.

3.1   SNDPF Architecture

The aim of SNDPF is to relieve the user from entering repeated set of data into different SNs. Instead, the SNs will be responsible for collecting the users' information from a single repository, owned and controlled by the user. It is quite obvious that collection of information in a manual fashion is not reasonable. Therefore automation should have been built into SNDPF to remove human and SN intervention. Implementing such framework not only makes life easier for the users, but also creates a more uniform set of user information on different SN services.

Information can pass the communication channel through two distinct strategies: *Push* or *Pull*. There are always at least two sides when dealing with communication over the internet. On one side we have the information provider and on the other side, the information consumer. Should the information provider processes the information without the recipient's request, an *information push* has occurred. Conversely, when a specific set of information is requested by the recipient and the provider prepares the information based on the request received, an *information pull* is occurred.

A choice has to be made between push and pull strategies. Although both frameworks were taken into consideration, the pull strategy was chosen for further development. Let's review the reasons for this choice. To better illustrate the distinction, the reader is encouraged to have a look at the architecture of SNDPF based on "push" and "pull" strategies, depicted in figures 1 and 2, respectively.

Let's consider "push" strategy. The user populates the personal knowledge base with her information. All the security measures and access control constraints are set by the user in the policy base, which has interaction with the personal knowledge base. The user in this architecture has a two-fold role: inputting the data to the knowledge base and extracting data out of the knowledge base in order to push it into the target SN. In order to push the data to the target SN, first the user should be authenticated. Afterwards the SQL Query Generator receives the data in the form of a result set produced by RDQL[1] and generates an SQL query to run on the server of the SN. Finally the SQL Query Engine runs the query on the SN's server and the appropriate changes actually affect the database.

Through "push" strategy, the user is required to log in to every SN, to which the information is to be pushed. This requires multiple authentications carried out and time wasted accordingly. This contradicts the primary aim of SNDPF. Therefore, we developed the framework taking advantage of "pull" strategy, depicted in figure 2, forming the core of our discussion.

---

[1] RDQL (RDF Data Query Language) is a query language which queries RDF documents using a SQL-alike syntax.

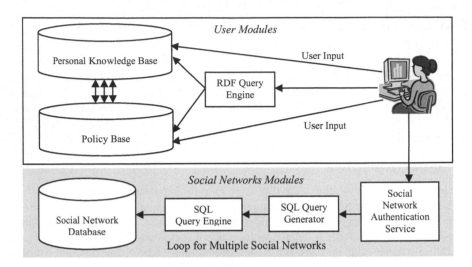

Fig. 1. SNDPF based on Push Strategy

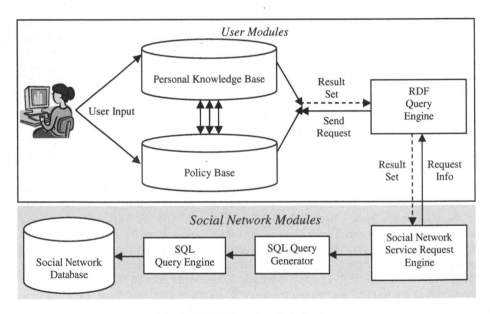

Fig. 2. SNDPF based on Pull Strategy

The architecture of SNDPF shown in figure 2 above is a SOA-based solution. The advantages of SOA influence the reusability as well as the maintainability of internet services; especially the property of reusability, if realized appropriately, can lead to a considerable cost reduction for software development. [2]

In SNDPF architecture based on "pull" strategy, the user is only involved in one single action and that is populating the personal knowledge base, along with the security and privacy measures. Social Network Service Request Engine is the module, from which all the scenario is triggered. This engine is responsible of submitting requests to the user's knowledge base at predefined time intervals. This engine requests specific pieces of information from the user's knowledge base by generating a query string. This query string is then passed to the RDF Query Engine which is responsible to run the query on the user's knowledge base in case such permission is given. Having successfully run the query on the Personal Knowledge Base, a result set is given back to the engine which is, in turn, passed on to the Social Network Service Request Engine. At this point, we have the requested data which is in fact supposed to affect the user's profile on the SN's database.

An important issue is that we have to actually map the data at hand with the database available on the SN. Most of the SNs today are making use of relational database management systems. Therefore we have chosen MemoRythm [3] (mentioned earlier in this paper) as our sample SN, which uses MySQL 5.0 as its Database Management System.

Referring to the Social Network Modules shown in figure 2, some predefined query templates exist in the SQL Query Generator module. These templates are complemented by a result set provided by the Social Network Service Request Engine. The query strings are formed in the SQL Query Generator and passed on to the SQL Query Engine in turn. SQL Query Engine will ultimately run the queries on the server and the actual updates will take place in the Social Network Database.

As it may occur to the reader, there are some security measures to be taken into account while passing the queries from one module to another. It is up to the SNDPF to handle the risks involved in the knowledge base, but as soon as the queries are generated and passed to the SQL Query Engine, it will be up to the SN to take the appropriate action upon securing the server for running of the received queries.

3.2 Tools and Technologies Used

- *Data Model*: The selected data model should be suitable for interoperable social networking. It should be capable of mapping with all existing APIs and being extensible. The only data model that fulfills these constraints is the Resource Description Framework (RDF). RDF is considered as the foundational language of the Semantic Web. RDF is a graph-based data format and secondly, it uses URIs as globally unique identifiers. Thirdly, it is defined primarily on the level of semantics, allowing the data itself to take any form (XML, JSON, APIs).

- *Ontology Editor*: Protégé 3.3
- *Sample Social Network*: MemoRythm[2]
- *Query Language*: RDQL
- *Server-Side Scripting Language*: PHP 5.0
- *IDE*: Macromedia Dreamweaver

4 Developed Prototype and Actual Findings

In order to measure the feasibility of SNDPF based on pull strategy, a prototype has been developed. In doing so, we use the tools and technologies as mentioned in

---

[2] MemoRythm is a social network with the aim of sharing memories in the form of text, image, audio and video. It is primarily designed to partly emulate the human brain in terms of memory storage and retrieval. (http://www.memorythm.com) [4].

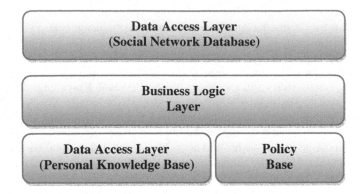

Fig. 3. SNDPF Implementation Layers

section 3.2. Let's take a look at the implementation layers of SNDFP displayed in figure 3 below.

Three layers are involved in the developed prototype. Unlike most systems, this prototype includes two different Data Access Layers (DALs). The personal knowledge base of the user serves as one DAL and the SN's database forms the other. The personal knowledge base is in RDF and is extracted from *people.rdf*[3]. Due to the simplicity of this ontology, we have partially modified it in order to make it more expressive. This ontology expresses the information of a person including name, position, age, email and website addresses, friends of a person and etc. This ontology has been filled with test data for the purpose of this prototype. The Policy Base is embedded within the same layer as the Personal Knowledge Base. A portion of the *people* ontology describing a person is shown below.

```
<?xml version="1.0" encoding="UTF-8" ?>
<rdf:RDF xmlns:rdf="http://www.w3.org/1999/02/22-rdf-syntax-ns#"
xmlns:rdfs="http://www.w3.org/2000/01/rdf-schema#"
xmlns:dt="http://foo.org#">
<rdf:Description about="http://foo.org/persons/pooyan">
  <dt:name>Pooyan</dt:name>
  <dt:email>pooyan.balouchian@emu.edu.tr</dt:position>
  <dt:age>26</dt:age>
  <dt:friend rdf:resource="http://foo.org/persons/emad" />
  <dt:friend rdf:resource="http://foo.org/persons/pejman" />
  </rdf:Description>
</rdf:RDF>
```

No user interface has been designed for this prototype. All the personal data have been given through the RDF file. Designing the user interface will be carried out as a future extension. The framework has two different types of users; individual user and

---

[3] http://phpxmlclasses.sourceforge.net/rdql.html

an SN. The individual users input their profile information. The SN is not directly in touch with the user's personal knowledge base. It is in fact communicating with the personal knowledge base through the Business Logic Layer. Referring to figure 2, the Social Network Request Engine and the SQL Query Generator form the Business Logic Layer modules. These two modules reside on the SN's server. The starting point of the scenario begins by submission of a request from the Social Network Service Request Engine. This engine submits requests to the user's knowledge base once per day at a pre-determined time. Having the retrieved information, queries matching the target SN's schema are generated automatically and passed to the Social Network Data Access Layer. This layer runs the generated queries on the SN's database server. Figure 4 depicts the part of MemoRythm's Data Model, on which the tests have been performed.

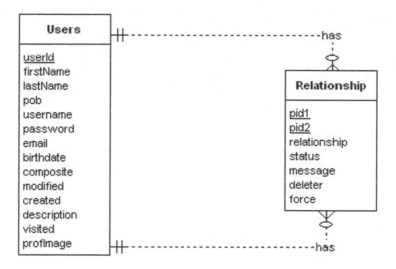

Fig. 4. Part of MemoRythm Data Model

The public profile of a user is shown in figure 5. As shown, the email of the user is *pooyan1982@yahoo.com* on MemoRythm SN. Let's consider the case that the user intends to change her profile email to *pooyan.balouchian@emu.edu.tr*. In the conventional approach, the user has to first sign in to MemoRythm's SN using her username and password. Afterwards she has to navigate to her profile page and click on Edit Profile. Finally she changes her email to *pooyan.balouchian@emu.edu.tr* and the change will affect MemoRythm's database. If such change is needed in all the SNs, in which the user is a member, then all these actions must be repeated in every single SN. But this is not the case with SNDPF. The user simply changes her email in her personal knowledge base and all the SNs, to which the user gives access, will update her email address accordingly in an automatic fashion even without the need to sign in to the user's account.

Fig. 5. Public Profile of a user on MemoRythm before Information Pull

The following RDQL query retrieves the new email from the user's knowledge base. The result set given back by the following query is simply *pooyan.balouchian@emu.edu.tr*. This email is passed on to the Business Logic Layer where the appropriate SQL Query is generated for further processing. Having generated the SQL Query, matching the target SN, being MemoRythm in this case, the database is affected by the query. The resulting profile is shown in Figure 6.

```
SELECT ?email
FROM <people.rdf>
WHERE (?user,<dt:name>,?name), (?user,<dt:email>,?email)
AND ?name == 'pooyan balouchian'
USING dt for <http://foo.org#>,
rdf for <http://www.w3.org/1999/02/22-rdf-syntax-ns#>
```

Fig. 6. Public Profile of a user on MemoRythm after Information Pull

In conventional approach, the sequence of actions that the user would have to take in order to update her profile must be repeated in every SN service. The sequence includes signing in to the user's account, followed by a sequence of links to click on and finally updating the profile and ultimately repetition of the same action for every single SN service, in which the change is required. In SNDPF, however the time taken to take these steps and the level of convenience of the user in this scenario is not even comparable.

The test case discussed above is a very simple test that merely measures the feasibility of SNDPF. In order to make the system fully functional, advancements including improvement of the personal knowledge base (ontology) and attaching appropriate security and privacy measures are required. The following section discusses the recent works in relation with our work.

5 Related Works

The first effort towards making a semantic SN was FOAF (Friend-Of-A-Friend) developed using RDF. FOAF is a machine-readable ontology describing persons, their activities and their relations to other people and objects. Anyone can use FOAF to describe him or herself. FOAF allows group of people to describe social networks without the need for a centralized database.

The Data Portability Workgroup (DPWG) has been trying to coordinate and formalize the process of developing data-portability technologies. The organization is attempting to define the nature of data portability among social-networking sites, promote best data-portability practices, and advocate open standards for implementing these practices [5].

Six Apart has announced that it will let people reuse their own social graph data elsewhere. Six Apart believes in openness of the users' social data on the web and has made use of existing technologies such as OpenID[4], the Microformats hCard[5] and XFN[6], and FOAF[7] in order to reach its objectives.

"The Decentralized Social Graph" by Brad Fitzpatrick, "Bill of Rights for Users of the Social Web" by Joseph Smarr et al., "The World is Now Closed" by Dan Brickley and the Social Network Aggregation Protocol (SNAP) and the Extensible Messaging and Presence Protocol (XMPP) are all on-going research carried out in this area of knowledge [1].

The major difference between our approach and those mentioned above is that on the subject of data portability across SNs, they concentrate on specific areas such as friend finding or reusing the username and password across different SNs or websites requiring authentication. The Social Network Aggregation Protocol (SNAP) deals with aggregation of SN data. These may be termed as SN-centric, whereas SNDPF is user-centric. Another difference is that we do not intend to force the current SNs to follow a specific standard, but rather standardize the way the information is entered to their databases and retrieved accordingly.

In SNDPF, the user has all her social data at hand and manages only one single personal knowledge base. The SNs are the ones to pull data out of the user's knowledge base. As a matter of fact, we are relieving the user from updating information in different SNs. Besides, this approach is far more secure, since the social data is

---

[4] http://openid.net
[5] http://microformats.org/wiki/hcard
[6] http://gmpg.org/xfn/
[7] http://esw.w3.org/topic/FoafDowntime

handled individually and not controlled by any other external party, except for cases when the users transfer such access to an SN.

6 Conclusion

In this paper, we first reviewed the history of social networking on the web from its emergence till the present time. Existing problems in the field of social networking were discussed thereafter. The problems mainly affect the users and not the SNs. The users provide every new SN service with information already given to other SNs. All the friendship relations, photos, videos, comments and etc. must be re-entered into each SN by the user. This is against the reusability of data on the web. The reason is the SN-centric approach that is currently dominating the social networking arena. The SN services own and control the social data of users and do not return them.

As a solution to this problem, we introduced Social Network Data Portability Framework (SNDPF), which we developed based on information pull strategy. This framework provides the user with a personal knowledge base to be owned and controlled by her. The user applies the security measures and access controls on the knowledge-base. The idea is not to replace the existing social networks, but creating means toward portability as well as reusability of social data on the web. Therefore while still keeping all the existing accounts in different social networks; the user only updates her profile on her own knowledge-base. The social network, in turn, would pull the data out of the user's knowledge base if such access is given. Finally, a prototype developed to measure the feasibility of SNDPF was presented.

As the future work towards improvement of SNDPF in order to make the system fully functional, we can point out some features to be added, namely inserting a friendly user interface, improving the personal knowledge base (ontology) and attaching security, trust and privacy measures.

References

1. Breslin, J., Decker, S.: The Future of Social Networks on the Internet: The Need for Semantics. Internet Computing IEEE 11(6), 86–90 (2007)
2. Thies, G., Vossen, G.: Web-Oriented Architectures: On the Impact of Web 2.0 on Service-Oriented Architectures. In: Asia-Pacific Services Computing Conference, 2008, APSCC 2008, December 9-12, 2008, pp. 1075–1082. IEEE, Los Alamitos (2008)
3. MemoRythm Social Network, http://www.memorythm.com
4. Balouchian, P., Safaei, M., Goudarzi, A.: Enforcing Security & Privacy Measures on Semantic Networks. In: Elci, A., Ors, B., Preneel, B. (eds.) Proc. The First International Conference on Security of Information and Networks (SIN 2007), pp. 328–335. Traffort Publishing, Canada (2008)
5. Heyman, K.: The Move to Make Social Data Portable, April 2008, vol. 41(4), pp. 13–15. IEEE Computer Society, Los Alamitos (2008)

Part IV
Agent and Multiagent Systems

A Novel Formalism to Represent Collective Intelligence in Multi-agent Systems[*]

Juan José Pardo[1], Manuel Núñez[2], and M. Carmen Ruiz[1]

[1] Departmento de Sistemas Informáticos
Universidad de Castilla-La Mancha, Spain
jpardo@dsi.uclm.es, MCarmen.Ruiz@uclm.es
[2] Departamento de Sistemas Informáticos y Computación
Universidad Complutense de Madrid, Spain
mn@sip.ucm.es

Abstract. In this paper we introduce a new formalism to represent multi-agent systems where resources can be exchanged among different agents by maximizing the utility of the agents conforming the systems. In addition to introduce a formalism to specify agents, we provide a formal framework to test whether an implementation *conforms* to the specification of an agent of the system.

1 Introduction

Computational collective intelligence has as its main objective to extract information from the knowledge of the components of a group that could not be extracted by taking this knowledge in an isolated way. A very simple example, extracted from [15], is that if one agent knows that $a \leq b$ and another one knows that $b \leq a$, then we can extract from the system a new piece of information $a = b$. We can use a similar approach in multi-agent systems where agents can compete to obtain resources: If we do this in a collaborative, collective way, all the agents can profit. For example, let us consider that a seller is willing to sell a certain item by at least n money units while a buyer is willing to pay at most m money units for this item. By putting together this information, we know that a deal can be reached as long as the selling price is between n and m money units. Multi-agent systems have been clearly identified as one of the fields of application of computational collective intelligence. Actually, they have been used in different application domains, in particular, outside purely computer science problems. One of the areas where research and development activities have been particularly increasing, maybe due to its financial applications, is in e-commerce systems where agents are in charge of some of the computations that users would have to perform otherwise (see, for example, [5,10,21,12,22,3]).

Formal methods are a powerful tool that allows the analysis, validation and verification of systems in general and of e-commerce systems in particular. In fact, due to

[*] This research was partially supported by the WEST project (TIC2006-15578-C02) and the Junta de Castilla-la Mancha project "Aplicación de metodos formales al diseño y análisis de procesos de negocio" (PEII09-0232-7745).

N.T. Nguyen et al. (Eds.): New Challenges in Compu. Collective Intelligence, SCI 244, pp. 193–204.
springerlink.com   © Springer-Verlag Berlin Heidelberg 2009

the complexity of current systems, it is very important to use a formal approach already in the early development stages since the sooner the errors of the system under development are detected the less harm, in particular in monetary terms, is done. In the context of multi-agent systems, formal methods can be used to express the high-level requirements of agents. These requirements can be defined in economic terms. Basically, the high-level objective of an e-commerce agent is *"get what the user said he wants and when he wants it."* Let us note that *what the user wants* includes not only what goods or services he wants but also other conditions, such as when and how to pay for the obtained goods and services, and these other conditions must be included in the specification of the system. There have been already several proposals to formalize multi-agent systems and to use existing formal methods within their scope (see, for example, [19,18,8,1,16,11]).

We have emphasized the importance of formal methods to specify the behavior of the system. However, it is even more important to ensure that the current implementation of the system is correct. In this line, testing [14,2], is one of the most extended techniques to critically evaluate the quality of systems. Traditionally, since the famous Dijkstra aphorism *"Program testing can be used to show the presence of bugs, but never to show their absence,"* testing and formal methods have been seen as rivals. However, during the last years there is a trend to consider them as complimentary techniques that can profit from each other. In fact, work on formal testing is currently very active (see, for example, [20,7,9,23,6]). The idea is that we have a formal model of the system (a specification) and we check the correctness of the system under test by applying experiments: We match the results of these experiments with what the specification says and decide whether we have found an error. Fortunately, and this is where formal methods play an important role, having a formal description of the system allows to automatize most of the testing phases (see [24] for an overview of different tools for formal testing).

The initial point of the work reported in this paper can be found in one formalism previously developed within our research group [17,13]. The idea underlying the original framework, called *utility state machines*, is to specify the high-level behavior of autonomous e-commerce agents participating in a multi-agent system and formally test the implemented agents with respect to the existing specifications. In this formalism, the user's preferences are defined by means of *utility functions* associating a numerical value to each possible set of resources that the system can trade. After using utility state machines to describe several existing systems we found that their internal inherent structure sometimes complicates the task of formally defining some types of specification. The problem that we found is that utility state machines are based on *finite state machines*, where a strict alternation between inputs and outputs must be kept. However, there are frequent situations where several inputs can be sequentially applied without receiving an output, or where an output can be spontaneously produced without needing a preceding input. Therefore, we have decided to consider a more expressive formalism where the alternation between inputs and outputs is not enforced. This slightly complicates the semantic framework. In particular, we need to include the notion of *quiescence* to characterize states of the systems that cannot produce outputs. We have also to redefine the notion of test and how to apply tests to systems. On the contrary, we have reduced some of the complexity associated with our previous formalism. Most notably, we formerly considered that agents could have *debts* that can be compensated with future exchanges. For example, an agent could offer a value greater than the one given by its utility function for an item if he could include this item as part of a future deal that would compensate the transitory lost. Even though this is a very interesting characteristic, it complicated too much the underlying semantic model and our experience shows that this feature was not used very often. In this paper we introduce a new formalism, called *Utility Input-Output Labeled Transition System* (in short, UIOLTS) that includes these enhancements. We have already used our new framework to formally specify an agent participating in the 2008 edition of the Supply Chain Management Game [4] but, due to space limitations, we could not include it in the paper.

In order to formally establish the conformance of a system under test with respect to a specification, we define two different implementation relations. The first one takes into account only the sequence of inputs and outputs produced by the system. While this notion would be enough to establish what a correct system is in terms of what the system does, it can overlook some faults due to the way resources are exchanged. Thus, we introduce a second implementation relation that also considers the set of resources that the system has after an action is executed.

The rest of the paper is structured as follows. In Section 2 we introduce our model. In Section 3 we define our implementation relations. In Section 4 we give the notion of test and how to apply tests to implementations under test. Finally, in Section 5 se present our conclusions and some lines for future work.

2   Modelling E-Commerce Agents by Using UIOLTSs

In this section we present our language to formally define agents and introduce some notions that can be used for their ulterior analysis. Intuitively, a UIOLTS is a labeled transition system where we introduce some new features to define agent behaviors in an appropriate way. The first new element that we introduce is a set of variables, where each variable represents the amount of the resource that the system owns. In addition, we associate a utility function to each state of the system. This utility function can be used to decide whether the agent accepts an exchange of resources proposed by another agent. Intuitively, given a utility function u we have that $u(\bar{x}) < u(\bar{y})$ means that the basket of resources represented by $\bar{y}$ is preferred to $\bar{x}$.

Definition 1. *We consider* $\mathbb{R}_+ = \{x \in \mathbb{R} \mid x \geq 0\}$. *We will usually denote vectors in* $\mathbb{R}^n$ *(for* $n \geq 2$*) by* $\bar{x}, \bar{y}, \bar{v} \ldots$ *Given* $\bar{x} \in \mathbb{R}^n$, x_i *denotes its i-th component. We extend to vectors some usual arithmetic operations. Let* $\bar{x}, \bar{y} \in \mathbb{R}^n$. *We define the addition of vectors* $\bar{x}$ *and* $\bar{y}$, *denoted by* $\bar{x} + \bar{y}$, *simply as* $(x_1 + y_1, \ldots, x_n + y_n)$. *We write* $\bar{x} \leq \bar{y}$ *if for all* $1 \leq i \leq n$ *we have* $x_i \leq y_i$.

We will suppose that there exist $n > 0$ *different kinds of resources. Baskets of resources are defined as vectors* $\bar{x} \in \mathbb{R}^n_+$. *Therefore,* $x_i = r$ *denotes that we own r units of the i-th resource. A utility function is a function* $u : \mathbb{R}^n_+ \longrightarrow \mathbb{R}$. *In microeconomic theory there are some restrictions that are usually imposed on utility functions (mainly, strict monotonicity, convexity, and continuity).*

As we indicated in the introduction, we consider two different types of actions that a system can perform. On the one hand, output actions are initiated by the system and cannot be refused by the environment. We consider that the performance of an output action can cost resources to the system. In addition, the performance of an output action will usually have an associated condition to decide whether the system performs it or not. On the other hand, input actions are initiated by the environment and cannot be refused by the system (that is, we consider that our systems are *input-enabled*). In contrast with output actions, the performance of an input action can increase the resources of the agent that performs it (that is, it receives a transfer of resources from the environment to *pay* the agent for the performance of the action). Let us remark that what we call in this informal description the *environment* can refer to a centralizer overviewing the activities of the different agents or to another agent that sends/receives actions to/from our agent. In addition to these two types of actions we need a third type that we introduce for technical reasons. It is possible to have a state where the system patiently waits for an input action and it cannot execute any output action for a certain combination of the existing resources. These states are called *quiescent*. In order to represent quiescence, we include a special action, denoted by δ, and special transitions labeled by this same δ action.

Definition 2. *A Utility Input Output Labeled Transition System, in short UIOLTS, is a tuple $M = (S, s_0, L, T, U, V)$ where*

- *S is the set of states, being $s_o \in S$ the initial state.*
- *V is an n-tuple of resources belonging to R_+. We denote by $\bar{v}_0$ the initial tuple of values associated with these resources.*
- *L is the set of actions. The set of actions is partitioned into three pairwise disjoint sets $L = L_I \cup L_O \cup \{\delta\}$ where*
  - *L_I is the set of inputs. Elements of L_I are preceded by the symbol ?.*
  - *L_O is the set of outputs. Elements of L_O are preceded by the symbol !.*
  - *δ is a special action that represents quiescence.*
- *T is the set of transitions. The set of transitons is partitioned into three pairwise disjoint sets $T = T_I \cup T_O \cup T_\delta$ where*
  - *T_I is the set of input transitions. An input transition is given by a tuple $(s, ?i, \bar{x}, s_1)$ where $s \in S$ is the initial state, $s_1 \in S$ is the final state, $?i \in L_I$ is an input action, and $\bar{x} \in \mathbb{R}^n_+$ is the increase in the set of resources. Since the system has to be input-enabled, we require that for all $s \in S$ and $?i \in L_I$ there exist $\bar{x}$ and s_1 such that $(s, ?i, \bar{x}, s_1) \in T_I$.*
  - *T_O is the set of output transitions. An output transition is given by a tuple $(s, !o, \bar{z}, C, s_1)$ where $s \in S$ is the initial state, $s_1 \in S$ is the final state, $!o \in L_O$, $\bar{z} \in \mathbb{R}^n_+$ is the decrease in the set of resources, and $C \in \mathbb{R}^n_+ \longrightarrow \{\texttt{true}, \texttt{false}\}$ is a predicate on the set of resources.*
  - *In addition, for each situation where a state cannot perform an output action we add a transition representing quiescence. This transition is a loop and does not need/receive resources to be performed. That is, $T_\delta = \{(s, \delta, \bar{0}, C_s, s) | s \in S \wedge C_s = \bigwedge_{(s, !o, \bar{z}, C, s_1) \in T_O} \neg C \wedge C_s \not\sim \texttt{false}\}$. Let us remark that $\bigwedge \emptyset = \texttt{true}$.*
- *$U : S \longrightarrow (\mathbb{R}^n_+ \longrightarrow R_+)$ is a function associating a utility function to each state in S.*

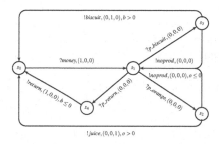

Fig. 1. Example of UIOLTS: A vending machine

In order to record the current situation of an agent we use *configurations*, that is, pairs where we keep the current state of the system and the current amount of available resources.

Definition 3. *Let M be a UIOLTS. A configuration of M is a pair $(s, \bar{v})$ where s is the current state and $\bar{v}$ is the current value of V. We denote by Conf(M) the set of possible configurations of M.*

In order to explain the main concepts of our framework, we use a simple running example.

Example 1. We consider a simple vending machine that sells biscuits and orange juice. The price of each of the items is one coin. This machine accepts four actions: Insert coin and press three different buttons, one for biscuits, one for orange juice and one for returning the money. In the initial state, after receiving a coin the machine waits for the user to press a button. If the pressed button is the return button the coin is returned and the machine returns to the initial state. If the user presses other button then if the machine has units of this product left, then it provides it; otherwise, it returns to the state to wait the user presses other button. Thus, the set of input actions for this machine is $L_i = \{?money, ?p\_orange, ?p\_biscuit, ?p\_return\}$ while the set of output actions is $L_O = \{!orange, !biscuit, !noprod, !return\}$. The tuple of resources has three variables $V = (m, b, o)$ that contain the amount of money (m), biscuits(b), and bottles of orange juice (o) currently in the machine. The utility function in each state takes a very simple form: $U(s_i)(V) = m + b + o$ that represents that every resource in the machine has the same value, meaning that the machine is *equally happy* with one coin or with one unit of a product.

If we consider that at the beginning of the day the machine has 10 bottles of orange juice, 10 packages of biscuits and it does not have money, the initial configuration of the system will be $c_0 = (s_0, (0, 10, 10))$.

In Figure 1 we graphically describe the system transitions. In order to increase readability, we have omitted trailing transitions labeled by δ and transitions that have to be added to obtain an input-enabled machine (for example, the transition $(s_0, ?p\_biscuit, (0,0,0), s_0)$).

Now we can define the concatenation of several transitions of an agent to capture the different evolutions of a system from a configuration to another one. These evolutions can be produced either by executing an input or an output action or by offering a *exchange of resources*. As we will see, exchanges of resources have low priority and will be allowed only if no output can be performed. The idea is that if we can perform an output with the existing resources, then we do not need to exchange resources.

Definition 4. *Let* $M = (S, s_0, L, T, U, V)$ *be a UIOLTS. M can evolve from the configuration* $c = (s, \bar{v})$ *to the configuration* $c' = (s', \bar{v'})$ *according to one of these options*

1. *If there is an input transition* $(s, ?i, \bar{x}, s_1)$ *then this transition can be executed. The new configuration is* $s' = s_1$ *and* $v' = v + \bar{x}$.
2. *If there is an output transition* $(s, !o, C, \bar{z}, s_1)$ *such that* $C(\bar{v})$ *holds then the transition can be executed. The new configuration is* $s' = s_1$ *and* $\bar{v'} = \bar{v} - \bar{z}$.
3. *Let us consider the transition associated with quiescence at s:* $(s, \delta, C_s, \bar{0}, s)$. *If* $C_s(\bar{v})$ *holds, that is, no output transition is currently available, then this transition can be executed. The new configuration is* $s' = s$ *and* $\bar{v'} = \bar{v}$.
4. *Let us consider again the transition associated with quiescence at s, that is,* $(s, \delta, C_s, \bar{0}, s)$. *If* $C_s(\bar{v})$ *holds, then we can offer an exchange. We represent an exchange by a pair* $(\xi, \bar{y})$ *where* $\bar{y} = (y_1, y_2, \ldots y_n)$ *is the variation of the set of resources. Therefore,* $y_i < 0$ *indicates a decrease of the resource i while* $y_i > 0$ *represents an increase of the resource i. M can do an exchange* $(\xi, \bar{x})$ *if* $U(s, v) < U(s, v + \bar{x})$. *If another agent is accepting the exchange, the new configuration is* $s' = s$ *and* $\bar{v'} = \bar{v} + \bar{y}$.

We denote an evolution from the configuration c to the configuration c' by the tuple $(c, (a, \bar{y}), c')$ where $a \in L \cup \{\xi\}$ and $\bar{y} \in \mathbb{R}^n$. We denote by $Evolutions(M, c)$ the set of evolutions of M from the configuration c and by $Evolutions(M)$ the set of evolutions of M from (s_0, v_0), the initial configuration.

A trace of M is a finite list of evolutions. $Traces(M, c)$ denotes the set of traces of M from the configuration c and $Traces(M)$ denotes the set of traces of M from the initial configuration c_0. Let $l = e_1, e_2, \ldots, e_m$ be a trace of M with $e_i = (c_i, (a_i, \bar{x}_i), c_{i+1})$. The observable trace associated to l is a triple (c_1, σ, c_{n+1}) where σ is the sequence of actions obtained from $a_1, a_2, \ldots, a_m$ by removing all occurrence of ξ. We sometimes represent this observable trace as $c_1 \stackrel{\sigma}{\Longrightarrow} c_{n+1}$.

Example 2. When the machine described in Example 1 is in its initial configuration, a user can insert one coin and the machine accept this action. After this action, the new configuration of the machine is $(s_1, (1, 10, 10))$. Now the user, who wants a pack of biscuits, press the biscuit button and the new configuration is $(s_3, (1, 10, 10))$ because by performing this action the set or resources does not change. In this configuration, the machine performs the output action *!biscuit* and the user can take his pack. The new configuration of the machine is $(s_0, (1, 9, 10))$.

Let us suppose that after a set of interactions our machine has reached the configuration $(s_0, (15, 5, 0))$. In this situation, the person in charge of the machine would like to refill it and take as many coins as new items he adds. This action can be represented as a exchange of resources. The tuple that represents the variation of the resources is $(-15, 5, 10)$. After this exchange the new configuration is $(s_0, (0, 10, 10))$.

One of the traces corresponding to our vending machine is

$$(c_0, (?money, (1,0,0)), c_1), (c_1, (?p\_biscuit, (0,0,0)), c_2), (c_2, (!biscuit, (0,1,0)), c_3),$$
$$(c_3, (\xi, (-1,1,0)), c_4).$$

while the associated observable trace is $(c_0, (?money, ?p\_biscuit, !biscuit), c_4)$.

3   Implementation Relations for UIOLTSs

In this section we introduce our implementation relations that formally establish when an implementation is correct with respect to a specification. In our context, the notion of correctness has several possible definitions. For example a person may consider that an implementation I of a system S is good if the number of resources always increases while another one considers that the number of resources can decrease sometimes.

We define two different implementation relations. The first one is close to the classical **ioco** implementation relation [23] where an implementation I is correct with respect to a specification S if the output actions executed by I after a sequence of actions is performed are a subset of the ones that can be executed by S. Intuitively, this means that the implementation does not *invent* actions that the specification did not contemplate. First, we introduce some auxiliary notation

Definition 5. *Let* $M = (S, s_0, L, T, U, V)$ *be a UIOLTS,* $c = (s, \bar{x}) \in Conf(M)$ *a configuration of* M, *and* $\sigma \in L^*$ *be a sequence of actions. Then,*

$$c \text{ after } \sigma = \{c' \in Conf(M) | c \overset{\sigma}{\Longrightarrow} c'\}$$

$$out(c) = \{!o \in L_O | \exists s_1, \bar{z}, C : (s, !o, C, \bar{z}, s_1) \in T \wedge C(\bar{x})\}$$

$$\cup \{\delta | \exists C_s : (s, \delta, C_s, \bar{0}, s) \in T \wedge C_s(\bar{x})\}$$

Intuitively, c after σ returns the configuration reached from the configuration c by the execution of the trace σ while $out(c)$ contains the output actions that the system can execute from the configuration c.

Definition 6. *Let* I, S *be two UIOLTSs with the same set of actions* L. *We write* I **ioco** S *if for all sequence of actions* $\sigma \in L^*$ *we have that* $out(I \text{ after } \sigma) \subseteq out(S \text{ after } \sigma)$.

Example 3. In figure 2 we show three UIOLTSs that model three possible implementations of our vending machine where, as before, we have removed trailing occurrences of input actions and of δ transitions.

Let us consider first the implementation I_1. In this case, the specification specifies that after $?money$ the machine does not execute any output actions.

So, $out(S \text{ after } ?money) = \{\delta\}$. The implementation I_1 indicates that the machine can execute the output action $!return$ after the input action $?money$. Therefore, $out(I \text{ after } ?money) = \{!return\}$. Since, $out(I \text{ after } ?money) \nsubseteq out(S \text{ after } ?money)$ we conclude that I_1 **ioco** S does not hold.

If we consider the implementation I_2 we can check that I_2 **ioco** S because the requested sets containment hold. In fact, this machine is *very good*, from the owner's

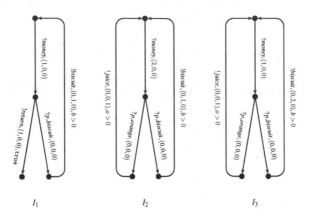

Fig. 2. Possible implementations of our vending machine

point of view, since it requests two coins for each of the items. Similarly, I_3 **ioco** S, but this is a bad implementation because it provides two packets of biscuits per each coin. This shows the weakness in our framework of an implementation relation concentrating only on the performed actions.

Our second implementation relation is also based on the **ioco** mechanism but we take into account both the resources that the system has and the actions that the system can execute. In order to define the new relation we need to redefine again the set out of outputs. In this case, our set of outputs has two components: The output action that can be executed and the set of resources that the system has. We also introduce an operator to compare sets of pairs (output, resources).

Definition 7. *Let $M = (S, s_0, L, T, U, V)$ be a UIOLTS and $c = (s, \bar{x}) \in Conf(M)$ be a configuration of M. Then*

$$\text{out}'(c) = \{(!o, \bar{y}) \in L_O \times \mathbb{R}_+^n \,|\, \exists s_1, \bar{z}, C : (s, !o, C, \bar{z}, s_1) \in T \wedge C(\bar{x}) \wedge \bar{y} = \bar{x} - \bar{z}\}$$

$$\cup \{(\delta, \bar{x}) \,|\, \exists C_s : (s, \delta, C_s, \bar{0}, s) \in T \wedge C_s(\bar{x})\}$$

Given two sets $A = \{(o_1, \bar{y}_1), \ldots, (o_n, \bar{y}_n)\}$ and $B = \{(o_1, \bar{x}_1), \ldots, (o_n, \bar{x}_n)\}$, we write $A \sqsubseteq B$ if $act(A) \subseteq act(B)$ and $\min(rec(A, o)) \geq max(rec(B, o))$, where $Act(A) = \{a \,|\, (a, \bar{y}) \in A\}$ and $rec(A, o) = \{r \,|\, (o, r) \in A\}$.

The set out$'(c)$ contains those actions (outputs or quiescence) that can be performed when the system is in configuration c as well as the set of resources obtained after their performance. Next, we introduce our new implementation relation. We consider that an implementation I is correct with respect to a specification S if the output actions performed by the implementation in a state are a subset of those that can be performed by the specification in this state and the set of resources of implementation I is *better* than the set of resources in the specification.

Definition 8. *Let I, S be two UIOLTSs with the same set of actions L. We write I ioco$_{\text{r}}$ S if for all sequence of actions $\sigma \in L^*$ we have that out$'(I$ after $\sigma) \sqsubseteq$ out$'(S$ after $\sigma)$.*

Example 4. Let us consider the specification S given in Figure 1 and the implementation I_3 given in Figure 2. We have, according to the specification, that

$$\text{out}'(S \text{ after } ?money?p\_biscuit) = \{(!biscuit, (1, 9, 10)\}$$

while, according to I_3, we have that

$$\text{out}'(I \text{ after } ?money?p\_biscuit) = \{(!biscuit, (1, 8, 10)\}$$

We can observe that the set of output actions is the same in both cases but the resources of the implementation are smaller than the ones of the specification. Then, we conclude that I_3 **ioco$_r$** S does not hold.

4   Tests: Definition and Application

Essentially, a test is the description of the behavior of a tester in an experiment carried out on an implementation under test. In this experiment, the tester serves as a kind of artificial environment of the implementation. This tester can do four different things: It can accept an output action started by the implementation, it can provide an input action to the implementation, it can propose a exchange of resources, or it can observe the absence of output actions, so that it can detect quiescence. When the tester receives an output action it checks whether the action belongs to the set of expected ones (according to its description); if the action does not belong to this set then the tester will produce a fail signal. In addition, each state of a test saves information about the set of resources that the tested system has if the test reaches this state.

In our framework, a test for a system is modeled by a UIOLTS, where its set of input actions is the set of output actions of the specification and its set of output actions is the set of input actions of the specification. Also, we include a new action θ that represents the observation of quiescence. In order to be able to accept any output from the tested agent, we consider that tests are *input-enabled*, since its inputs correspond to outputs of the tested agent.

Definition 9. *Let* $M = (S, s_0, L, T, V)$, *with* $L = L_I \cup L_O \cup \{\delta\}$. *A test for* M *is a UIOLTS* $t = (S^t, s_0^t, L^t, T^t, V)$ *where*

- S^t *is the set of states where* $s_0^t \in S^t$ *is the initial state and there are two special states called* fail *and* pass *with* fail $\neq$ pass. *We represent by* $res(s)$ *the information about resources saved in state s.*
- L^t *is the set of actions.* $L = L_O \cup L_I \cup \{\theta, \xi\}$ *where*
  - L_O *is the set of inputs (these are outputs of M).*
  - L_I *is the set of outputs (these are inputs of M).*
  - θ *is a special action that represents the detection of quiescence.*
  - ξ *is an special action that represents the proposal of an exchange.*
- T^t *is the set of transitions. Each transition is a tuple* $(s, a, \bar{x}, s_1)$ *where s is the initial state,* s_1 *is the final state,* $a \in L_O \cup L_I \cup \{\theta, \xi\}$ *is the label of the transition and* $\bar{x} \in \mathbb{R}^n$ *is the variation in the set of resources.*

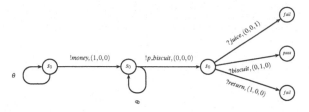

Fig. 3. Test for the vending machine.

We define configurations of a test in the same way that we defined them for UIOLTSs, and we thus omit the definition. In Figure 3 we present a test for the system described in Figure 1. Given an implementation I and a test t, running t with I is the parallel execution of both taking into account the peculiarities of the special actions δ, θ and ξ. Let $c = (s, \bar{v})$ and $c' = (s', \bar{v}')$ be configurations of I, and $c_t = (q_t, \bar{v}_t)$ and $c'_t = (q'_t, \bar{v}'_t)$ be configurations of t. The rules for the composition operator, denoted by $\|\lceil$, are the following

$$A) \quad \frac{c \xrightarrow{(?a,\bar{x})} c', c_t \xrightarrow{(!a,\bar{x})} c'_t, a \in L_i}{c\|\lceil c_t \xrightarrow{(a,\bar{x})} c'\|\lceil c'_t}$$

$$B) \quad \frac{c \xrightarrow{(!a,\bar{x})} c', c_t \xrightarrow{(?a,\bar{x})} c'_t, a \in L_O}{c\|\lceil c_t \xrightarrow{(a,\bar{x})} c'\|\lceil c'_t}$$

$$C) \quad \frac{c \xrightarrow{\delta}, c_t \xrightarrow{\theta} c'_t}{c\|\lceil c_t \xrightarrow{\theta} c\|\lceil c'_t}$$

$$D) \quad \frac{c \xrightarrow{(\xi,\bar{x})} c', c_t \xrightarrow{(\xi,-\bar{x})} c'_t}{c\|\lceil t \xrightarrow{(\xi,\bar{x})} c'\|\lceil c'_t}$$

Definition 10. *A test execution of the test t with an implementation I is a trace of $i\|\lceil t$ leading to one of the states* pass *or* fail *of t.*

We say that an implementation I passes a test t if all test execution of t with I go to a pass *state of t.*

Another definition for passing a test, considering the resources administered by the system, is the following.

Definition 11. *An implementation I passes$_r$ a test t if all test execution of t with I go to a* pass *state s of t and the set $\bar{x}$ of resources in the configuration of I is greater than or equal to the set or resources $rec(s)$ that the test says the implementation may have if the test reaches the state s.*

The final step is to prove that the application of a certain set of tests has the same discriminatory power as the implementation relations previously defined. Due to space limitations we cannot include this proof. The idea is to derive tests from specifications using an adaption to our new framework of the algorithm given in [17].

5   Conclusions and Future Work

In this paper we have introduced a new formalism, called Utility Input Output Labeled Transition Systems, to specify the behavior of e-commerce agents taking part in a multi-agent system. We have also defined a testing methodology, based in this formalism, to test whether an implementation of a specified agent behaves as the specification says that it behaves. In this theory we have defined two different implementation relations and a notion of test.

This paper is the first step in a long road that we expect to continue in the future. We currently focus on two research lines. The first one is based on theoretical aspects and we would like to extend our formalism in order to specify the behavior of agents that are influenced by the passing of time. The second line is more practical since we would like to apply our formalism to real complex agents. In this line, we have already developed a prototype tool that allows us to do automatic generation of tests, but we have to *stress test* the tool to check how it scales.

References

1. Adi, K., Debbabi, M., Mejri, M.: A new logic for electronic commerce protocols. Theoretical Computer Science 291(3), 223–283 (2003)
2. Ammann, P., Offutt, J.: Introduction to Software Testing. Cambridge University Press, Cambridge (2008)
3. Balachandran, B.M., Enkhsaikhan, M.: Developing multi-agent E-commerce applications with JADE. In: Apolloni, B., Howlett, R.J., Jain, L. (eds.) KES 2007, Part III. LNCS (LNAI), vol. 4694, pp. 941–949. Springer, Heidelberg (2007)
4. Collins, J., Rogers, A., Arunachalam, R., Sadeh, N., Eriksson, J., Finne, N., Janson, S.: The supply chain management game for 2008 trading agent competition. Technical Report CMU-ISRI-08-117, Carnegie Mellon University (2008)
5. Guttman, R., Moukas, A., Maes, P.: Agent-mediated electronic commerce: A survey. The Knowledge Engineering Review 13(2), 147–159 (1998)
6. Hierons, R.M., Bogdanov, K., Bowen, J.P., Cleaveland, R., Derrick, J., Dick, J., Gheorghe, M., Harman, M., Kapoor, K., Krause, P., Luettgen, G., Simons, A.J.H., Vilkomir, S., Woodward, M.R., Zedan, H.: Using formal methods to support testing. ACM Computing Surveys 41(2) (2009)
7. Hierons, R.M., Bowen, J.P., Harman, M. (eds.): FORTEST. LNCS, vol. 4949. Springer, Heidelberg (2008)
8. Hindriks, K.V., de Boer, F.S., van der Hoek, W., Meyer, J.-J.C.: Formal semantics for an abstract agent programming language. In: Rao, A., Singh, M.P., Wooldridge, M.J. (eds.) ATAL 1997. LNCS (LNAI), vol. 1365, pp. 215–229. Springer, Heidelberg (1998)
9. Jacky, J., Veanes, M., Campbell, C., Schulte, W.: Model-Based Software Testing and Analysis with C#. Cambridge University Press, Cambridge (2008)
10. Jennings, N.R.: On agent-based software engineering. Artificial intelligence 117(2), 357–401 (1999)
11. Lomazova, I.A.: Nested Petri Nets for adaptive process modeling. In: Avron, A., Dershowitz, N., Rabinovich, A. (eds.) Pillars of Computer Science. LNCS, vol. 4800, pp. 460–474. Springer, Heidelberg (2008)
12. Ma, M.: Agents in e-commerce. Communications of the ACM 42(3), 79–80 (1999)
13. Merayo, M.G., Núñez, M., Rodríguez, I.: Formal specification of multi-agent systems by using EUSMs. In: Arbab, F., Sirjani, M. (eds.) FSEN 2007. LNCS, vol. 4767, pp. 318–333. Springer, Heidelberg (2007)
14. Myers, G.J.: The Art of Software Testing, 2nd edn. John Wiley and Sons, Chichester (2004)
15. Nguyen, N.T.: Computational collective intelligence and knowledge inconsistency in multi-agent environments. In: Bui, T.D., Ho, T.V., Ha, Q.T. (eds.) PRIMA 2008. LNCS (LNAI), vol. 5357, pp. 2–3. Springer, Heidelberg (2008)

16. Núñez, M., Rodríguez, I., Rubio, F.: Formal specification of multi-agent e-barter systems. Science of Computer Programming 57(2), 187–216 (2005)
17. Núñez, M., Rodríguez, I., Rubio, F.: Specification and testing of autonomous agents in e-commerce systems. Software Testing, Verification and Reliability 15(4), 211–233 (2005)
18. Padget, J.A., Bradford, R.J.: A pi-calculus model of a spanish fish market - preliminary report. In: Noriega, P., Sierra, C. (eds.) AMET 1998 and AMEC 1998. LNCS (LNAI), vol. 1571, pp. 166–188. Springer, Heidelberg (1999)
19. Rao, A.S.: AgentSpeak(L): BDI agents speak out in a logical computable language. In: Perram, J., Van de Velde, W. (eds.) MAAMAW 1996. LNCS (LNAI), vol. 1038, pp. 42–55. Springer, Heidelberg (1996)
20. Rodríguez, I., Merayo, M.G., Núñez, M.: $\mathcal{HOTL}$: Hypotheses and observations testing logic. Journal of Logic and Algebraic Programming 74(2), 57–93 (2008)
21. Sandholm, T.: Agents in electronic commerce: Component technologies for automated negotiation and coalition formation. In: Klusch, M., Weiss, G. (eds.) CIA 1998. LNCS (LNAI), vol. 1435, pp. 113–134. Springer, Heidelberg (1998)
22. Sierra, C.: Agent-mediated electronic commerce. Autonomous Agents and Multi-Agent Systems 9(3), 285–301 (2004)
23. Tretmans, J.: Model based testing with labelled transition systems. In: Hierons, R.M., Bowen, J.P., Harman, M. (eds.) FORTEST. LNCS, vol. 4949, pp. 1–38. Springer, Heidelberg (2008)
24. Utting, M., Legeard, B.: Practical Model-Based Testing: A Tools Approach. Morgan Kaufmann, San Francisco (2007)

A Formal Model for Epistemic Interactions

Paweł Garbacz, Piotr Kulicki, Marek Lechniak, and Robert Trypuz

John Paul II Catholic University of Lublin,
al. Racławickie 14, 20-950 Lublin, Poland
{garbacz,kulicki,lechniak,trypuz}@kul.pl
http://www.l3g.pl

Abstract. The conceptual world of AI is inhabited by a number of epistemic puzzles whose role is to provide a test harness environment for various methods and algorithms. In our paper we focus on those puzzles in which agents either collaborate or compete with one another in order to adopt their epistemological situations to their environment. Our goal is to devise a formal model for epistemic interactions and a family of reasoning mechanisms that would solve those puzzles. Once specified in the abstract manner, they are implemented in the Prolog environment.

Keywords: knowledge representation, ontology, epistemic change.

1 Introduction

The conceptual world of Artificial Intelligence (AI) is inhabited by a number of epistemic puzzles whose role is provide a test harness environment for various methods and algorithms. In our paper we will focus on those puzzles in which agents either collaborate or compete with one another in order to adopt their epistemological situations to their environment. In the puzzles at stake they exhibit certain behaviours by means of which they attempt to reach certain epistemic goals. Here is an exemplary list of such puzzles: the hats puzzle ([2]), the wisemen puzzle ([8]), Mr Sum and Mr Product puzzle ([8]), the muddy children puzzle ([1]). Our goal is to devise a family of reasoning mechanisms that would solve the puzzles. Once specified in the abstract manner, they will be implemented in the Prolog environment. As a working example we will use a well-known *hats puzzle*:

> Three people Adam, Ben and Clark sit in a row in such a way that Adam can see Ben and Clark, Ben can see Clark and Clark cannot see anybody. They are shown five hats, three of which are red and two are black. The light goes off and each of them receives one of the hats on his head. When the light is back on they are asked whether they know what the colours of their hats are. Adam answers that he doesn't know. Then Ben answers that he doesn't know either. Finally Clark says that he knows the colour of his hat. What colour is Clark's hat?

John McCarthy in [7] points out that AI methods of solving problems suffer unbalance between their two parts: epistemological and heuristic. The first one is a representation

N.T. Nguyen et al. (Eds.): New Challenges in Compu. Collective Intelligence, SCI 244, pp. 205–216.
springerlink.com

of the world which enables one to solve problems while the second one is a mechanism of finding solutions. McCarthy notes that most of the work is devoted to heuristic part.

In this paper we will pick up the task of developing the epistemological part of intelligence. Accepting the tenets of the Knowledge Representation paradigm, we believe that developing a general formal model of agents' epistemologies will provide a firm and universal basis for the algorithms we are up to.

In section 2 a general model of epistemic interactions is introduced and described. In section 3 we present an implementation of our model in Prolog. Then in section 4 a comparison of our model with other frameworks is given. At the end we discuss some directions of evolution of our model and its applications.

2   Towards a General Model of Epistemic Interactions

Our model of epistemic interactions has two components: ontological and epistemological. The ontological part represents, in rather rough and ready way, the world our knowledge concerns. The epistemological part of the model represents the phenomenon of knowledge in its static and dynamic aspects.

We start to discuss the ontological component of the model with an analysis of the notion of "situation". A belief, as an intentional entity, refers to an external chunk of reality, which we call a *situation* (or state of affairs). So when I believe that Warsaw is the capital of Poland then this belief of mine concerns the situation *that Warsaw is the capital of Poland*, which situation is somehow part of the real world. In a general, the situation at stake may have any ontic structure. Thus, there are situations "in which" certain objects possess certain properties, situations "in which" certain objects participate in certain relations or processes, etc.

Let $ElemSit$ be a set of elementary ontic (possible) situations. Briefly speaking, a situation is *elementary* if no situation is part of it. For instance *that Adam has a red hat* would be an elementary situation and *that both Adam and Ben have red hats* would not be an elementary situation.

In the set $ElemSit$ we define the relation of compossibility. Intuitively, $x \parallel y$ means that a situation x may co-occur with situation y.[1] For example, if $x = $ *that Adam has a red hat*, $y = $ *that Ben has a red hat* and $z = $ *that Adam has a white hat*, then $x \parallel y$, $y \parallel z$ and $x \nparallel z$. For obvious reasons, the relation $\parallel$ is reflexive and symmetric in $ElemSit$, but is not transitive.

In what follows we will mainly deal with non-elementary situations (situations, for short), which will be represented as sets of elementary situations. We use the notion of situation instead of the notion of elementary situation because a lot of our beliefs do not concern elementary situations. Let $\emptyset \notin Sit \subseteq \wp(ElemSit)$ be a set of (possible) ontic situations.[2] In our example, that both Adam and Ben have red hats might be repre-

---

[1] It is our intention that the meaning of "may" remains within the domain of ontology. In other words, if $x \nparallel y$, then x cannot co-occur with y because of some ontological principle, where ontology is understood broadly enough to include laws of logic. However, the concept of ontology used here has more to do with the engineering understanding than with the philosophical tradition (see, for instance, [5]). So, effectively, relation $\parallel$ is relative to *an ontology*.

[2] As before, the specific content of Sit is determined by the ontology presupposed in a given domain (or puzzle).

sented as the set $\{x, y\} \in Sit$. Given our understanding of the relation $\|$, the following condition is the case:

$$X \in Sit \rightarrow \forall y, z \in Xy \parallel z. \tag{1}$$

We do not accept the opposite implication because we do not want to commit our model to such entities as the situation that Warsaw is the capital of Poland and π is an irrational number.

We can now define the notion of possible world:

$$X \in PossWorld \triangleq X \in Sit \wedge \forall Y(X \subset Y \rightarrow Y \notin Sit). \tag{2}$$

Let $Agent$ be a set of epistemic agents, i.e. those agents that are capable of having beliefs and $Time = (t_1, t_2, \ldots)$ be a sequence of moments. The *actual epistemic state* of an agent at a given moment will be representeded by a subset of $PossWorld$: $epist(a, t_n) \subseteq PossWorld$. Any such state collectively, so to speak, represents both the agent's knowledge and his or her ignorance. Due to its actual epistemic state, which is represented by a set $epist(a, t_n)$, and for every $X \in Sit$, agent a may be described (at t_n) according to the following three aspects:[3]

Definition 1. *Agent a knows at moment t_n that situation X holds (written: $K_{a,t_n}(X)$) iff $X \subseteq \bigcap epist(a, t_n)$.*

Definition 2. *Agent a knows at moment t_n that situation X does not hold (written: $\overline{K}_{a,t_n}(X)$) iff $X \cap (\bigcup epist(a, t_n)) = \emptyset$.*

Definition 3. *Agent a does not have any knowledge at moment t_n about situation X iff $\neg K_{a,t_n}(X) \wedge \neg \overline{K}_{a,t_n}(X)$.*

However, the puzzles we are dealing with do not presuppose that we know the actual epistemic state of a given agent. Thus, we extend the notion of actual epistemic state to the notion of possible epistemic state. A *possible epistemic state* of an agent represents a body of knowledge (resp. of ignorance) that the agent may exhibit given the ontic situation and epistemic capabilities of this agent. In our case, the possible epistemic states arc determined by the relation of seeing (perceiving), other agents' announcements, and the agent's deductive capabilities.

A possible epistemic state of an agent at a given moment will be represented by the set $epist_i(a, t_n) \subseteq PossWorld$. One of the possible epistemic states is the actual epistemic state, i.e. there exists i such that $epist(a, t_n) = epist_i(a, t_n)$. As before (for $Y \in Sit$), an agent may be described (at t_n) according to the following three aspects:

Definition 4. *Agent a knows in a possible epistemic state $X = epist_i(a, t_n)$ that (ontic) situation Y holds (written : $K_X(Y)$) iff $Y \subseteq \bigcap X$.*

Definition 5. *Agent a knows in a possible epistemic state $X = epist_i(a, t_n)$ that (ontic) situation Y does not hold (written : $\overline{K}_X(Y)$) iff $Y \cap (\bigcup X) = \emptyset$.*

---

[3] The definitions, and their extensions 4, 5, and 6 below, presuppose that the set Sit is closed under intersections: $X, Y \in Sit \wedge X \cap Y \neq \emptyset \rightarrow X \cap Y \in Sit$.

Definition 6. *Agent a does not have any knowledge in a possible epistemic state* $X = epist_i(a, t_n)$ *about ontic situation* Y *iff* $\neg K_X(Y) \wedge \neg \overline{K}_X(Y)$.

We will use later the following auxiliary notions: $Epist(a, t_n)$ – a set of all possible epistemic states of agent a at t_n, $Epist(t_n)$ – a set of sets of possible epistemic states of all agents at t_n, $Epist$ – a set of sets of all possible epistemic states of all agents (from $Agent$) at all moments (from $Time$).

When $a \in Agent$, then "$\sim_a$" will represent the relation of epistemological indiscerniblity, which we treat as an equivalence relation. In general, the epistemological indiscernibility covers a number of epistemic constraints of agents. For example:

- (Because Adam does not see his head), he does not discern the situation in which he has a red hat from the situation in which has a white hat,
- (Because of his daltonism), Ben does not discern the situation in which the traffic signal is red from the situation in which the traffic signal is green,
- (Because of the thermostat's failure), the central heating system does not "discern" the situation in which the temperature in this room is higher than 285 K from the situation in which the temperature is higher than 290 K.

As usual, the relation $\sim_a$ may be defined by means of the set of equivalence classes in $PossWorld$.

In a special case, the relation of epistemological indiscerniblity depends on the knowledge obtained thanks to the behaviour of an agent. For example, if Adam says that he does not know what hat he has, then this utterance may lead Ben and Clark (because of their current epistemic state) to the discernment between the situation that Adam has a red hat and they have white ones and all other (relevant) situations.

The relation of epistemological indiscerniblity is, as the notion of knowledge itself, relative to time: $\sim_{a,t_n}$ is the relation of epistemological indiscerniblity for agent a at time t_n.

It is assumed in our approach that possible epistemic states coincide with the abstraction classes of the epistemological indiscerniblity relation:

$$Epist(a, t_n) = PossWorld/\sim_{a,t_n} \tag{3}$$

We assume that all changes of epistemic states are caused by the behaviours of the agents—in particular by their utterances, by means of which they expose their (current) epistemic states—and not by their inference processes.

We will now define a number of rules that govern the dynamics of epistemic states. In the hats puzzle the only rule that sets the epistemic states in motion is the following one: *if* agent a (says that) he or she does not know that X holds and in an epistemic state $epist_i(a, t_n)$ a knows that X holds, *then* after the aforementioned utterance that state (i.e. $epist_i(a, t_n)$) is effectively impossible, i.e. we remove its elements from all possible epistemic states of all agents. Formally,

Rule 1. *If (a says that)* $\neg K_{a,t_n}(X)$ *and* $Y \in Epist(a, t_n)$ *and* $K_Y(X)$, *then for every* $a' \in Agent$, $Epist(a', t_{n+1}) = \delta_0(Epist(a', t_n), Y)$, *where*

Definition 7. δ_0 *maps* $Epist \times \bigcup\bigcup Epist$ *into* $Epist$ *and satisfies the following condition:*

$$\delta_0(Epist(a,t_n), X) = \begin{cases} Epist(a,t_n)\setminus\{X\}, & \text{if } X \in Epist(a,t_n), \\ (Epist(a,t_n)\setminus\{Z\})\cup\{Z\setminus X\} & \text{if } Z \in Epist(a,t_n) \\ & \text{and } X \cap Z \neq \emptyset \\ Epist(a,t_n) & \text{otherwise.} \end{cases}$$

In our conceptual framework we may also define other rules:

Rule 2. *If (a says that)* $K_{a,t_n}(X)$ *and* $Y \in Epist(a,t_n)$ *and* $\neg K_Y(X)$, *then for every* $a' \in Agent$, $Epist(a',t_{n+1}) = \delta_0(Epist(a',t_n),Y)$.

Rule 3. *If (a says that)* $\neg\overline{K}_{a,t_n}(X)$ *and* $Y \in Epist(a,t_n)$ *and* $\overline{K}_Y(X)$, *then for every* $a' \in Agent$, $Epist(a',t_{n+1}) = \delta_0(Epist(a',t_n),Y)$.

Rule 4. *If (a says that)* $\overline{K}_{a,t_n}(X)$ *and* $Y \in Epist(a,t_n)$ *and* $\neg\overline{K}_Y(X)$, *then for every* $a' \in Agent$, $Epist(a',t_{n+1}) = \delta_0(Epist(a',t_n),Y)$.

As one can easily appreciate, these rules presuppose certain degree of rationality of epistemic agents. The rationality manifests itself in the fact that the epistemic condition of those agents, or rather their utterances about their condition, accords with their epistemic states (cf. 4, 5 and 6). Moreover, the rules presuppose that the agents are infallible and (epistemologically) sincere.

If we weaken those assumptions, we may arrive at another notion of rationality. Namely, we assume now that our agents are infallible and (epistemologically) sincere. We do not however assume that the agents' utterances are based on their knowledge of (their) epistemic states. Then when agent a says that:

- he or she knows that X *holds*, then we remove from every epistemic state of every agents all the possible situations that *do not belong* to X,
- he or she knows that X *does not hold*, then we remove from every epistemic state of every agents all the possible situations that *belong* to X.

Formally,

Rule 5. *If (a says that)* $K_{a,t_n}(X)$, *then for every* $a' \in Agent$, $Epist(a',t_{n+1}) = \delta_1(Epist(a',t_n),X)$.

Rule 6. *If (a says that)* $\overline{K}_{a,t_n}(X)$, *then for every* $a' \in Agent$, $Epist(a',t_{n+1}) = \delta_2(Epist(a',t_n),X)$, *where*

Definition 8. δ_1 *maps* $Epist \times \bigcup\bigcup Epist$ *into* $Epist$ *and satisfies the following condition:*

$$\delta_1(Epist(a,t_n), X) = \begin{cases} (Epist(a,t_n)\setminus\{Z\})\cup\{Z\cap X\}) & \text{if } Z \in Epist(a,t_n), \\ Epist(a,t_n) & \text{otherwise.} \end{cases}$$

Definition 9. δ_2 *maps* $Epist \times \bigcup\bigcup Epist$ *into* $Epist$ *and satisfies the following condition:*

$$\delta_2(Epist(a,t_n), X) = \begin{cases} (Epist(a,t_n)\setminus\{Z\})\cup\{Z\setminus X\}) & \text{if } Z \in Epist(a,t_n), \\ Epist(a,t_n) & \text{otherwise.} \end{cases}$$

It seems that the factors that trigger the process of epistemic change are of two kinds: ontological and epistemological. Consider rule 2 once more. The ontological condition of this rule is the fact that agent a says that he or she knows that a certain ontic situation holds (i.e. that $K_{a,t_n}(X)$). On the other hand, the epistemological condition is his or her epistemic state ($Y \in Epist(a, t_n)$) in which the agent cannot know that this situation holds ($\neg K_Y(X)$). We may represent the epistemological conditions of rules for epistemic changes by means of the notion of epistemic state. However, in order to account for the ontological conditions, we distinguish in the set $\wp(ElemSit)$ a subset $AgentBeh$ that collects types (here: sets) of ontic situations that are those conditions. An example of such type may be a set of situations in which agents say that they do not know what hats they have. In general, those conditions may be classified as agents' behaviours, which include also such "behaviours" as being silent (cf. the wisemen puzzle). It is worth noting that, as rules 5 and 6 attest, a rule might not have epistemological conditions.

Let $a \in Agent$. A *rule for epistemic change* ρ is either

1. a mapping $\rho : \bigcup Epist \times AgentBeh \times \bigcup Epist \to \bigcup Epist$ (this condition concerns rules with epistemological conditions) or
2. a mapping $\rho : \bigcup Epist \times AgentBeh \to \bigcup Epist$ (this condition concerns rules without epistemological conditions).

It should be obvious that

1. if $\rho(X, Y, Z) = V$ and $X, Y \in Epist(t_n)$, then $V \in Epist(t_{n+1})$ (for rules with epistemological conditions),
2. if $\rho(X, Y) = V$ i $X \in Epist(t_n)$, to $V \in Epist(t_{n+1})$ (for rules without epistemological conditions).

The set of all such rules will be denoted by "*Rule*". For the sake of brevity, from now on we will consider only rules with epistemological conditions.

Note that all our examples of rules (i.e. rules 1, 2, 3, 4, 5, and 6) actually define multi-agent interactions. Making an utterance of a specific kind, which is described by the relevant ontological trigger of each rules, an agent reveals his or her epistemic condition to other agents who update their own epistemic conditions accordingly. Then, in a sense, our rules describe mutli-agent epistemic actions.

Summarizing, we suggest that one should base his or her solution to a puzzle at stake on such dynamic model of knowledge. In order to obtain the solution, one needs the following input data:

- set Sit,
- temporal sequence $Time = (t_n)$,
- set of epistemic agents $Agent$,
- set of sets of epistemic states of any such agent at the initial moment t_1: $Epist_1 = \{Epist(a, t_1) : a \in Agent\}$,
- function $dist : Time \to \wp(ElemSit)$.

Our function $dist$ is to distribute the agents' behaviours over the set of moments. The behaviours may or may not belong to the elements of $AgentBeh$. When "nothing happens", the value of $dist$ is equal to the empty set.

The evolution of sets of epistemic states is triggered by the ontological conditions, which are determined by function $dist$, according to the accepted rules of epistemic change. This implies that the following condition holds[4]:

$$\exists \rho \in Rule \, \exists X \in AgentBeh \qquad (4)$$
$$[\rho(Epist(a, t_n), X, Epist(a', t_n)) = Z \wedge dist(t_n) \cap X \neq \emptyset] \rightarrow Epist(a, t_{n+1}) = Z.$$

We also assume that epistemological states change only when a certain rule is triggered:

$$Epist(a, t_{n+1}) \neq Epist(a, t_n) \equiv \qquad (5)$$
$$\exists \rho \in Rule \, \exists X \in AgentBeh \, \exists Y \in \bigcup Epist[\rho(Epist(a, t_n), X, Y) = Epist(a, t_{n+1})].$$

Of course, it might happen that a rule ρ is triggered vacuously, i.e. for certain a, t_n, X, and Y, it is the case that $\rho(Epist(a, t_n), X, Y) = Epist(a, t_n)$.

After applying the last rule, say at moment t_k ($k \leq n$), we will reach one of the four possible outcomes: (1) in every epistemic state at moment t_k agent a knows that situation X holds (or does not hold), (2) in one epistemic state at moment t_k agent a knows that situation X holds (or does not hold) and in another epistemic state he or she does not know that, (3) in no epistemic state at moment t_k agent a knows that situation X holds (or does not hold) and (4) set $Epist(a, t_{n+1})$ is empty. Only the first case represents the situation in which we (or any other agent, for that matter) are in a position to solve the puzzle at stake. On the other hand, the last situation implies that the puzzle was inconsistent.

The conceptual model, which supports automatic resolutions of a broad class of puzzles, may be seen as a dynamic epistemological model. In general, it is a triple $\langle Sit,$ $Time, Epist \rangle$, which is uniquely determined by the initial assumptions represented by the quadruple $\langle Sit, Epist_1, dist, Rule \rangle$.

3   Implementation in Prolog

The presented model can be implemented as a program in Prolog in a straightforward way.[5] The user has to introduce the set of parameters: the list of agents, possible values of the attributes (number and colours of hats) and a specification of the perception relation (cf. *sees* relation). On that basis, the initial set of sets of possible epistemic states for all agents is computed. The crucial issue for the puzzle is to find out when an agent knows the colour of his hat and how information about that fact changes other agents epistemic states. We use definition 4 and rule 1 of our model. Definition 4, in the case of the hats puzzle, takes the form of the predicate knows_his/3 finding all possible epistemic states in which the agent knows the colour of his hat, defined below.

---

[4] For the sake of simplicity, we assume that function $dist$ does not trigger more than one rule at a time, i.e. we assume that: $[\rho(X_1, Y_1, Z_1) = V_1 \wedge dist(t_n) \cap Y_1 \neq \emptyset] \wedge [\rho(X_2, Y_2, Z_2) = V_2 \wedge dist(t_n) \cap Y_2 \neq \emptyset] \rightarrow V_1 = V_2$.

[5] We present that elements of the program which directly corespond with our model of knowledge ommitting purely technical parts. Full text avialable at http://l3g. Program was tested using SWI-Prolog (ver. 5.6.61).

```
knows_his(_,[],[]).
knows_his(A,[H|T1],[H|T2]):-
  unique_in_partition(A,H), !,
  knows_his(A,T1,T2).
knows_his(A,[_|T1],T2):-
  knows_his(A,T1,T2).
unique_in_partition(Agent,Situations):-
  intersection(Situations,Knowledge),
  member((Agent,_),Knowledge).
```

Rule 1 is represented by the predicate change/4, in which the function δ_0 defined by the predicate delta/3 is applied to all epistemic states of all agents.

```
delta([],_,[]).
delta([X|T1],X,T2):-
  delta(T1,X,T2), !.
delta([H1|T1],X,[H2|T2]):-
  intersection(H1,X,I), I\=[],
  subtract(H1,X,H2),
  delta(T1,X,T2), !.
delta([H|T1],X,[H|T2]):-
  delta(T1,X,T2).
```

For example, if one introduces the parameters by the following predicates' definitions:

```
colour(red, 3).
colour(white, 2).
agents([Adam,Ben,Clark]).
sees(Adam,Ben).
sees(Adam,Clark).
sees(Ben,Clark).
```

and ask the following query:

```
?- all_agents_i_p_s(L), change(not_knows,Adam,L,M),
change(not_knows,Ben,M,N),get_result(Clark,N,HatList).
```

the program gives the answer:

```
HatList = [red]
```

The presented model can be applied to a wide range of puzzles concerning knowledge. Just by changing parameters of the program one can solve the hats puzzle with any number of agents and different perception relations. Two of the special cases here are known from literature puzzles: *three wise men* and *muddy children*. With minor changes the program can be also applied to the puzzle *Mr P. and Mr S.* and some puzzles about knights and knaves presented by R. Smullyan.

4   Comparison with Other Approaches

Knowledge is formally studied in the following frameworks:

(DEL) Dynamic Epistemic Logic
(EBA) Event-Based Approach
(SC) Situation Calculus

DEL has two components, Kripke epistemic structures and dynamic epistemic logic languages to make assertions about them. Kripke structures considered here are of the form $\mathcal{K} = \langle S, Agent, \sim \rangle$, where S is a set of situations (or possible worlds), $Agent$ is a set of rational agents and $\sim$ is a function assigning a set of pairs of situations about which it is said that they are indiscernible to every element of $Agent$. Usually $\sim_a$ are assumed to be equivalence relations. In this approach, knowledge is not expressed directly in $\mathcal{K}$ structures, but syntactically, by formulas with modal operator of epistemic logic of the form $K_a \varphi$— agent a knows that φ [15,6]. The satisfaction condition for the knowledge formulas in model $\mathcal{M} = \langle \mathcal{K}, v \rangle$ (where $\mathcal{K}$ is the described structure and v is a standard valuation function) and for $s \in S$ is the following:

$$\mathcal{M}, s \models K_a \varphi \iff \forall s' \in S \ (s \sim_a s' \implies \mathcal{M}, s' \models \varphi)$$

It says that the formula "agent a knows that φ" is true in situation s (of $\mathcal{M}$) if and only if for all situations which are indiscernible (for agent a) with s, φ is true.

Model of knowledge proposed by "pure" epistemic logic is static, i.e. does not allow for representing a change of agent's knowledge. In order to make it dynamic, dynamic epistemic logics have been created [14]. An example of such logic is a public announcement logic ([11,4]) which is built by adding to the language of epistemic logic a new operator which enables to express formulas of the form "$[\varphi]\psi$", which are read "after it was publicly and truthfully announced that φ, it is the case that ψ. That formula has the following satisfaction condition in $\mathcal{M}$:

$$\mathcal{M}, s \models [\varphi]\psi \iff \mathcal{M}, s \models \varphi \implies \mathcal{M}^\varphi, s \models \psi$$

where $\mathcal{M}^\varphi = \langle S', A, \sim', v' \rangle$ is characterized by the following conditions:

- $S' = \{s \in S : \mathcal{M}, s \models \varphi\}$,
- for every $a \in Agent$, $\sim'_a = \sim_a \cap (S' \times S')$
- and for every $p \in Atm$, $v'(p) = v(p) \cap S'$.

The second approach, **EBA**, called the event-based approach [3], is more typical for game theory and mathematical economics. Instead of Kripke structures the following structures are in use here: $\langle S, Agent, P \rangle$, where S and $Agent$ are, similarly as in Kripke structures, a set of situations and a set of agents respectively and P is a set of partitions of S for every agent from $Agent$. The event-based approach focuses only on events (and sometimes states of affairs), which are represented by sets of situations. It is said in the approach that event $\mathbf{X}$ takes place in situation s if and only if $s \in \mathbf{X}$.

Knowledge in this approach (cf. [3, Section 2.5]) is modeled by a function $\mathbf{K}_a :$ $2^S \longrightarrow 2^S$ defined as follows $\mathbf{K}_a(\mathbf{X}) = \{s \in S : P_a(s) \subseteq \mathbf{X}\}$, where

- P_a is partition of S for agent a and $P_a(s)$ is a cell of the partition P_a to which s belongs
- $\mathbf{X}$ is an event or state of affairs.

Thus $\mathbf{K}_a(\mathbf{X})$ is a set of situations in which agent a knows that event $\mathbf{X}$ occurs. It can be said that a knows that $\mathbf{X}$ in situation s if and only if $\mathbf{X}$ holds at every situation $s' \in P_a(s)$.

In [10] we can find another definition of knowledge which is understood as a union of set $\mathbf{K}_a(\mathbf{X})$ and set $\overline{\mathbf{K}}_a(\mathbf{X}) = \{s \in S : P_a(s) \cap \mathbf{X} = \emptyset\}$. $\overline{\mathbf{K}}_a(\mathbf{X})$ is a set of situations in which a knows that $\mathbf{X}$ does not occur.

There is an evident correspondence between the structures of **EBA** and Kripke structures. Every partition P_a can be defined in an obvious way by the indiscernibility relation $\sim_a$ and vice versa by definition:

$$s \sim_a s' \iff P_a(s) = P_a(s')$$

In **SC** [12], knowledge is modeled as a fluent, i.e. a predicate or a function whose truth value may vary. In [9,13] a fluent $Knows(P, s)$ (P is known in situation s) is defined as follows:

$$Knows(P, s) \iff \forall s'(K(s, s') \implies P(s'))$$

where K is accessibility relation between situations. In the scope of this approach, the following issues are considered:

- which actions are knowledge-producing actions and which are not,
- under which conditions carrying out a knowledge-producing action in some situation leads to the successor situation in which something is known (General Positive Effect Axioms for Knowledge) or is not known (General Negative Effect Axioms for Knowledge),
- *successor state axiom* combining General Positive and Negative Effect Axioms for Knowledge, completeness assumption (i.e. an assumption that General Positive and Negative Effect Axioms for Knowledge characterize all the conditions under which an action leads to knowing or not knowing) and unique name axioms,
- frame problem (typically solved by successor state axioms),
- question of knowledge persistence.

Our model share many intuitions which lie behind the frameworks shortly described above and at the same time is distinct form each of them in some aspects. Below we shortly describe some similarities and differences.

- In our approach a set of possible situations (or worlds) is defined, whereas in all listed above frameworks it is taken to be primitive. Possible situation/world in our model is a maximal set of elementary ontic (possible) situations.
- The concept of event/state of affair $\mathbf{X}$ in **EBA** and **PAL**, for $X \in Sit$, corresponds to the set: $X^{PW} = \{Y \in PossWorld : X \cap Y \neq \emptyset\}$
- For fixed time t_n, $Epist(a, t_n)$ and $epist_i(a, t_n)$ correspond to partition P_a and to some cell of P_a in **EBA** respectively.
- Indiscernibility relations in **PAL** and in our model are isomorphic

– Correspondence between the concept of knowledge in our model and in **EBA** (in static aspect, i.e. without taking into account the interaction between agents) is established by the formula:

$$K_{epist_i(a,t_n)}(X) \iff epist_i(a,t_n) \subseteq \mathbf{K}_a(X^{PW})$$

$$\overline{K}_{epist_i(a,t_n)}(X) \iff epist_i(a,t_n) \subseteq \overline{\mathbf{K}}_a(X^{PW})$$

– In contrast to **PAL** in our model we do not have a distinction between a structure and a language in which the assertions about the structure are made. In our model we express all kinds of intuitions in one framework.

– The concepts of knowledge in our model and in **PAL** are similar. Having in mind the satisfaction condition for knowledge operator in **PAL** we can define knowledge in our model in **PAL**-like style

$$K_{epist_i(a,t_n)}(X) \iff \forall Y \in PossWorld \ (Y \in epist_i(a,t_n) \implies Y \in X^{PW})$$

$$\overline{K}_{epist_i(a,t_n)}(X) \iff \forall Y \in PossWorld \ (Y \in epist_i(a,t_n) \implies Y \notin X^{PW})$$

– The idea of reducing the agent's ignorance by making proper changes in the sets of indiscernible situations as a result of receiving new information is shared by **PAL** and our model. However, what makes a difference is the generality of our approach. In our model we consider many kinds of ontological and epistemological factors that trigger the process of epistemic change, whereas in **PAL** we just have announcements as triggers.

– We share with **SC** the interest in the issues which were listed above. In our model we have the condition of knowledge persistence (see formula 5) and our rules imitate the General Positive Effect Axioms for Knowledge. Our concept of (maximal) situation is very similar to McCarthy's view on situations as snapshots and is far different from Reiter's situation as a sequence of actions.

5   Further Work

Our plan for future work includes:

– Development of our model of epistemic interactions by incorporating more sophisticated phenomena such as: shared and common knowledge or non-verbal behavior. Another line of development would take into account more features of the real-world knowledge. For example, the relation of indiscernability, in most of the realistic scenarios, is not transitive.

– Investigating computational properties of our model when applied in more complex epistemic situations.

– More detailed work concerning the relation of our model to other frameworks dealing with knowledge.

References

1. Barwise, J.: Scenes and other situations. Journal of Philosophy 78(7), 369–397 (1981)
2. Bovens, L., Rabinowicz, W.: The puzzle of the hats. Synthese (2009) doi: 10.1007/s11229-009-9476-1
3. Fagin, R., Halpern, J.Y., Moses, Y., Vardi, M.Y.: Reasoning About Knowledge. MIT Press, Cambridge (2003)
4. Gerbrandy, J.D., Groeneveld, W.: Reasoning about information change. Journal of Logic, Language and Information 6, 147–169 (1997)
5. Gomez-Perez, A., Corcho, O., Fernandez-Lopez, M.: Ontological Engineering. Springer, London (2001)
6. Hintikka, J.: Knowledge and Belief: An Introduction to the Logic of The Two Notions. Cornell University Press, Ithaca (1962)
7. McCarthy, J., Hayes, P.: Some philosophical problems from the standpoint of artificial intelligence. In: Meltzer, B., Michie, D. (eds.) Machine Intelligence, pp. 463–502. Edinburgh University Press (1969)
8. McCarthy, J.: Formalizing Common Sense. Ablex Publishing, Greenwich (1990)
9. Moore, R.C.: Reasoning about knowledge and action. Technical report, SRI International (1980)
10. Orlowska, E.: Logic for reasoning about knowledge. Bulletin of the Section of Logic 16(1), 26–36 (1987)
11. Plaza, J.A.: Logic of public communications. In: Emrich, M.L., Pfeifer, M.S., Hadzikadic, M., Ras, M., W., Z. (eds.) Proceeddings of the 4th International Symposium on Methodologies for Intelligent Systems, pp. 201–216 (1989)
12. Reiter, R.: Knowledge in Action: Logical Foundations for Specifying and Implementing Dynamical Systems. MIT Press, Cambridge (2001)
13. Scherl, R.B., Levesque, H.J.: The frame problem and knowledge-producing actions. In: Proceedings of the Eleventh National Conference on Artificial Intelligence, pp. 689–695. AAAI Press/The MIT Press, Menlo Park (1993)
14. van Ditmarsch, H., van der Hoek, W., Kooi, B.: Dynamic Epistemic Logic. Synthese Library Series, vol. 337. Springer, Heidelberg (2007)
15. von Wright, G.H.: An Essay in Modal Logic. North Holland, Amsterdam (1951)

Issues on Aligning the Meaning of Symbols in Multiagent Systems

Wojciech Lorkiewicz and Radosław P. Katarzyniak

Institute of Informatics
Wrocław University of Technology
Wybrzeże Wyspiańskiego 27
{wojciech.lorkiewicz,radoslaw.katarzyniak}@pwr.wroc.pl

Abstract. The autonomy of a multiagent system in relation to external environment can be greatly extended thorough the incorporation of a language emergence mechanism. In such a system the population of agents autonomously learn, adapt and optimize their semantics to the available mechanisms of perception and the external environment, i.e. it dynamically adapts the used language to suit the shape of external world, assumed perception mechanism and intra-population interactions. For instance, used symbols should denote only the directly available states of the external world, as otherwise the symbols have no meaning to the agents. Further, the incorporated language sign, denoting certain meaning, representation can be adapted to suit the demands of communication, e.g. by lowering the energy utilization – shorter signs should denote more frequent symbols. Additionally, the proposed approach to language emergence is applied in the area of tagging systems, where it helps to solve and automate several problems.

Keywords: symbol grounding, meaning development, autonomous agent.

1 Introduction

Communication is a necessity in a highly distributed systems, as the lack of central coordination processes demands intensive interaction between individuals. In most cases the process of communication is treated as a purely symbolic task, i.e. narrowed to just simple symbol exchange and/or manipulation. However, the importance of grounding the language symbols in agent perception is commonly neglected and the meaning of language symbols is externally imposed. This seems natural for systems that require direct human-computer interaction, whilst in the opposite cases such assumption seems to be against the common sense. Simply because the nature of individuals and their perception apparatus is often different from the one that humans use and therefore external imposing conceptualization can lead to many problems. Further, the issue of symbol grounding is crucial for highly autonomous systems, i.e. the multiagent systems. Such a system consists of numerous highly autonomous entities, where the autonomy of an individual should be present both in the area of agent's inner organization and both in the area of agent's language competence. Harnad in [1]

N.T. Nguyen et al. (Eds.): New Challenges in Compu. Collective Intelligence, SCI 244, pp. 217–229.
springerlink.com © Springer-Verlag Berlin Heidelberg 2009

proposed an idea, rather a simple mechanism of grounding, i.e. reaching the coupling between the real world experience and the language symbols, that was further applied by Vogt in [7] allowing agents to reference the incorporated language symbols with certain real world perceptions. As Luciano Floridi has stated "*the grounding problem is the most important problem in the philosophy of information*".

In the case of highly autonomous multiagent systems the problem of aligning the meaning of language symbols in a group of agents is of even greater importance, i.e. semiosis - process of shaping the meaning of language symbols in a given population. The term *semiosis*, as introduced by C.S. Pierce, describes the process that interprets signs as referring to their real world objects. Further, as noted by de Saussure (See [9]), the meaning of a symbol in a given population results from a certain convention and is a result of a common agreement. Roy Harris states that "*The essential feature of Saussure's linguistic sign is that, being intrinsically arbitrary, it can be identified only by contrast with coexisting signs of the same nature, which together constitute a structured system*". As such the language is strictly correlated with the process of conceptualization (See [13]) that depends on the environment, empirical experience, and internal organization of an individual. Moreover, a sign cannot function until the audience distinguishes it, as it only then triggers the cognitive activity to interpret the data input and so to convert it into meaningful information: "*A sign is not a link between a thing and a name, but between a concept and a sound pattern*". It should be underlined that the problem of semiosis is a multidisciplinary problem that has its origins in the study of human language acquisition and it seems to be neglected in the area of technical science.

This paper is organized as follows: in the following chapter the basic notions, such as: agent, environment, object, index and label, are defined and briefly described. Next, the problem of aligning the meaning of symbols is presented and decomposed to three basic sub problems. All pinpointed situations are further sketched, modelled and discussed in the context of possible application area, i.e. applying the proposed mechanism in tagging systems. Further, short summary with future research plans is presented.

2 Objects, Indexes and Labels

It is assumed that the multiagent system is located in a given dynamic environment, external to the system. The environment is treated as a collection of individually identifiable objects, where each object is represented by a unique "*id*" number that is prior known to the agents. In particular, the set of all objects is denoted as $O=\{id_1, id_2, ..., id_N\}$. Further, it is assumed that the set of all objects is finite and static. The environment is dynamic as at any given time point the object is in a certain state, can be ascribed by representing a certain property, and these property-object relation change over time. Nevertheless, the precise determination of object's state may be impossible even for the system itself, this is a crucial and fundamental assumption to the introduced approach. It allows to describe the state of the external world by the means of properties, that are reflected by the existing objects.

Proposed mechanisms of aligning the meaning of symbols can be incorporated in a wide range of real world systems in order to improve their operability. Most of the

application areas are rather obvious and straightforward, as they are a direct consequence of the introduced formalism, i.e. smart sensors, sensor networks, robotics, etc.
However, the application in tagging systems seems to be very promising (See [10]
and [11]). Further, due to their current popularity and great usefulness we will adopt it
and use as an example throughout this paper. In particular, the focus will be on systems that allow the users to append certain symbols to objects existing in the system,
i.e. based on their strictly private and autonomous decisions. Each such act of labelling, or tagging, can be perceived as a language symbol utterance. It is a common
situation in nowadays systems that the assigned tag is adequate only to a certain object by the means of external user, as a consequence all tags are just purely symbolic
to the system itself, i.e. their meaning is intrinsic to the system. Introduction of the
mechanisms from SGP (symbol grounding problem) could provide a really useful tool
for automating several tasks. In the tagging system metaphor the set of objects O
should be understood as a set of elements that exist in the system, i.e. the subject.

2.1 The Agent

The agent is an autonomous entity that is capable of performing individual observations on the given environment and can communicate with other agents. It is also
competent of differentiating objects in the environment from the background [8] and
determining their current state. Through the empirical interaction with the environment the agent collects sensory data, that are further available to its internal processes.
However, in order to prevent agent's sensory overload, the collected data should be
filtered and only the salient ones should receive full cognitive attention. As such the
inner representation of observation should be based on a simplified model of the real
world, i.e. allowing the ranking of data elements in terms of their significance and
filtering out the irrelevant ones. Each observation is made upon a single object and is
further represented by an internal index $I(id_j)$ (See point 2.2), i.e. the filtering mechanism. This mechanism is strictly dependant on agent's perception capabilities, i.e.
defined by the set of available sensors, the external environment and the assumed
intrinsic model. In short, the agent formulates an internal projection of the objects
ideal state on it's perception space.

Secondly, each agent is capable of uttering a language symbol $L=\{L_1,L_2,...,L_W\}$.
Each symbol describes a certain aspect of the current state of the environment (See
point 2.3) observed by the agent. Resultant the agent can attribute a label (language
symbol) to the observed state of a given object in the environment (internal index).
Moreover, the agent based on the internal representation is able to categorise the current state of an object to one of already known and differentiated labels or can further
create a new label and therefore expand the set of used language symbols.

Third, the relation S between label L and internal index I forms (See point 2.4)
agent's model of understanding of L. Generally, all such models assemble to form
agent's embodied ontology, that represents agent's autonomous and individual understanding of available language symbols. What is crucial in this approach is the fact,
that the meaning is coded on the very basic level of empirical experience of the environment, as it reflects agent's direct perception. Moreover, this triadic relation (See
Figure 3) defines the very fundamental conceptualization that can be perceived as a

certain form of embodied ontology, i.e. the direct connection between internal meaning and the real world.

2.2 Index

As stated, it is assumed that there exists a certain mechanism that applied on the object o is able to create an internal representation of its current state. This process can be perceived as an act of observation, where an individual is determining, through an array of sensors, the surrounding environment. For the sake of this paper the notion of index will be further used to describe this resultant internal representation.

Index can be perceived as an internal representation of an element, existing in the system, that captures its individual aspects and characteristic features. As the direct manipulation on the object can be hard and even impossible, it allows automatic and convenient way of manipulating the objects. Moreover, it defines the perceptual space of the system.

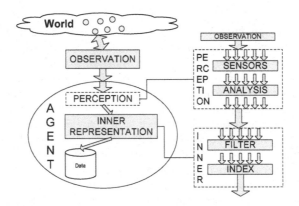

Fig. 1. Perception

Three basic cases of index types are discussed:

Case 1. Binary vector

$I(id_j)=<I(id_j)^1, I(id_j)^2, ..., I(id_j)^K>$ element of the set $\{0,1\}^K=I_s$.

Represents the simplest type of observation in which a certain sensor might be activate(1) or inactive(0). As such the object state is denoted by a binary vector that can be further processed through the internal processes. This case is treated as the basic one throughout this paper.

Case 2. Fuzzy vector

$I(id_j)=<I(id_j)^1, I(id_j)^2, ..., I(id_j)^K>$ element of the set $[0,1]^K=I_f$.

Represents the general case, assuming that the sensor has a certain intensity of the percept, i.e. 0 is total inactive state, 1 total active state and value between 0 and 1 are the intensity. As such the state of the external object is denoted as a fuzzy vector that can be further used in the internal processes.

Case 3. Set of terms

$I(id_j)$ is a subset of the set of terms $T=\{T_1, T_2, ..., T_L\}=I_T$.

Represents a certain mixture of the earlier cases, where there are numerous sensors available to the agent and still only few are activated at the same time. In such a case the vector representation is inappropriate, as only few coordinates are non-zero, whilst the majority can be neglected. Therefore it is much more convenient to only denote the non-zero coordinates as a certain elements T_k from the set of all coordinates T.

In tagging systems each object is present in the system, for instance it can be a picture, a short note, an article, etc. Further, each object is automatically processed by an indexing mechanism and it's state is projected from the real world object onto the agent's perceptual space. Depending on the type of objects this mechanism can be rather simple and straightforward, or very complex and multidimensional. The resultant vector, defined as an element of the perceptual space, is treated as the inner representation of the object – index.

2.3 Labels

A label $L_k(id_j)$ is a language symbol that can be uttered by an agent. It is assumed that there is a finite and static set of all possible language symbols $L=\{L_1, L_2, ..., L_W\}$. Each label can be perceived as a form of external, external to the agent and internal to the agent population, representation of shared meaning. In short, for each agent the uttered label denotes its internal index, i.e. the meaning applied by the agent. As underlined, this direct relation between label and certain form of internal index constitutes the meaning of a given language symbol. This meaning is therefore directly correlated with a physical experience of an entity, and is grounded in this experience. In practice the label is a pre assumed pattern, like voice pattern, graphical pattern, light pattern, etc., depending on the medium used for internal communication.

The labels are just tags assigned to an object that represent certain categorise that are autonomously applied and defined by the agent upon inner representation, i.e. each category defines the meaning of the label/symbol relation. Moreover, the set of all inner categories can be perceived as agents embodied ontology, as it relates the inner representation, the direct reflection of a real world state, with a certain language sign, a certain label.

Importantly, each label has its energy demands, that is the amount of energy required to utter a given language symbol. Basic interpretation is that each symbol must be transported from one agent to the other, and as such simple labels require less effort from the agent, whilst the more complicated ones require more effort from the agent. Therefore the utilization function $U:L \rightarrow [0,1]$ represents the required effort to utter a given label, i.e. the lower the value is the less energy is required, whilst the higher the value is the more energy is required. It should be underlined that a common sense trend tends to assign less energy consuming symbols to more frequent messages, and more consuming symbols to less frequent messages.

In the tagging system the tags are usually assigned by external users, where they tend to capture their individual conceptualization of the environment. However, the system itself is capable of performing the procedure of tag assignment, i.e. based on certain clustering method or due to learnt symbol/meaning mapping. Both cases are important in real life systems, as it will be further discussed.

2.4   The Sense

As stated the meaning of a certain label $L_k(id_j)$ is hidden in the correlation between a language symbol L_k and the possible inner representations that are adequate to this label. Therefore the meaning S divides the space of all possible observations, i.e. the space I (either space I_s,I_f or I_T), into M clusters, i.e. each language symbol L_k has an adequate group S_k. Therefore, the sense can be treated as a classifier (See [11]) that for a certain inner representation $I(id_j,t)$ of an object id_j in time point t assigns a set of language symbols $S(I(id_j,t))$, or more generally as a certain fuzzy mapping between these spaces (See [2])

Definition 1. The sense S is the fuzzy mapping between the inner representation space I (either I_s, I_f or I_T) and the language symbol space L, that is $S(I(id_j,t))=\{<s,v>:s=1,M,v=[0,1]\}$, where s is the language symbol and v is the correlation between observed inner representation $I(id_j,t)$ and language symbol s.

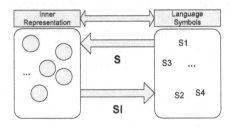

Fig. 2. Sense representation

For a given pair (s,v) *from* $S(I(id_j,t))$ the higher the correlation v is the more adequate the language symbol s is to the given inner representation $I(id_j,t)$. Further, the symbol with maximum correlation can be uttered in the time point t.

The inverse mapping from the space of language symbols L to the space of inner representation I can be further used to decode the received utterances to perceptions, and as such perform the process of understanding. For instance, let us assume that the agent has received a language symbol s concerning object id_j from the population of agents. Based on the developed sense mapping S, this agent can define the fuzzy set $SI(s(id_j))=\{<I(id_j),v>:I(id_j)=I,v=[0,1]\}$, where v represents the correlation value, representing the sense. That is the higher the correlation is the more adequate the inner representation $I(id_j)$ is to the language symbol s.

3 Aligning the Meaning of Symbols - Semiosis

In order to reach a successful communication the sense of symbols should be shared among the population, i.e. the meaning should be consistent inside the group of communicating agents. In particular, the language symbol should result from the process of semiosis, and only in such a case it is convenient to introduce the notion of shared understanding of language symbols among agent population. Further, this understanding should be grounded in agents experience by the means of triadic relation, that is the relation between object, index and a label. It should be stressed that this idea in not new as it is just a certain instance of the Piercean semiotic triangle [6] (See Figure 2).

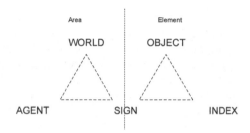

Fig. 3. Semiotic Triangle (a) and Tag relation (b)

This relation can be interpreted as follows. First of all, the real world experience through agents perception evokes agent's thought. That is the real element of the environment, for instance a certain object, is represented by an index – thought. Secondly the agent refers it to a certain symbol that is also evoked by it. That is the index is correlated with a tag assigned to the object by the sens relation R. Finally, through the relation between experience, thought and symbol agent refers the symbol back to the reflection of its current perception, i.e. the relation between object in the environment and a tag - thought. Formally, object id_j, its index $I(id_j,t)$ - as perceived by an agent in the time point t - and a certain symbol L_k – assigned by the agent form the internal relation between $I(id_j,t)$ and L_k. The latter, can be generalized in order to form the model representing the grounded, i.e. related to the external world experience, meaning of the language symbol L_k.

Developing the complete mechanisms of language symbol alignment requires to deal with three basic problems, i.e. **individual language emergence**, where a single agent is aligning with the general population semantics, **population language emergence**, where each single agent is forming the general population semantics, and **multi population language emergence**, where two or more populations align their own semantics. As this mechanism is highly dynamic and very complex it requires the introduction of consistency measures, i.e. a measure stating the popularity of a certain language symbol sense spread among the population. In particular, such measure would support the study of convergence of the whole learning process in all introduced cases.

3.1 Individual Semiosis

This is the basic situation, as it is straightforward and corresponds directly with the problem of language acquisition. In short, an individual is shaping and aligning it's language with a distinguished external source of meaning. In particular, a single agent A_0 is interacting with a group of mature agents A (with a predefined meaning of language symbols) in order to learn how to correlate the language symbols - imposed by the mature population - with empirically perceived external states of the environment - the grounded meaning of the language symbols. It should be underlined that this is a common case in MAS where an individual is introduced to a mature population of agents and where the premature agent should learn the sense of symbols used by the mature population.

It is also a very typical situation in tagging systems. It resembles the situation in which a group of objects from the environment has a pre-assigned set of tags. Most commonly these tags are assigned by human users or by the system itself. Further, the system should be able to acquire the relation between tag and the inner representation of the objects. Of course this is possible only under the assumption that all incorporated tagging sources, either a human operator or/and predefined automatic mechanisms, use consistent senses of symbols.

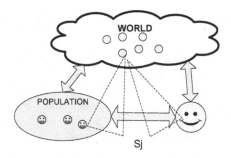

Fig. 4. Individual Semiosis

In general, individual semiosis can be modelled as a simple learning task, where an agent observes the interaction between mature population and the shape of external world, and correlates the acquired language symbols (tags) with current perception (inner representation - index). As such the learning agent is able to form the learning set $LS(id_j,t) = \{<I_j(t),L_j(t)>\}$ of observed past, i.e. to the time point t, semiotic relations.

Definition 2. The learning set for an agent A_0 is the set $LS(t) = \{LS(id_1,t),LS(id_2,t),...,LS(id_N,t)\}$, where each $LS(id_j,t)=\{<I_j(t),L_j(t)>\}$ is the set of all past semiotic relations concerning object id_j, where $I_j(t)=\{I(id_j,t_k):t_k<t\}$ is the set of all past observations of object id_j and $L_j(t)=\{L(id_j,t_k):t_k<t\}$ is the set of all correlated (with observations) language symbols uttered in context of object id_j.

Further, based on the learning set the agent is able to formulate the sense mapping S that represents the correlation between the space of language symbols L and the space of inner representations I, and as such forms agents embodied ontology.

In the simplest case of binary representation the agent has the knowledge about past and current tagging relations represented in the form of binary vectors. Further, it may correlate a certain label $L(id_j)$ with a set of activated attributes of the index $<I(id_j)^1, I(id_j)^2, ..., I(id_j)^K>$ to formulated the relation S. This basic case of learning algorithm is rather straightforward and has been already proposed and discussed in [3]. As shown, this short procedure seems to generate the needed and reasonable results. However, one may notice that the proposed approach has several drawbacks that prevent it from direct application and further usage. The precise analysis of this mechanism is outside of the scope of this paper.

Returning back to the tagging system example we may stress that the application of proposed mechanism is straightforward and requires only to neglect the time dependency, as the objects in such a system are rather purely static and do not change over time. The individual, that is the system itself, has a finite learning set LS, i.e. the set of objects with assigned tags. Resultant the system itself has the ability to formulate the mapping S and the inverse mapping SI, representing the embodied ontology storing the information about the sense of used tags. Further, using the learned relation the system may automatically assign tags to new objects or identify tags that are inappropriate for a given object. What is crucial in this attempt is the fact that the resultant tag verification is based on the learned embodied ontology.

3.2 Group Semiosis

More generally the problem of shaping the meaning of language symbols can be deliberated without the need of a distinguished and a predefined population of mature agents. As such the process of developing the agreed language semantics is distributed among individual agents, where each is developing and adapting it's personal semantics, the process of individual grounding, and continuously aligns it with the whole population, the process of semiosis. In particular, the agents are indirectly negotiating the meaning of used language symbols in order to reach a successful and optimal communication. Of course the agent is a part of a larger population, where each agent may be in a different stage of language development, as may already have a developed semantics or may be in the early stage. This semantic inequality of population allows to introduce the mechanisms of cross learning between the agents. Further, due to the fact that uttering a language symbol requires a given amount of energy, the process of adaptation should be biased by the frequency of incoming external world states. In short the more frequent signals should be denoted by less consuming symbols, whilst the rare signals should be denoted by more consuming symbols.

In the case of tagging system this situation does not necessary involves the existence of a set of objects with already pre-assigned tags, i.e. the meaning consistency among the population is minimal. Further, this consistency should be developed over time through the direct or/and indirect interaction between system entities and its environment. Consequently, the system itself is trying to somewhat categorize the existing objects and further assign precise labels to each of the resultant category. Moreover due to incorporated optimization process, the system itself is trying to analyse the frequency of indexes and further adapts the labels.

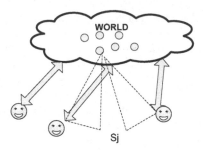

Fig. 5. Group Semiosis

It is assumed that the agents only share the same set of possible utterances, i.e. language symbols. The goal of the whole process, i.e. group semiosis, is to develop a shared fuzzy mapping S among the population. This is realised by constant shaping the individual processes of meaning determination and adapting with through the interaction with the population. In the case where all agents share the same perception mechanisms and the same perception space, i.e. have identical inner representations, the task seems to be quite straightforward, as the whole process can be narrowed to a simple task of labelling, i.e. identifying the mapping between index and label.

Two strategies are proposed, more chaotic - inspired by the studies on language emergence in children - and more formal - directly taking into account the energy utilization.

The chaotic approach is developed over very simple mechanisms and a random factor. In the begging an agent has no semantics imposed, i.e. there is no mapping between inner representations and language symbols (the correlation in S and SI is zero for all possible mappings). As the agent interacts with the environment it gathers knowledge about it through constant observations, however the agent is unable to communicate its individual perceptions to the population. In the early development stage the agent may just randomly choose the label to a given observation, i.e. modify the appropriate correlation in the mapping S and SI. This is analogous to early stages of language development in children, when they try to create something in the form of personal language. However, parallel Resultant the agent must balance between its personal biases and the population biases, i.e. both the random internal process and the process of adaptation to the population modify the correlation between inner representation and language symbol in the S and SI mapping.

The formal approach to the whole process can be decomposed into two basic stages. In the beginning each agent has only the ability to collect internal representations, i.e. stored in the form of vector of perception – inner index. Simple analysis of this data may lead to the determination of frequency $f: I \rightarrow [0,1]$ of received signals I. Parallel, the agent still receives the utterances from the population and as such updates the mapping S and SI accordingly. Further, after reaching a certain border value of intensity of the received signals, i.e. after a certain time window, after reaching a given number of registered states, after reaching a given frequency, etc., the agent enters the second stage of development. Based on the frequency the agent may order the inner representations from the most common f_{max} to the least common f_{min}. As such the agent in order to reach the minimal energy utilization will individually tend to assign the label L_k, where $U(L_k)=\max_j U(L_j)$, to the inner representation with f_{max} frequency, and label L_m, where $U(L_m)=\min_j U(L_j)$, to the inner representation with f_{min} frequency. This first bias modifies the mapping S and SI, accordingly. Nevertheless, the agent exists in a given population that it communicates with. As such the agent receives language symbols and correlates them with a certain inner representations, i.e. analogous to the case of individual semiosis. This second bias also modifies the mapping S and SI, accordingly. Throughout the evolution of the system each agent tends to balance both individual and population biases.

Learning procedure in this case is not so obvious, as in the previous situation. First of all an individual is still incorporating the individual semiosis mechanism, as the agent learns from the population. Secondly, the agent is individually shaping the meaning of language symbols. As in the random stance the agent randomly assigns a unused symbol to newly observed state of external world and as such creates a correlation between index and label. In the formal stance, the agent first collects information from the environment and then correlates the most frequent states with the shortest, in terms of energy utilization, symbols.

3.3 Multi-group Semiosis

The third case is when two mature populations, with differently predefined semantics (or differentially developed as of the Group Semiosis), are set together in a common environment. Treated as a whole the collated populations have various meanings assigned to the incorporated language symbols, and in order to communicate their semantics should be aligned, or may have different language symbols, and both dictionaries should be related to each other. As such, it can be perceived as a more general, a broader case of the first situation. Further, it should be underlined that it is strictly correlated with the fundamental problem of ontology alignment.

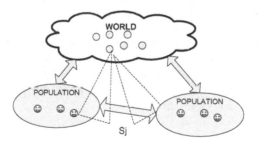

Fig. 6. Multi-group Semiosis

However, in the traditional approach the ontologies are not related to the real world experience of an entity, in short are not grounded. In consequence the task of aligning ontologies seems to be very complex or even impossible to solve. Therefore, in the proposed approach the ontologies are grounded in agents perception and as such the meaning assigned to each concept can be directly addressed through the space of real world experience – the observed state of the environment. The novelty of this approach lies in the fact that all of the aligned ontologies share the same environment, that is common to all populations. This common background of reference can lead to a proper correlation between ontologies, as the existence and state of the external world is objective rather than subjective. It should be also noted that the problem of ontology alignment, on the formal level, has been broadly studied in the literature, whilst the latter case, with grounded ontologies on the perception level, is still neglected, despite its great practical importance and possible applications.

The third case represents a typical situation in tagging systems, where groups of similar users are set together in a common environment. Each of such group uses a specific semantics and as such the system should be able to address each separated population in an individual fashion. Further, in case of searching task the system should be able to automatically enrich user queries with the labels/tags that are equivalent in other groups, as such increasing the completeness and the accuracy of the search results. Moreover the aligning of ontologies could result in the creation of a single common ontology for the whole system, and as such would allow users to easily communicate in this general fashion between individual groups.

In general, the multigroup case of semiosis can be modeled as a task of determining the relation between meaning fuzzy mappings in each population and the relation between inverse meaning fuzzy mappings in each population. More formally, let L_1 be the set of labels used by population P_1 and S_1, SI_1 be the fuzzy mapping of meaning and the inverse mapping in population 1. Analogous, let L_2, S_2 and SI_2 be labels, fuzzy mapping and inverse fuzzy mapping incorporated by population P_2. Each such group of agents can acquire, through the direct linguistic interaction and the real world reference, the reflection of meaning assigned by the other group. Further, this reflection can be formalized in the form of external mapping ES_i and external inverse mapping ESI_i, for an example of such a case please find [2]. In particular, the population P_1 would be able to identify S_1, SI_1, ES_2 and ESI_2, and on this basis define the relations $R_{1,2}$ and $RI_{1,2}$ between $S_1 \leftrightarrow ES_2$ ($R_{1,2}$) and $SI_1 \leftrightarrow ESI_2$ ($RI_{1,2}$). Based on this relations the agent can map it's individual semantics with its own reflection of the external semantics of the other population. Resultant the task of this case of language symbol alignment is to develop the aforementioned relations, that allow proper communication between populations. However, one may go even further and integrate both groups in order to form a single population with both semantics aligned. That is the resultant semantics would incorporate language having symbols from both L1 and L2, and the resultant mappings Sc and SIc. Nevertheless, the precise formulation of postulates for the proper alignment is outside of the scope of this paper, as here only the synopsis are presented and briefly discussed.

4 Future Work

As briefly sketched, the problem of aligning the meaning of language symbols in an artificial population is fundamental to the field of multiagent systems, especially to embodied multiagent systems, e.g. robotics or smart sensors networks. The superior goal of language symbols meaning alignment in a multiagent systems is the resultant formulation of consistent and common substance of symbols, i.e. symbol semantics and associated signs, among a given agent population. It is important to stress that the problem should not be treated only at the theoretical level of deliberation, as it has several significant application areas. Further the proposed models and mechanisms can be easily adapted to other, then the one presented above, areas of application, such as robotics and smart sensor networks.

Currently the simplest case of individual semiosis has been implemented in the JADE framework and basic study has been performed. This work focused mainly on the aspect of creating an environment for the assumed multiagent system, communicating about the external environment and handling the aforementioned process of perception. In this system, two algorithms for acquiring language semantics were implemented, i.e. one directly from work [3] and its iterative modification. What should be stressed and underlined is that the early results seem to be very promising, as the whole procedure is performing well as desired.

Future work will focus mainly on the development of effective algorithms that could perform all three cases of semiosis. Further, the aspect of additional assumptions will be studied, along with the quest for other effective solutions. As a result it is assumed that the whole system will be implemented and experimental analysis performed.

References

1. Harnad, S.: The Symbol Grounding Problem. Physica D 42, 335–346 (1990)
2. Lorkiewicz, W., Katarzyniak, R.P.: Representing the meaning of symbols in autonomous agents. In: ACIIDS 2009, pp. 183–189 (2009)
3. Katarzyniak, R.P.: Grounding Crisp and Fuzzy Ontological Concepts in Artificial Cognitive Agents. In: Gabrys, B., Howlett, R.J., Jain, L.C. (eds.) KES 2006. LNCS (LNAI), vol. 4253, pp. 1027–1034. Springer, Heidelberg (2006)
4. Katarzyniak, R., Nguyen, N.T., Jain, L.C.: A Model for Fuzzy Grounding of Modal Conjunctions in Artificial Cognitive Agents. In: Nguyen, N.T., Jo, G.-S., Howlett, R.J., Jain, L.C. (eds.) KES-AMSTA 2008. LNCS (LNAI), vol. 4953, pp. 341–350. Springer, Heidelberg (2008)
5. Katarzyniak, R.: The Language Grounding Problem and its Relation to the Internal Structure of Cognitive Agents. J. UCS 11(2), 357–374 (2005)
6. Pierce, C.S.: Collected Papers, vol. I-VIII. Harvard University Press, Cambridge (1932-1958)
7. Vogt, P.: The emergence of compositional structures in perceptually grounded language games. Artificial Intelligence 167(1-2), 206–242 (2005)
8. Spelke, E.S.: Innateness, learning, and the development of object representation. Developmental Science 2, 145–148 (1999)
9. de Saussure, F.: Course in General Linguistics, trans. Roy Harris (La Salle, Ill.: Open Court, 1983)
10. Steels, L.: Fifty Years of AI: From Symbols to Embodiment - and Back. In: 50 Years of Artificial Intelligence 2006, pp. 18–28 (2006)
11. Steels, L., Hanappe, P.: Interoperability through Emergent Semantics. A Semiotic Dynamics Approach. Journal on Data Semantics (2006)
12. Ogden, C.K., Richards, I.A.: The Meaning of Meaning, 8th edn. Brace & World, Inc., New York (1923)
13. Whorf, B.: Language, thought, and reality. MIT Press, Cambridge (1956)
14. ASL (Applied Science Laboratories), Eye Tracking System Instructions ASL Eye-Trac 6000 Pan/Tilt Optics, EyeTracPanTiltManual.pdf, ASL Version 1.04 01/17/2006

Artificial Evolution and the EVM Architecture

Mariusz Nowostawski

Information Science Department
The University of Otago
PO BOX 56, Dunedin, New Zealand
mariusz@nowostawski.org

Abstract. This article presents the concepts behind the Evolvable Virtual Machine architecture (EVM). We focus on the main features, its biological inspirations and the main characteristics of the implementation. EVM has been designed from ground up with the automated program generation in mind and utilises a modern stack-based virtual machine design. It uses auto-catalytic cycles and biological symbiosis as the underlying mechanisms for building complexity in a multi-agent systems in an autonomous fashion.

1 Overview

Designing, developing and controlling massively concurrent asynchronous computing systems is an important and difficult area in both, computer science in general, and software engineering in particular. Massively parallel systems become ubiquitous, and at the same time we need to control and use their emergent and complex behaviour [11]. They range from multi-chip and multi-core systems on server farms, through personal computers utilising global inter-connectivity of Internet through to portable wireless devices that provide the means of establishing ad-hoc networks [13].

The complexity of parallel and distributed systems in the area of bio-inspired, intelligent and evolutionary computation has been long known as difficult to manage [9]. One of the fields that addresses the issue of information integration and distributed intelligent information processing is the field of multi-agent systems (MAS). MAS systems utilise decomposition and well-established intuitive notions from the realms of the physical and social sciences to facilitate and help with the inherent complexity of such systems. The Evolvable Virtual Machine architecture (EVM) [12] is a computing architecture based on the notion of distributed interactive asynchronously communicating agents. The EVM provides a massively decentralised and distributed asynchronous framework for experimenting with, and studying the properties of open multi-agent systems that can be evolved. EVM can be used for distributed multi-task learning [2, 14] and for automated program discovery [16]. The main emphasis is on collaborative behaviour, self-organisation and spontaneous integration of computation, formed by a large number of asynchronously communicating processing agents.

2 Motivation

Traditional human-centred computational computing languages are not best suited for fully autonomous and automated programming. Existing frameworks are almost

N.T. Nguyen et al. (Eds.): New Challenges in Compu. Collective Intelligence, SCI 244, pp. 231–242.
springerlink.com   © Springer-Verlag Berlin Heidelberg 2009

exclusively designed for human operators in such a way, that it makes it easy for humans to manage the complexity of the software systems. In recent years we can observe two important phenomena. First, there is a rapid growth of complexity of software projects. This, together with the fast evolution of specifications and change of requirements makes large monolithic designs impractical. Second, there is visible paradigm shift in the way software is designed and deployed. This is driven by the change in hardware architectures: from fast single CPU mainframes, to slower, but integrated and highly parallel multi-chip and multi-core based cluster computing facilities. To address these changing needs a paradigm shift is needed. The changes in the technology world are changing the way we think about computation. On one hand, there is a need for much more flexible, parallel and asynchronously communicating system architectures. On the other hand, there is an interest to adapt bio-inspired computing models that could help us to manage, design and deploy such systems. Bio-inspired models can potentially facilitate rapid evolution of software systems.

To address these and other challenges and opportunities of the modern computing architectures, we decided to design a prototype and test a new model of massively parallel asynchronous computational architecture. Our model has been built from bottom up with automatic programming in mind and it tries to benefit from the modern views on software and hardware architectures.

When developing our computational framework, we focused on two main objectives: (1) The EVM computational model must be well suited for automatic generation of massively parallel, autonomously interacting computational units, without the need for human programmer; in other words, architecture that can be used in a fully automated way, by non-human programmers, to generate distributed programs and computation deployed on multi-core and multi-chip parallel architectures; (2) The EVM computational framework should intuitively model naturally occurring phenomena; in particular, we are interested in modelling biological processes of life and evolution through a massively distributed systems of autonomous collaborating entities.

3   The EVM Architecture

The Evolvable Virtual Machine consists of two parts. The abstract layer, called EVMA (Evolvable Virtual Machine Architecture) and the concrete instantiation, called EVMI (Evolvable Virtual Machine Instantiation). The full description of the EVMA/EVMI has been presented in [10]. Here we focus only on few selected components.

The current EVMI provides a framework to experiment with and instantiate EVMA models. The current EVMI supports most but not all of the EVMA features. Originally, we sought to implement EVM directly on top of the Java Virtual Machine. However, the Java bytecode capabilities turned out to be too limiting in terms of the required reflective mechanisms. Other virtual machines (Lisp or .Net) provided enough capabilities in terms of tail-recursion and list manipulations, however, the representation and redefinition of individual instructions proved another obstacle. One of the primary reasons to implement our own VM for the EVM bytecode is the ability of the VM to redefine its own base instruction set at runtime. This is the fundamental feature that distinguishes EVM from any other Virtual Machine implementations. EVMI provides a general platform for investigating different architectural aspects of evolvability and computation on massively parallel asynchronous computing frameworks.

The architecture consists of independently operating computational *cells*. The cells operate in a local environment where they provide their results and obtain the feedback. A computational entity, a cell, acts (operates) in an unknown environment, trying to maximise its own reward intake. The cell continuously adjusts its actions in such a way as to collect more positive and avoid negative feedback. The rewards (positive reinforcement) and punishments (negative reinforcement) can be delayed in time and do not necessarily correlate directly with the last actions of the cell. As in Q-learning [8, p.367–387], the actual real reinforcement mechanism is unknown and can be subject to dynamic changes.

The cell computations consume the resources that the cell has been initiated with. As long as the cell's reward mechanisms are refuelled and the cell continues to perform its activities, the *task* is considered solved within the EVMI context. Note, this is different from the ideal solution of the task in the EVMA[10]. The EVMI approximates an unknown, infinitely long, computational solution to the task by a temporal snapshot that has been computed up to a specified time mark. The task is considered as failed or unsolved when the activities of the cell do not bring the required reward intake for the cell, and cell's computational expenditure cannot be balanced out with the cell's reward intake. In our experiments we have used two types of situations:

- there is a periodic feedback to the cell, with rewards, or punishments, or both (positive and negative reinforcement accordingly). The cell operates in an internal loop mode for an extended amount of time. In this case the cell's state trajectory is evolving in the state space in a more or less continuous fashion.
- there is only one reward feedback provided after all the activities of the cell have ceased. In this case the cell's state trajectory has been clearly divided into generations. After each activity-reward feedback cycle, the cell's state has been reset to the initial state. Note, that the environmental state is not being reset to its initial state, but continues its own evolution in a continuous fashion. This results in a complex dynamics between internal and external activities of the cells.

Cells. Computational cells are organised into a regular lattice (or grid). Neighbourhood is typically 4 or 8 adjacent cells. This can be programmatically controlled. Various topologies are possible, although in this article we only focus on regular grids. EVMI cells are constrained in such a way that an externally provided resource (reward) is necessary for the cells to continue to exist (or to be allowed to perform their activities). In other words the cells' computational resources (memory and CPU cycles) are abstracted into a single parameter, called a *resource*, which again, is balanced by a single parameter *reward*. This is an external system-level constraint that may not have any direct linkage with the problem domain of any of the tasks. It can be treated simply as an *organising principle* [5]. In other words, it is a domain-independent artifact that models certain constraints that are conceptually (metaphorically) equivalent to physical constraints. For example, two distinct material objects cannot occupy the same physical space, or a material object cannot emit energy indefinitely, etc. Such physical constraints and more complex organising principles are modelled through the abstract notions of a *resource* and *reward*.

Intra-cell computing. A cell's program takes data from and produces output to the *environment*. In the abstract architecture these mechanisms can be arbitrary, and the EVMA does not mandate any particular mechanism. For efficiency reasons however, we have primarily designed and used certain fixed global properties of the cell processing. Thus, though each cell specialisation and program execution mechanism was built-in, and could not be changed by a given cell itself, in the abstract sense, these could have been reified and provided as mechanisms on the base level, subject to direct manipulation, degradation and improvement when needed.

At every iteration, the execution engine tries to solve some tasks provided by the environment. For each task, it runs the cell's program (possibly calling some neighbours' programs) in interaction with the task's resource. If the output is correct, the cell receives some rewards. A specialisation mechanism uses these rewards to modify its state and provide a new program for the next iterative step. The *specialisation mechanism* is an umbrella term that relates to the ability of the cell to trim its computational capabilities (that means some of the computational capabilities are removed from the cell repertoire). Generally, each cell starts the search process as a general computing machine, capable of computing any computable function. This capability however is not needed for a particular one-off task – what a cell needs is only a subset of the general capabilities. In other EC systems the decision to trim the capabilities is done outside of the search process through the appropriate selection of machine instructions used. This is augmented in EVMI by the mechanisms that allow a cell to trim the instruction set during the runtime too. This trimming process we generally refer to as *specialisation*. The general instruction set is being trimmed and specialised for a narrow task at hand. Individual cells work in a distributed and asynchronous fashion. The processing of an individual cell is sequential and consists of sets of operations, also called *iterations*. The iterations can be imposed by the environment (as was the case with some of our experiments), can be managed by the cell and cell program alone, or they can be in an arbitrary order (subject to a mixture of environmental and cellular influences). At every iteration, the specialisation mechanism of the cell must generate one program. This program will try to solve some tasks from the environment, and this process may yield some rewards coming from the environment. From an artificial life perspective, the program attempts to maximise the reward and maintain its existence by trying to achieve homeostasis with regard to the external resources (rewards).

Inter-cell communication. A key feature of the cellular system is the interactions among the basic components: the cells. From these interactions emerges the global behaviour of the system. It has been assumed that symbiosis among the cells is a crucial/necessary property of the system to reach higher levels of complexity. From a machine learning perspective, symbiosis facilitates knowledge diffusion and capability reuse. Knowledge reuse, on the other hand, facilitates multitask learning. Any cell can use any other cell's program (as long as they are connected) to solve a given task. Programs of a machine can use neighbours' programs (linked cells) as their own instructions (Figures 1 and 3). If rewards are gained and these rewards are shared between all of the involved cells so that they all benefit from the relationship, we model this as a simple symbiotic mechanism based on mutualism. As a consequence, two (or more) cells can effectively *merge* (operate as a single organisational unit), in the reward-driven dynamics. The reward sharing is distributed proportionally to all the merged cells.

Other cellular systems, like classical cellular automata have behavioural properties that are sensitive to synchronisation issues. Indeed, some authors argue that some of the apparent self-organisation of CAs is an artifact of the synchronisation of the global CA clock [17]. Others claim that information is less easily propagated in asynchronous CAs and that such CAs may be harder to analyse [19]. That would be also the case with an asynchronously executing and communicating multi-cell EVM system. With the current implementation, even though individual cells work concurrently, each cell's own *view* of the execution is synchronised and straightforward to track and analyse. Each cell operates independently of the other, in an asynchronous fashion. However, each individual cell represents only a single thread of computation and is therefore traceable and easier to analyse. In principle, in the abstract architecture, each individual cell may be composed of many asynchronously executing cells. Therefore, each individual cell would be capable of modelling a parallel distributed system on its own, keeping detailed track of all the executed sequences and data. Even though it is possible to show that this is feasible in principle, it would complicate the storage and execution for practical implementations at the present time.

Cell merging. There are two basic mechanisms to implement simple merging in a cellular system. One way is to have a single program per cell and provide a mechanism to merge individual cells into higher-order cells (collaboration). The other possibility is to provide the architectural mechanisms for a cell to have the ability to contain multiple programs (containment). We have implemented and tried both of these variants. In the case where a single cell is capable of containing several programs (we also refer to such a collection of programs as a *machine*), our primary goal was to specialise a single cell for solving a set of tasks. It is a basic model for multi-task learning – several programs per cell together with a decision mechanism within each cell which decides when to use which routine. It is easier to have several programs (or several building blocks) solving certain specialised tasks and having a management layer that can decide which program is appropriate and to be used for what situation. In the extreme case, we could imagine having as many programs per cell as tasks in the environment and a good discrimination mechanism. In the experimental section we will discuss some issues regarding the co-evolution of a cell's programs. With only one program per cell, locality of interconnections between the cells plays an important role, and our EVM cellular system exhibits interesting artificial life features and structural tendencies. For instance, since knowledge sharing is not as straightforward as with several programs per machine[1], the system must find some artifacts to propagate solutions to other cells that need it (e.g. through parasitism). Moreover, because each cell remains quite simple (just one sequential stack-based program), the implementation of the specialisation mechanism is facilitated more easily (it has to specialise only one program). This is especially well-suited for genetic algorithms and probabilistic search methods. Thus because of

---

[1] In the case of multiple programs per cell, programs naturally interlink and use each other within a cell to solve more complex tasks. With only a single program in a cell, the information and capabilities sharing must be conducted across the cellular level.

its simplicity and generality, in most experiments we chose to use a single program per cell model.

4   Environment and Resources

Every cell maintains a program[2]. The cell's *goal* is to find a successful program: one that, by solving a task, yields enough rewards for the cell to survive. Programs can call other programs (Figure 1 and 3).

When a given program obtains a reward, the reward will be shared proportionally with any programs used as *assistants* to compute the solution. All of the participants will benefit from their relationship. In other words, symbiotic (mutualistic or parasitic) relationships will appear between programs. This ability to access other programs has thus facilitated complex hierarchical organisation and self-assembly. As a consequence, cells are able to collaborate to solve complex problems. Problems that none of the cells would be able to solve on their own can thereby be solved through cell collaboration.

Figure 1 depicts a simple example of cell C_1 using other cells during the course of C_1's computation. The computation starts with C_1 executing its own instruction number 1. The second instruction of C_1 calls its right neighbour, the cell C_3. Therefore the computation follows to the first instruction of the cell C_3. C_3's program is executed up to instruction number 2, and the third instruction of C_3 calls the program of upper neighbour, cell C_5. And so on. On the bottom of the figure you may see the actual order of instructions being executed and where they belong across all cells. This type of hierarchical assembly between the cells is distinct from the hierarchy on the virtual machine level, which is supported by the EVM assembly language. This multi-level machine hierarchy (as opposed to cell's hierarchy) is being depicted on Figure 2. Note that the Current Machine (CM) index and Higher Machine (HM) index may point to various parts of the HM list.

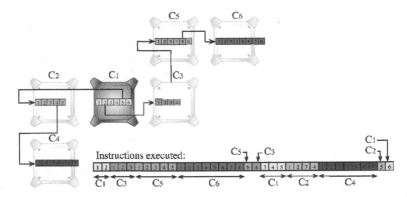

Fig. 1. The dark cell executes its program. Arrows show instructions that call neighbours' programs.

---

[2] A program in this case means simply a sequence of instructions in the EVM assembly language.

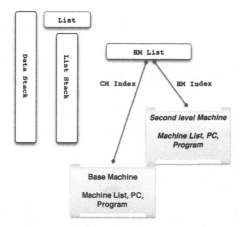

Fig. 2. EVM hierarchy. The HM (Higher Machine) List stores indices to all created and managed machines. The CM (Current Machine) pointer points to the currently executing machine, and the HM index points to the higher machine that can be manipulated by the EVM instructions.

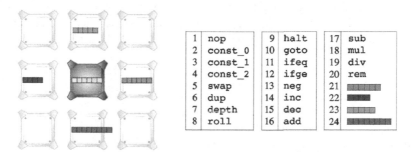

1	nop	9	halt	17	sub
2	const_0	10	goto	18	mul
3	const_1	11	ifeq	19	div
4	const_2	12	ifge	20	rem
5	swap	13	neg	21	
6	dup	14	inc	22	
7	depth	15	dec	23	
8	roll	16	add	24	

Fig. 3. Instruction set I for the dark cell's program. Instructions 21 to 24 execute the programs of its neighbours. The von Neumann neighbourhood (on a regular grid: up, right, down, left neighbours) is used for this example. A program can thus look like the following: add dup *leftNeighbourProgram* mul *rightNeighbourProgram*.

The environment represents the external constraints on our system. Its role is to keep the system under pressure to force it to solve the tasks specified from the outside. We have designed the environment as a set of *resources*. There is a one-to-one mapping between the resources and the tasks to solve (every resource corresponds to a task). The purpose of these resources is to give rewards to the cells when they solve their task. *Cell specialisation* consists of finding a successful program for the cell. In other words, the cell self-adapts to a particular task in the environment. Several specialisation mechanisms have been studied by the author: classic genetic algorithms, ad hoc stochastic search (maintaining a tree of probabilities of potential building blocks), or an adaptation of an environment-independent reinforcement learning method (proposed by [15]). We have experimented with different methods of learning, including random search, stochastic search and genetic algorithms. The results of our experiments have been published [11]. Some argue that many multi-task problems share a similar internal structure, therefore reusing common properties and regularities aids the solution search process [18]. Multi-task learning is an area of machine learning which studies methods that can take advantage of previously learnt knowledge by generalising and reusing it while solving a set of possibly related tasks [1]. Our EVM architecture benefits from those properties, by concurrently solving multiple tasks and by recombining the already obtained partial solutions or solutions to sub-problems.

5   Static and Dynamic Aspects of the EVM Cell

During the course of execution, an initial program can create, use and dispose of an arbitrary number of intermediate virtual machines running programs in appropriate languages. The trace of such an execution can be investigated and analysed, leading to a better automatic and autonomic generation of suitable virtual machines for given tasks. The implemented EVM architecture helps to model and keep track explicitly of dynamical properties of a running program, which otherwise may be completely implicit and often, intractable. Many cells may be involved in a single thread of computation and it may become complicated to move data between all of them. A fast parallel implementation (in hardware) is difficult to design, but not impossible. We envision, in the future, special self-adapting operating systems executing on re-configurable hardware implementations, such as an FPGA-based TTA (Transfer Triggered Architerture) for example [3].

6   Hypercycles

Let us consider a sequence of reactions in which products, with or without the help of additional reactants, undergo further transformations. The *reaction cycle* or *cycle* is such a sequence of reactions such that some of the products are identical with the reactant of any previous step of the sequence. The most basic is a three-members cycle, with a substrate, an enzyme, and a product. The enzyme transforms the substrate into an enzyme-substrate, and then enzyme-product complexes, which in turn are transformed into the product and a free enzyme. See Figure 4.

In such a case, the cycle as a whole works as a *catalyst*. Unidirectional cyclic restoration of the intermediates presumes a system far from equilibrium. This can be associated with a dissipation of energy into the environment. Equilibration occurring in a

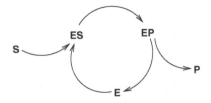

Fig. 4. An example of a three-members catalytic cycle: the free enzyme (E), the enzyme-substrate (ES) and the enzyme-product (EP) complexes all demonstrate a catalytic cyclic restoration of the intermediates in the turnover of the substrate (S) to the product (P).

closed system will cause each individual step to be in balance: catalytic action in such a closed system will be microscopically reversible. Let us now consider a reaction cycle in which at least one of the intermediates themselves is a catalyst (see the work of Kaufmann on autocatalytic nets, e.g. [6]). The simplest representative of this category is a single *autocatalyst* (or a self-replicative unit). A system which connects autocatalytic or self-replicative units through a cyclic linkage is called a *hypercycle*. One of the goals of the EVMA is to provide a computational environment where hypercycles can be created, simulated and observed.

Hierarchies. In Darwinian systems all self-replicative units competing for selection are non-coupled. In other words, the selection forces operate purely on a single level. It can be a level of individuals, of species, or of genes. We have here a conservation of a limited amount of information, which cannot pass a specified threshold (see the discussion above). Hypercyclic systems also deal with similar evolutionary selective pressures. However, in this case we also deal with integrating properties, and this allows for cooperation of otherwise competing units to develop: hypercycles are able of establishing higher-order linkages. When inter-cyclic coupling is established, individual hypercycles may form hierarchies. In other words the basic unit of selection is not a single hypercycle; it is a whole chain of interrelated hypercycles [4]. Similarly, in the EVMA, although lower level selective pressure may lead to survival or extinction of certain individual programs (cells), the basic unit of selection is not a single cell. The successful units will invariably form long interlinked chains. It will be a network of many individual cells contributing to the overall success. Only the entire network can then survive, as a whole.

7   Autopoiesis

The theory of autopoiesis has been developed by two Chilean biologists, Humberto Maturana and Francisco Varela, in the early 1970s. In recent years the theory has been re-shaped slightly and used as a basis for discussions on new notions of the concepts of biological information, information processing and complexity. The original assumptions on the physical embodiment, central to autopoiesis, can be then considered to be the shortcomings of the theory. In the following sections we provide the introduction to the general theory of autopoiesis.

Machines. Machines are *unities* which are made out of components. All components are characterised by certain properties capable of satisfying certain relations that determine within the machine (in the unity) the interactions and transformations of these components. The actual nature of components and their particular properties, other than those participating in the interactions and transformations which constitute the machine, are irrelevant and can be arbitrary. In the context of the EVM machines, all the side-effects that do not contribute to the relations characteristic of the unity are not considered to be important and can be any. That is, several different programs can be considered as the same machine, as long as they satisfy all the necessary relationships. This is inherently simple and intuitive in the context of computation and virtual machines. For example any Java interpreter is considered a Java interpreter, no matter what extra functionality that interpreter may posses, or how internally it has been implemented. Thus there is a certain level of granularity that is mapped to a given notion of a machine.

The *organisation* of a machine is defined as all the relations which define a machine as a unity and determine the dynamics of interactions and transformations which it may undergo as such a unity. The organisation of a machine does not specify the properties of the components which realise a concrete machine. The organisation of a machine is independent of the arbitrary properties of its components, and a given machine can be realised in many different manners by many different kinds of components. In other words, the organisation is the functional abstraction over the actual physical (or computational) realisation of a given machine.

The *structure* of a machine is defined as the actual relations which hold among the components which integrate a concrete machine in a given space. This is the actual realisation, or implementation, of a given machine. This represents the actual program of the machine, with all its properties and functionality. Note that a given machine (machine with fixed organisation) can be realised by many different structures. For example, an organisation may remain constant by being static, by maintaining its components constant, or by maintaining constant certain relations between components which are otherwise in continuous flow or change.

An *autopoietic machine* is defined as a unity by a network of production, transformation, and destruction of components which: (i) through their interactions and transformation continuously regenerate and realise the network of relations that produced them, and (ii) constitute the machine as a concrete unity in the space in which the components exist, by specifying the topological domain of its realisation as such a network [7]. This somewhat abstract definition, in the context of the EVM architecture, means that an autopoietic machine is such a machine that continuously regenerates and realises all the necessary components of the interlinked network of dependencies, and by doing so, maintains all the necessary dependencies and computations. In other words, it maintains itself, it achieves homeostasis.

A living system is considered to be a unity in physical space. It is an entity topologically and operationally separable from the physical background. It is defined by an organisation that consists of a network of processes of production and transformation of components, molecular and otherwise, that through their interactions: a) recursively generate the same network of processes of production of components that generated them; and b) constitute the system as a physical unity by determining its boundaries in the physical space. As defined above, this organisation is called an autopoietic system. An autopoietic machine is an homeostatic, or rather a relations-static, system which has its own organisation as the fundamental variable which it maintains constant. A machine whose organisation is not autopoietic does not produce the components that constitute it. The product of such a machine is different from the machine itself. The physical unity of such a machine is determined by processes that do not enter into its organisation. Such a machine is called *allopoietic* [7]. Allopoietic machines have input and output relations as a characteristic of their organisation: their output is the product of their operation, and their input is what they transform to produce this product. The phenomenology of an allopoietic machine is the phenomenology of its input-output relations. The realisation of allopoietic machines is determined by external processes, and these external processes do not enter into the machine's organisation.

8 EVM and the Theory of Autopoiesis

In the EVM architecture we follow the Maturana's model [7] in the sense that evolution is a side-effect, a consequence, not the prerequisite of life. We always deal with limited resources and restricted computations, thus we have the selection and evolutionary processes naturally occurring within our computational models. Our model departs from Maturana's work in some important points. First, we do not stress the notion of "physical space" (which, by the way, Maturana does not define formally in his work [7]) as a prerequisite of autopoiesis. For the theory to work, it does not actually matter what sort of "space" is being used for synthesis and research of autopoietic processes. It is just like the situation with different branches of science: some are more abstract than others, but the aim is always the same: to make predictions about physical reality. For some fields it makes sense to ground the theory in a real physical reality, for example in case of biology; but for some it does not help at all, for example in mathematics. We believe, that a general abstract theory of life which could be applied easily to both virtual and physical spaces is of great interest to progress the research, not only in theoretical biology, but also in evolutionary biology, computer science and information science. Hence, we do not require the "space" to be of a physical nature. Second, we do not require the "machines" from Maturana's original autopoietic theory to be Turing-like. We believe that this is too restrictive, and we allow any computing machine to be used as a component in the autopoietic system.

9 Summary and Future Work

This article presents a general description of the EVM architecture together with a detailed discussion of its biological inspirations. We have discussed how the multi-cellular computational EVM model mimics theoretical models of biological systems. For detailed description of the model together with the experimental results please refer to [10]. Future work will consists of re-implementing the model on a modern stack-based virtual machine running on massively multicore servers and modern GPUs. This will enhance performance and allow more complex studies. We hope to achieve sufficient level of computational power so that the EVM model can be used in real-world systems and tackle complex optimisation and machine learning problems.

References

[1] Baxter, J.: A model of inductive bias learning. Journal of Artificial Intelligence Research 12, 149–198 (2000)
[2] Caruana, R.: Multitask learning. Machine Learning 28(1), 41–75 (1997)
[3] Corporaal, H.: Microprocessor Architectures: From VLIW to TTA. John Wiley & Sons, Inc., New York (1997)

[4] Eigen, M., Schuster, P.: The Hypercycle: A Principle of Natural Self-Organization. Springer, Heidelberg (1979)

[5] Haken, H.: Synergetics, An Introduction: Nonequilibrium Phase Transitions and Self-Organization in Physics, Chemistry, and Biology, 3rd revised and enlarged edn. Springer, Berlin (1983)

[6] Kauffman, S.A.: The origins of order: self-organization and selection in evolution. Oxford Press, New York (1993)

[7] Maturana, H.R., Varela, F.J.: Autopoiesis: The organization of the living. In: Cohen, R.S., Wartofsky, M.W. (eds.) Autopoiesis and Cognition: The Realization of the Living. Boston Studies in the Philosophy of Science, vol. 42. D. Reidel Publishing Company, Dordrech (1980); With a preface to 'Autopoiesis' by Sir Stafford Beer. Originally published in Chile in 1972 under the title De maquinas y Seres Vivos, by Editorial Univesitaria S.A

[8] Mitchell, T., Utgoff, P., Banerji, R.: Learning by experimentation: Acquiring and refining problem-solving heuristics. In: Michalski, R., Carbonell, J., Mitchell, T. (eds.) Machine Learning: An Artificial Intelligence Approach, ch. 6, pp. 163–190. Springer, Heidelberg (1984)

[9] Nowostawski, M.: Parallel genetic algorithms in sequential optimisation. Master's thesis, School of Computer Science, University of Birmingham, Birmingham, UK (September 1998)

[10] Nowostawski, M.: Evolvable Virtual Machines. PhD thesis, Information Science Department, University of Otago, Dunedin, New Zealand, 12 (2008)

[11] Nowostawski, M., Epiney, L., Purvis, M.: Self-Adaptation and Dynamic Environment Experiments with Evolvable Virtual Machines. In: Brueckner, S.A., Di Marzo Serugendo, G., Hales, D., Zambonelli, F. (eds.) ESOA 2005. LNCS (LNAI), vol. 3910, pp. 46–60. Springer, Heidelberg (2006)

[12] Nowostawski, M., Purvis, M.K., Cranefield, S.: An architecture for self-organising evolvable virtual machines. In: Brueckner, S.A., Di Marzo Serugendo, G., Karageorgos, A., Nagpal, R. (eds.) ESOA 2005. LNCS (LNAI), vol. 3464, pp. 100–122. Springer, Heidelberg (2005)

[13] Perkins, C.E., et al.: Ad hoc networking. Addison-Wesley, Reading (2001)

[14] Schmidhuber, J.: Self-referential learning, or on learning how to learn: The meta-meta-. hook. Diploma thesis, Institut fuer Informatik, Technische Universitaet Muenchen (1987), http://www.idsia.ch/juergen/diploma.html

[15] Schmidhuber, J.: Environment-independent reinforcement acceleration. Technical Note IDSIA-59-95, IDSIA, Lugano (1995)

[16] Schmidhuber, J.: A general method for incremental self-improvement and multiagent learning. In: Yao, X. (ed.) Evolutionary Computation: Theory and Applications, ch. 3, pp. 81–123. Scientific Publishers Co., Singapore (1999)

[17] Sipper, M., Tomassini, M.: An introduction to cellular automata. In: Mange, D., Tomassini, M. (eds.) Bio-Inspired Computing Machines, vol. 3, pp. 49–58. Presses Polytechniques et Universitaires Romandes, Lausanne (1998)

[18] Thrun, S.: Is learning the n-th thing any easier than learning the first. In: Touretzky, D., Mozer, M. (eds.) Advances in Neural Information Processing Systems (NIPS), vol. 8, pp. 640–646. MIT Press, Cambridge (1996)

[19] Wolfram, S.: A New Kind of Science, 1st edn. Wolfram Media, Inc. (May 2002)

On Multi-agent Petri Net Models for Computing Extensive Finite Games

Rustam Tagiew

Institute for Computer Science of TU Bergakademie Freiberg, Germany
tagiew@informatik.tu-freiberg.de

Abstract. Modelling multi-agent strategic interactions by using Petri nets is the addressed issue. Strategic interactions are represented as games in extensive form. Representations in extensive form are known in artificial intelligence as game trees. We use transition systems of Petri nets to build game trees. Representable games are restricted to be finite and of complete information. A language for representation of these games is created and is expected to represent also time dependent aspects. Two perspectives of application are considered - game server definition and calculation of equilibria. The approach is compared with related works and its advantages are discussed. The most important advantages are the graphical representation in comparison to logic based approaches and the slenderness in comparison to default game representation in extensive form. Formal definition, algorithms and examples are given. An implementation is already tested.

1 Introduction

If increasing of subjective utility needs interaction with other agents, a rational agent has to interact [1, p.161]. If an interaction between rational agents takes place, then it means that at least one agent wants to increase his utility. If agents want to achieve same goals, they cooperate. If their goals are contrary, they compete. Most cases of such interactions are between these two extrema. A good advise for an interacting agent is to anticipate the goals of other agents. A better advised agent can partially predict the behavior of other agents as a consequence of their goals. The reasoning needed in such interactions is strategic. A strategic interaction (SI) is an interaction, where strategic reasoning takes place.

SI or also games are investigated in game theory [2]. Game theory considers rationality and the possibility of predicting rational behavior. The presumption is existence of common knowledge of rationality. That means that every interacting participant believes in rationality of the others and believes that they believe in his rationality and so on. Predicted behavior of interacting interaction participants or players is the equilibrium. None of players benefits, if he or she deviates from equilibrium. That is, why it is called equilibrium. There is at least one equilibrium in finite games [3]. Finite means finite number of players, states and actions. Solving a game is to find the equilibria of this game. Games can be classified in games of perfect, imperfect and incomplete information. Games of imperfect information are interactions, where at least one player is uncertain in at least one step about the state of interaction. In games of incomplete

N.T. Nguyen et al. (Eds.): New Challenges in Compu. Collective Intelligence, SCI 244, pp. 243–254.
springerlink.com

information, players are uncertain about the structure of the game. In this work, we concentrate on perfect and imperfect information.

Games of imperfect information can be represented in strategic or in extensive form. Fig. 1 shows the game Matching Pennies in strategic and in extensive form. It is a zero sum game. Alice and Bob simultaneously choose one of two available actions - head or tail. If both chosen actions are the same, Alice wins otherwise Bob. This game can be presented as double matrix (left) or as a game tree (right). Lower left entries of the matrix are the payoffs of Alice and other entries are Bob's. In the game tree representation simultaneous decisions have to be sequential. The dashed line in the game tree means that Bob is uncertain about Alice's decision and hence about the tree node for the current game state. So connected nodes are called an information set. After Alice's action, Bob can choose between H_ an T_ (_ is variable). A player can be also replaced by so called *nature*. Nature is a chance element in the game. It is an unpaid player with known distribution over his actions. In following text we use only extensive representation of games (or also game tree representation).

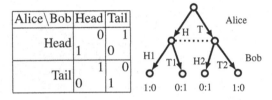

Alice\Bob	Head	Tail
Head	0 / 1	1 / 0
Tail	1 / 0	0 / 1

Fig. 1. Strategic vs. extensive form.

For artificial intelligence, games have at least two application - agent design and mechanism design [1, p.632]. In agent design, we have a game and must compute adequate behavior. In mechanism design, we have an expectation about the behavior and must invent game rules. These two tasks can be solved theoretically by running algorithms over a game tree or in practical way by constructing an environment where different real players can interact. In most cases games are defined in low level code [4]. Our goal is to develop a high-level language for definition of general games, which is applicable in both cases. For this goal, we model a game tree as a transition system of an enhanced Petri net (PN). A high level language for the definition of games has a couple of advantages. (1) Representation of a game is clear and convertible. Convertible means that it can be compiled to other representation formats and formalisms. Such representation can be transfered between systems - from a game server to a player e.g.. (2) The game server and the game solver use the same file. It reduces redundancy. It also guarantees that both run on the same game. (3) Game rules can be edited easier. Algorithms can be developed, which manipulate game representation in any possible way - 'reduce number of players' or 'remove simultaneous turns' e.g.. Game representations can be also used for evolutionary mechanism design [5]. It can be easier to define mutation on a high level domain specific language. (4) Techniques for defining behaviors of players can be developed on the basis of a game description. Players which are defined for playing multiple kinds of games can be constructed and benchmarked. (5) Protocoling of a game during an execution can be better defined. (6) One can consider automatic verbalisation algorithms of games rules and game rules editors for users who are not experienced in programming.

The next chapter tells about related work in developing a language for game definition. Chapter 3 explains details of PNSI[1](Petri net for strategic interaction) formalism. In chapter 4, we present an example for modelling games in PNSI. Chapter 5 evinces the appropriate game server algorithm [6]. Chapter 6 shows the way of constructing of a game tree using PNSI. Discussion follows thereafter. Then we present future work and conclude the results.

2   Related Work

Creating a formalism for representing a general class of games is not the main intention of game theory. Extensive form is a common used game represention formalism. However, there are approaches for solving games in slight definition of some subclasses - graphical games [7], local-effect games [8], congestion games [9] and action-graph games [10]. This field of research is called game representation in compact form. In these game theory motivated approaches, compact form can be achieved through losing generality. Then, there is an approach for modelling two player zero sum games called as shortest path games using PN [11]. Kanovich [12] considered non-deterministic PN for modelling vector games. Vector games are games of perfect information. Westergaard [13] used colored PN for modelling network protocols which can be seen as games. These approaches are not general enough to consider it in our work.

In artificial intelligence, there are two well-known independent approaches for modelling general games. The first is the GALA language [14]. The Abbreviation GALA means game-theoretic analysis for a large class of games. GALA is logic-based and has been developed for general representation and solving of games. The GALA language represents games as branching programs. Every branching node in a game tree is a call of functions and also a logic proposition, which can be satisfied in a couple of ways (branches). The supporting system for GALA is prolog. The GALA system generates a game tree using a definition of a game and forwards it to GAMBIT [15]. GAMBIT is state-of-art open source game solving system. GALA can also solve games itself using commercial linear programming libraries like CPLEX or MATLAB. GALA can represent finite games of imperfect information. No approach is known to define a game server based on GALA. GALA can not represent time dependent elements of a game like delays, timeouts or sudden events.

The other work is general game playing (GGP) [16]. It considers finite games of perfect information, which are called deterministic in game theory. The main idea of GGP is providing an environment for conducting artificial intelligence programming contestes between different artificial agents. GGP provides a game model. The game model is a graph consisting of game states connected by actions. Actions are the transitions between states. The game model allows circles. States are explained to be not monolithic. That means that they consist of a couple of separately changeable items like a database. GGP provides a logic-based language GDL for definition of the game model. GDL is based on situation calculus. It uses a vocabulary of predicates. A transition is performed as an update of the dynamic knowledge base. The GGP environment is constructed using webservices. It is still not discussed, how to represent time dependent elements with GDL.

---

[1] [ˈpɛnˈzaɪ]

3   Definition

We use transition systems (TS) of PN for modelling games [6]. PNSI is a combination of two elements - PN [17] and SI (strategic interaction) - $PNSI = (P, Q, F, W, M, I, C, N, D, A, O, H, B)$. P - set of places; Q - set of transitions, where $P \cap Q = \emptyset$ holds; $F \subseteq (P \times Q) \cup (Q \times P)$ - set of directed arcs; $W : F \to \mathbb{N}_1^+$ - function for weights at the arcs; $M \in \mathbb{N}^{|P|}$ - current assignment of places; I - set of agents, empty element ε stands for environment or also nature; $C \subset (Q^*)^*$ - subset of sequences of transitions, called choice sets; Every transition is a member of only one element of C; $N : C \to \mathbb{N}$ - numbering function, which is not injective; $D : \mathbb{N} \to (\mathbb{R}_0^1)^n$ - function for firing probability distribution in a choice set, where $\sum(D(\_)) = 1$ and n is number of elements of the related choice sets; $O : \mathbb{N} \to I \cup \varepsilon$ denotes ownership; $A : Q \to \mathbb{R}^{|I|}$ - payoff vector of a transition, if it fires; $H : P \to I^*$ provides for every place a subset of agents for which it is hidden. Agents can alter D for own numbers and see all unhidden places; $B : I \to \mathbb{R}$ - current account balance of agents.

Regarding the deficiencies of the previous approaches, we decided to develop a particular intuitive philosophy for game computing. PNSI implements this philosophy. A strategic interaction is a kind of a running engine. The engine runs in discrete 'steps'. The time period between two 'steps' is considered to be always the same. After a time period is expired, the engine makes some changes in its internal state. This engine has a couple of modulators. The set of modulators is constant. Every modulator has a current state and a single owner, who can alter its state. Manipulating the modulators can impact the running of the engine. A player acts by manipulating a modulator. Some player are able to see some of details of the internal state (M - the state, H - visibility) of the engine. In PNSI, we implemented the concept of numbered choice sets (C - choice sets, N - numbers). A choice set is a set of transitions, in which only one transition can be fired exclusively in a step. An owned number (O) is a modulator. A number without an owner models actions of the nature. Altering the distributions of numbers causes firing or not firing of transitions in corresponding choice sets.

Based on this idea, for the transition system (TS) of PNSI

$$TS = (S, Q, \to) \tag{1}$$

S - states, where $S = (M, B)$, $\to \subseteq S \times Q^* \times S$ - directed arcs between states,

we can formulate the rule for choice sets in following way

$$\forall c \in C : \forall t, t' \in c : \forall s, s' \in S : \neg(s \xrightarrow{tt'} s'). \tag{2}$$

Every choice set has a number. Multiple choice sets can have the same number. This is important for modeling that actions cause different consequences in different states. As we already said, a number can have an owner and a firing probability distribution over the transitions in corresponding choice sets. If a number has an owner then the owner can alter the distribution of his number. Further, we can calculate the probability for every arc in TS.

$$probability(\overset{t_1...t_n}{\rightarrow}) = \prod_{i=1}^{n} D(N(c))[position(t_i)], \quad \text{where } c[position(t_i)] = t_i,$$

and as a consequence of the equation (2) $n \leq |C|$ (3)

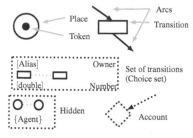

Fig. 2. Graphical representation of PNSI.

Hidding places is usefull for modelling imperfect information. Payoffs can not be hidden. To construct for a concrete game a default PNSI structure, one has to create for every action a transition and for every state a place. Every arc of this PN is weighted with 1. States are created as marking of the net. In this default modeling case, all places of a state have zero tokens, except of the place with one token, which corresponds to this state. Outgoing and incoming arcs for a transition can be derived on the basis of both connected states. But one can easily find a game, where one can construct more than one PNSI structure.

Fig.2 shows elements for the graphical representation of PNSI. These graphical elements are derived from the common representation of PN. All transitions and places are labeled. Arcs have weights. The set '[double]' is a distribution. The set '[Alias]' represents aliases for transitions in a choice set.

4   Modeling Games

As an example for the creation of a PNSI structure, we introduce the two player zero-sum game Nim for one heap. This game is of perfect information. At the initial state, one has a non empty set of items. The players perform successively turns. Every player can remove at least one but not more than three items. The player, who removes the last item, loses. PNSI for this game is presented in fig.3. At start, Bob has the turn. Initial account balances of Alice and Bob are   1 and 1 as. Consequences of actions are coded

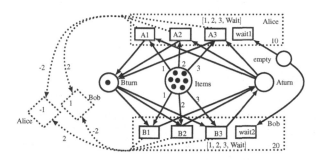

Fig. 3. PNSI structure of Nim for one heap [6].

by weights at arcs. The account balances swaps, if a player performs a turn. The game ends, when *Items* is empty. To underline the power of PNSI, one must say that it can be also used for coding time dependent rules. It is possible due to the fact that it is a Petri net based model. But, here is not enough space to discuss discrete time representation in PNSI.

The syntax of PNSI is based on 'Yet Another Markup Language' (YAML^{TM}). It is a human-friendly definition and serialization language. In YAML, one can represent nested collections. The advantage of YAML compared to other languages is its slight and highly readable syntax.

5   Game Server Implementation

The algorithm for the game server is a loop. Based on PNSI, a single iteration of the game server algorithm is defined (Alg.1). In line 2 the algorithm receives and implements the altering commands of the players in distributions of owned numbers. Line 7 means that a dead PNSI is a finished game. In lines 15-20, a transition in a choice set is chosen independently of proving sufficiency of incoming tokens. This detail enables representing of pause action like waiting in the Nim example. In the Nim example, if a player sets 1 as probability for waiting, the game stops for an infinite period of time. The same lines show that in identically numbered choice sets transitions can be chosen from different positions. This is done to avoid useless operations. Because of practical considerations, the distributions of owned numbers are restricted to be $\in \mathbb{N}_0^1$. That means that every player can only choose the position for his number, where to set 1. Then, there is no need to run the loop on numbers instead of choice sets. Every position of an owned number gets additionally an alias. An altering command of a player consists of a number and an action alias. Choice sets are not ordered. Line 24 implements this fact through choosing randomly transitions, each of them is already chosen from a choice set. Lines 36-46 send visible changes of the game state to the players. Sending only the changes of the game state reduces communication traffic.

PNSI represents games as running Petri nets. A PNSI representation itself is also a state of a game. Each in PNSI modelled running game can be stopped, saved to a file or data base and then loaded and executed again. This makes game computing in the game server case persistent. Changes of relations *places*, *accounts* and *distribution* are recorded in a protocol. *places* and *accounts* represent the state of the game and *distribution* represents players decisions.

6   Game Tree Generation

How to solve a game represented in PNSI? Our solution is to construct a game tree as a transition system of PNSI and forward this game tree to a program which can handle with game trees. Here, we use GAMBIT[15] as such program. GAMBIT accepts two kinds of game representations - extensive and strategic form. GAMBIT's output is a list of equilibria for the input game. At time, GAMBIT is the best known open source game solving library. The game representation in extensive form used by GAMBIT is a game

---

Algorithm 1. Game server iteration.

   Data: PNSI

 1 **while** *not time_period_expired* **do**

 2 | PNSI.implement(receive_altering_commands_from_players)

 3 **end**

 4 create_set(active)

 5 **foreach** *t in PNSI.transitions* **do**

 6 | **if** *PNSI.enough_incoming_tokens(t)* **then**

 7 | | active.add(t)

 8 | **end**

 9 **end**

10 **if** *active.empty* **then**

11 | complete_game

12 **end**

13 create_list(tobefired)

14 **foreach** *c in PNSI.choice_sets* **do**

15 | th = c.choose_randomly_transition_according_to_distribution

16 | **if** *active.contains(th)* **then**

17 | | tobefired.add(th)

18 | **end**

19 **end**

20 create_list(fired)

21 create_set(changed)

22 **while** *not tobefired.empty* **do**

23 | ta = tobefired.remove_at_index(random_value)

24 | **if** *PNSI.enough_incoming_tokens(ta)* **then**

25 | | PNSI.abolish_incoming_tokens(ta)

26 | | changed.add(ta.incoming)

27 | | fired.add(ta)

28 | **end**

29 **end**

30 **while** *not fired.empty* **do**

31 | tp = fired.remove_first

32 | PNSI.produce_outgoing_tokens(tp)

33 | changed.add(ta.outgoing)

34 | PNSI.produce_payoffs(tp)

35 **end**

36 **foreach** *a in PNSI.agents* **do**

37 | **foreach** *p in changed* **do**

38 | | **if** *not PNSI.hidden(p, a)* **then**

39 | | | add2message(a, p.id, p.value)

40 | | **end**

41 | **end**

42 | **foreach** *p in amounts* **do**

43 | | add2message(a, p.id, p.value)

44 | **end**

45 | send_message(a)

46 **end**

---

tree with repeated states. This representation is called extended form game (EFG). EFG is a tree with three kinds of nodes - chance node, personal node and terminal node. Every node contains an outcome, which is a payoff vector sized according to number of players. Chance nodes contain addionally a vector of probabilities for outgoing nodes. Personal nodes contain the owner and a vector of names for actions. Personal nodes can be connected in case of imperfect information. EFG grows exponentially with number of turns. The approach, which is chosen for converting PNSI to EFG, is to bound the quantity of turns.

A single iteration or also step of game server algorithm (Alg.1) can produce different assignments of places and accounts depending on players commands and nature. One step of this algorithm can be seen as one (simultaneous) turn in a game. To produce EFG of a PNSI for a couple of turns, we explain the routine for a single turn first. Alg.2 shows main points of this transformation. This algorithm is based on equation 3. *Step* is a structure, which contains all information for generating a game tree for a single step. In lines 2–9, the algorithm finds all choice sets which contain active transitions and numbers altering of whose distributions can affect the result. In lines 10 and following, the algorithm finds all sets of partially active transitions, their probabilities and required player commands. Some of this sets contain transitions,

---

Algorithm 2. PNSI to EFG, single step

  Data: PNSI, EFG
1 create(step)
2 **foreach** *c in PNSI.choice_sets* **do**
3   **if** *PNSI.contains_any_active_transition(c)* **then**
4   step.choice_sets.add(c)
5   **if** *has_owner(c.number)* **then**
6   step.ownednumbers.add(c.number)
7   **end**
8   **end**
9 **end**
10 step.firing_alternatives = create_all_firing_alternatives(step.choice_sets)
11 **foreach** *transitions_set in step.firing_alternatives* **do**
12   alternative = calculate_probabilities_and_turns(transitions_set)
13   **if** *alternative.probability = 0* **then**
14   step.firing_alternatives.remove(transition_set)
15   **else**
16   remove_dead_transitions_from_alternative(transitions_set)
17   resolve_conflicts_if_needed(transitions_set)
18   **end**
19 **end**
20 **foreach** *number in step.ownednumbers* **do**
21   EFG.add_to_leafs(step.branches(number))
22 **end**
23 EFG.add_a_choice_node_to_each_leaf(step)
24 **return** EFG.get_leaves(step)

---

which can not be fired together, because they can have the same incoming places. Line 17 resolves these conflicts. The resolving is done by splitting the set of transitions in multiple non-conflicting ones. The probabilities of the resulting non-conflicting sets of transitions are calculated considering the fact of missing order of choice sets (Alg.1, line 24). Then, the algorithm constructs the game tree with leading personal nodes and terminating with choice nodes.

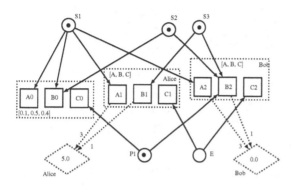

Fig. 4. Benchmark for Alg.2.

The game depicted on fig.4 is chosen as an example for this transformation. This game is a demonstration of what can be represented using PNSI and is a concrete case to demonstrate Alg.2. This figure models a situation, where Alice, Bob and nature make a simultaneous concurrent decision. GAMBIT has an ability to generate a colored image for given EFG file. Fig.5 shows a GAMBIT generated image of generated EFG on one step of PNSI structure (colors are removed). The root is a personal node of Bob. The nodes *nd2*, *nd20* and *nd37* are personal nodes of Alice. All other nodes are either chance nodes or terminal nodes. Chance nodes and actions of nature are gray. We see the resulting payoffs of the players as terminal nodes of the tree (Bob, then Alice). The root contains the initial account balances of players. We call further all nodes at the begin and at the end of a step like the root and the terminal nodes as border nodes. Every branch of choice nodes is denoted by a set of transitions and a probability of their firing. The depicted probabilities are not exact because of numerical errors. Personal nodes of Alice are connected, because it is a simultaneous turn. The game has only one equilibrium calculated by GAMBIT. The equilibrium is to take action A for both players. This equilibrium is depicted as probabilities at personal arcs.

The transformation of multiple steps needs to regard the hidden places. Every border node of resulting EFG complies to a marking of the net. For a player, two different nodes are in the same information set, if he can not see the places in which they differ. The algorithm (Alg.3) for transforming multiple states uses the algorithm for a single step (line 6). In line 3, it finds for every player sets of nodes, which he can not distinguish. Based on the result of previous steps, it extends EFG and connects personal nodes (line 7). As example for a transformation using multiple steps, we present fig.6(top) a representation of Matching Pennies in PNSI. Bob owns two choice sets with number 2. He can not alter the distribution for these choice sets separately. Our algorithm transforms it to EFG representation depicted on fig.6(bottom). Here, nodes *nd3* and *nd5* are connected. The explanation for fig.6(bottom) are the same as for fig.5. The difference is that the equilibrium is a mixed one.

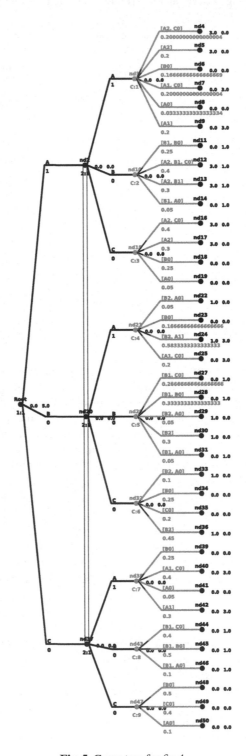

Fig. 5. Game tree for fig.4.

Algorithm 3. PNSI to EFG, multiple steps

Data: PNSI, EFG

```
1  initialize(EFG, border_nodes)
2  foreach i in steps do
3  │    information_sets = create_information_sets(border_nodes, PNSI, EFG)
4  │    clear(border_nodes)
5  │    foreach node in information_sets.flatten do
6  │    │    new_border_nodes = do_one_step(PNSI_state_of(node), EFG))
7  │    │    connect_personal_nodes(node, information_sets)
8  │    │    border_nodes.add(new_border_nodes)
9  │    end
10 end
```

---

7   Discussion

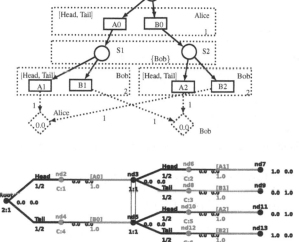

Fig. 6. Matching Pennies from PNSI to EFG.

We can figure out main advantages of PNSI as following. (1) PNSI representation is compacter than EFG. (2) It provides a graphical representation, which is not available using logic-based approaches. (3) It enables pro-bably representation of time dependent elements in a game. (4) It satisfies both game computing tasks - game solver and game server definition.

The only disadvantage is the size of representation, which is significantly bigger than in logic-based approaches. For instance, if one intends to model chess, one needs to create 13 places for every cell of the board (12 kinds of pieces plus empty). This makes 832 places for representing assignment of the board. The other case is representing of payoff matrixes. Every entry in a such matrix needs a transition. For a $M \times N$ two player payoff matrix, it means $M * N$ transitions. But some big payoff matrixes can be summarized by a couple of simple rules. Nevertheless, PNSI was already successfully used for conducting an experiment in behavioral game theory [18]. The scenario used in this application is a repeated zero sum two players game with imperfect information. The game rules are based on Roshambo. The PNSI definition of this scenario contains 161 transitions and 56 places.

8   Conclusion and Future Work

PNSI is a clear convertible game representation for both practical tasks of game computing - game server and game solver definition. PNSI is fully implemented and successfully applied. It is additionally expected to model time dependent elements like delays, timeouts, sudden events and so on. As an approach, which is based on Petri nets, it provides a possibility of graphical representation. But some games, which can be represented by logic-based approaches in a couple of simple rules, take too much space in PNSI. We propose a combination of logic-based game definition and PNSI, as a direction of future work. It can be also proposed to use PNSI for definition of multi-agent interaction protocols.

References

1. Russel, S., Norvig, P.: Artificial Intelligence. Pearson Education, London (2003)
2. Osborne, M.J., Rubinstein, A.: A course in game theory. MIT Press, Cambridge (1994)
3. Nash, J.: Non-cooperative games. Annals of Mathematics (54), 286–295 (1951)
4. Grosz, B., Pfeffer, A.: Colored trails, http://www.eecs.harvard.edu/ai/ct/
5. Phelps, S.G.: Evolutionary Mechanism Design. PhD thesis, University of Liverpool (2007)
6. Tagiew, R.: Multi-agent petri-games. In: Mohammadian, M. (ed.) CIMCA 2008 – IAWTIC 2008 – ISE 2008. IEEE Computer Society Press, Los Alamitos (2009)
7. Kearns, M., Littman, M., Singh, S.: Graphical models for game theory. In: Proceedings of UAI (2001)
8. Leyton, K., Tennenholtz, M.: Local-effect games. In: IJCAI (2003)
9. Rosenthal, R.: A class of games possessing pure-strategy nash equilibria. International Journal of Game Theory 2, 65–67 (1973)
10. Bhat, N., Leyton-Brown, K.: Computing nash equilibria of action-graph games. In: Proceedings of UAI (2004)
11. Clempner, J.: Modeling shortest path games with petri nets: a lyapunov based theory. Appl. Math. Comput. Sci. 16(3), 387–397 (2006)
12. Kanovich, I.M.: Petri nets, horn programs, linear logic and vector games. Annals of Pure and Applied Logic 75 (1995)
13. Westergaard, M.: Game coloured petri nets. In: Seventh Workshop on Practical Use of Coloured Petri Nets and the CPN Tools (2006)
14. Koller, D., Pfeffer, A.: Representations and solutions for game-theoretic problems. Artificial Intelligence 94(1-2), 167–215 (1997)
15. Turocy, T.L.: Toward a black-box solver for finite games. In: IMA Software for Algebraic Geometry Workshop. Springer, Heidelberg (2008)
16. Genesereth, M.R., Love, N., Pell, B.: General game playing: Overview of the aaai competition. AI Magazine 26(2), 62–72 (2005)
17. Priese, L., Wimmel, H.: Petri-Netze. Springer, Heidelberg (2008)
18. Tagiew, R.: Towards a framework for management of strategic interaction. In: First International Conference on Agents and Artificial Intelligence, INSTICC, pp. 587–590 (2009)

Multi-agent Verification of RFID System

Ali Selamat and Muhammad Tarmizi Lockman

Faculty of Computer Science and Information Systems,
Universiti Teknologi Malaysia,
81300 Skudai, Johor, Malaysia
aselamat@utm.my, mizis_mizis@yahoo.com

Abstract. In supply chain management and asset tracking, identifying an item and its' location is very important. RFID technology is suitable to perform the identification and tracking for better asset management. RFID tags can be embedded and attached to the items and their information can be classified for verification process. This is important to avoid problems such as data lost during data management as well as for time management. Multi-agent technology is useful to be implemented during the verification of RFID system architecture because intelligent agent has the capability to define specific verification process and can interact to each other in order to improve the efficiency of the RFID system. Multi-agent system, in this verification of RFID system architecture has a role in formulating the system taxonomy. Therefore, in this paper, the verification process of RFID system architecture has been discussed and successfully implemented in RFID shopping system. The implementation has been tested and evaluated. The results have been encouraging based on the investigation and verification done on a simulation platform.

1 Introduction

Radio frequency identification (RFID) is an automatic identification method, relying on storing and remotely retrieving data using devices called RFID tags or transponders. RFID tag can be applied to or incorporated into a product, animal, or person for the purpose of identification and tracking using radio waves. A technology called chipless RFID allows for discrete identification of tags without an integrated circuit, thereby allowing tags to be printed directly onto assets at a lower cost than traditional tags [7]. The basic building blocks of RFID environment are reader, tags, antenna, and air interface, together, these elements form RFID infrastructure that provides visibility to tagged items within field of view of the reader's antennas. A reader uses the air interface to transmit control parameter to tag and receive their unique serial number or user identity (UID). Fig 1 shows the illustration of the RFID building block that is similar to the RFID certification textbook [1].

Verification of RFID architecture is defined by listing all the steps and parts in the RFID layer system architecture that need to be verified. There are researches that have been done for RFID that are based on agent technology. One of the works uses agent in its RFID architecture [9] and another work is based on how multi agent architecture for RFID taxonomy can help the developer to implement a systematic

N.T. Nguyen et al. (Eds.): New Challenges in Compu. Collective Intelligence, SCI 244, pp. 255–268.
springerlink.com   © Springer-Verlag Berlin Heidelberg 2009

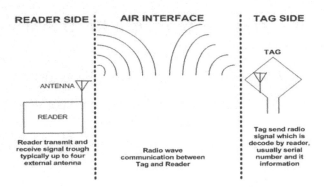

Fig. 1. The illustration of RFID building block

RFID classification scheme [2][10]. There is a need for a systematic, comprehensive, and robust classification scheme in order to enhance the verification process in RFID system. Verification in the system means that the system should conform to its specification [3]. Verification should establish confidence in which it is fit and is depending on the system's purpose, user expectations, and marketing environment. The verification processes need to define the best requirement to optimize the system in term of quality of service (QOS) for RFID system.

Real-time access to item's information is important for decision-making, identifying, and tracking the item in supply chain management. In a store where there are many products of different types, customers or the owner have difficulty in searching the exact location and finding the details information about the products. Lack of system management to manage the products at a specific section makes the items unmanageable and sometimes cannot be identified through user shopping system. Moreover, customers expect shops or stores to be able to provide up-to-date information of products that allows them to better monitor and respond in real-time to strategic objectives and operating constraints.

As a result, efficient management of real-time information becomes a focus and challenge to the supply chain management system. Information must be collected and transmitted in real time so that all the supply chain players have the latest information of product status. When tracking an item, a collected basic data from RFID readers is less valuable without effective and efficient software that can transfer the RFID data to the back-end server [6]. Usually, RFID middleware is implemented to be located between RFID hardware devices and back-end application. It translates requests from the back-end application into commands for RFID devices, receives the responses from RFID devices, and passes them back to the applications. In order to track item location and its environmental information, the RFID middleware in particular needs to solve the problems that occur during data collection from diverse RFID devices [7]. Since widely adopted standardization does not exist, the solution should have the ability to read data from diverse RFID products. Another issue is the reliability of the data transfer to the back-end system. It should be able to robustly transfer data under variably outdoor communication environment, which is very important when items are moving to other locations.

In this paper, we have proposed a multi-agent based RFID middleware system for the process of identifying and tracking items in RFID shopping system. The system promises better performance of multi-agent platform because collected raw data is processed locally by an agent before it is transmitted to the system application. Only useful information is sent back, thus reducing time and cost of data transfer. The structure of this paper is as follows. In section 2, we describe the related works. Next, in section 3, we present the RFID layer services and the taxonomy of RFID. Then, Section 4 describes the multi agent shopping system. Finally, in section 5, we conclude our findings.

2   Related Work

For RFID system, usually the features that the researchers provide are the solution to the services that are important to the system architecture. Hao He [5] gives a definition: "A service is a unit of work done by a service provider to achieve desired end results for a service consumer". Service oriented refers to a concept or approach for distributed computing and communication that think of computing and communication resources as services available on an information and communication technology (ICT) infrastructure [6]. Successful RFID system implementation is more than just technology and interconnection that can traditionally be satisfied by the use of the 7-layer Open System Interconnection (OSI) model. RFID systems are implemented to meet enterprise system needs. To help ensure RFID service alignment, a holistic enterprise architecture approach is required. The weaknesses in current IT architectures, for example, centralized data processing, delays in processing, and point solutions, make them less suitable to handle vast amount of RFID-generated data used in retail operations and real-time decision-making at the edges of the system. Successful RFID systems implementation requires service, multi-agent distributed architectures for verification middleware. The concept of service and agent are complementary because they need each other to perform their role [7].

RFID middleware is a middleware system that translates requests from back-end application into commands for RFID devices, receives the responses from the RFID devices, and passes them back to the applications. RFID middleware applies three-layer system architecture which includes 1) RFID hardware management (RHM) at the bottom layer, 2) RFID data management (RDM) at the middle layer, and 3) RFID application interface (RAI) at the top layer. For example, the former MIT Auto-ID Center mentioned the three layers of Savant, the most dominant RFID middleware infrastructure proposed by EPCglobal, that are named respectively as: the reader interface, Savant services, and the application interface [10][11]. **RHM** provides the connection between the middleware and RFID devices. Since there is no RFID standardization that is widely accepted, here, the bottom layer module must have the capability to hide the detail information of the heterogeneous RFID devices and present a universal data interface for RDM at the middle layer. **RDM** is designed to process collected RFID data from different RFID devices. It also responds to control commands from back-end applications, and events triggered by RFID data. RFIDStack filters and aggregates raw RFID data before disseminating them to utilize the restricted bandwidth efficiently [12]. **RAI** delivers the filtered information to back-end system, and shares the information with other applications. Considering its future market potential, many IT giants have combined RFID technology into their legacy enterprise system. IBM RFID solution is integrated into WebSphere RFID Premises Server V1.1 [14], Microsoft integrates RFID framework into their business process management-BizTalk Server 2006 [13], BEA into WebLogic server [16], Cisco into Application-Oriented Networking [15], Sybase into RFID Anywhere [17], and Sun into Java System RFID Software V3.0 [18].

Comparing with the above approaches, our proposed system is built on agent platform, which means that users can customize the information of tracked item flexibly. It shows benefits for distributed applications, e.g. more efficient bandwidth utilization and load balancing.

3   RFID Layer Services and Taxonomy of RFID System

There are several methods and algorithms used to verify data on the RFID. Issues such as anti-counterfeiting and authentication need to be handled toward verifying the tag on the reader and validating the original data, to confirm that the tag is valid on the system as the tag we want to read. Anti-counterfeiting and authentication technologies can be used to handle the verification of tag scanned by reader and the validation of original data, to ensure that the tag is valid and to confirm that it is really the requested tag. The basic RFID layer consists of physical, communication, services integration, and application layer. Fig 2 shows the RFID layer and the Open System Interconnection (OSI) system layer for data transition.

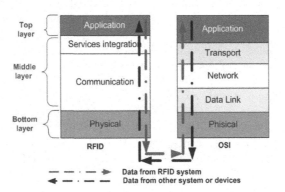

Fig. 2. RFID layer and OSI system layer

RFID physical layer usually consists of hardware such as reader, antenna, RFID tags sensor, and other wireless devices that assist RFID to operate in the air interface. It is equivalent to the physical layer in the OSI reference model in which verification in the system needs to specify type of reader, identification method used, and validation of the device detection at the backend of the system.

The communication service layer includes the networking of both wired and wireless readers and the communication link between readers and tags. The function of this layer is equivalent to the function of data link layer, network layer, and transport layer of the OSI reference. Multi-agent application could be developed within this layer to handle the interactions and coordination among a network of RFID readers [6][8]. This is a layer where the RFID efficiently handles a huge volume of raw RFID data in real-time. It also integrates the back-end enterprise applications with the RFID infrastructure. It is a service-oriented integration layer that becomes an important part of the extended enterprise RFID business solutions. Usually, RFID middleware can be implemented here to directly handle the data management and data filtering [4][9]. Finally, the last RFID layer is the application layer. This layer consists of various back-end enterprise applications that support the system including RFID applications. It requires information (data?) from RFID systems to process and display it as information or results.

The top view of each layer of the RFID system presented in Fig 3 illustrates the verification process done on each RFID system layer. The Physical layer is a key focal point for the RFID industry even though the focus has now moved up the protocol stack and other components in RFID architecture. New developments that include the "reader chip" like the Intel R1000 [9] would inevitably change the concept of an RFID reader. The ability of RFID connection for identification is needed to distinguish their length of coverage, especially the RFID reader whether it is passive RFID or active RFID. Agent based technology should be managed by different reader that has been used in the RFID system.

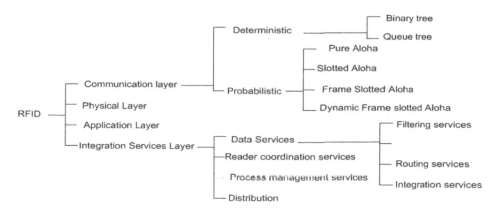

Fig. 3. RFID system layer and services

The communication service is important in any RFID system. It has attracted a large proportion of academics and researchers to improve the speed and accuracy of tag identification. In the communication services, the media access control protocol needs to specify verification and validation by using algorithm and protocol to communicate. Identification, authentication, and privacy are also included in this layer. Fig 3 shows several classifications of media access control protocols for RFID which take part in the verification and validation of the RFID devices and system. The integration services layer is usually referred to as the middleware of the system that provides data filtering, aggregation, routing, and relaying to the appropriate enterprise systems. Verification agent could be implemented here where it can reside on readers or servers. There are number of players in the integration layer market [20].

The application is depending on the system either for identification, tracking, censoring, or other functions. For identification, usually the system verifies the access control, anti-counterfeiting, surveillance, and distribution. When tracking a person or an item attached with RFID tag, the system usually only read the UID on the tag and compare it with a UID that is stored on the database for validation process. Information about the tag that has been installed at the specific target area can be read by the reader for tracking and identification.

4   Multi-agent Shopping System

An intelligent agent is a computer system that is capable of flexible autonomous actions in order to meet its design objectives. Multi-agent based approaches have been proven in many application scenarios to better reduce network traffic as compared to client-server based approaches simply by moving code close to the data (code-shipping) instead of moving large amount of data to the client (data-shipping). Other advantages include overcoming network latency, executing asynchronously and autonomously, and reacting to changes of environments autonomously [4].

4.1   System Overview

Fig 4 shows the architecture diagram of the RFID shopping system. The architecture consists of RFID fixed and mobile reader. The reader is connected to the RFID system through the wireless or wired environment as the physical layer interconnection. RFID middleware is located between the reader and system application because RFID middleware performs data management and filtering of collected information. Usually, system database contains information about the RFID tags attached to the items and unique identity numbers that can differentiate between each of the RFID tags.

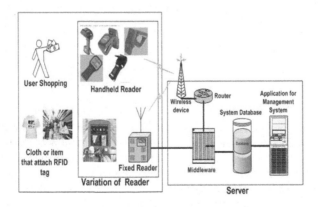

Fig. 4. Deployment diagram of shopping system

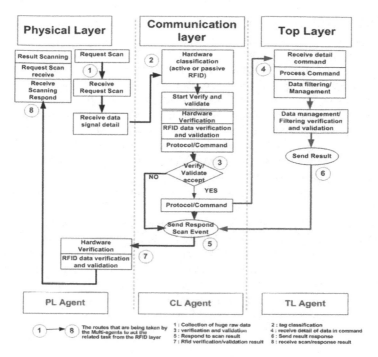

Fig. 5. Agent architecture RFID layers.

4.2 Agent System Architecture

Basically, in RFID system, there are several modules that are based on the RFID system process. The first module includes RFID layer agents that specify the verification process from RFID bottom layer (physical layer) to the upper layer (application layer). The physical layer consists of hardware information where tags or readers are handled by the agent. The second module is RFID communication services agent. This module is responsible to handle and manage interconnection and communication activities by using algorithm and protocol to communicate. Finally, the last module is the RFID top layer agent which is responsible to manage the system integration of data services, such as filtering, routing, integration, and displaying a result to the back end of the system. The information about the activity of item identification and tracking is handled by the agent to ensure the successful process in RFID system. Fig 5 shows the multi-agent interconnection for RFID system that can be implemented.

RFID reader requests information about the scanned item. The information that the reader scans from RFID tags includes the item's information and its serial number or UID. At the RFID physical layer, PL agent manages the scanned data by requesting the scanning event of that particular tag from the communication layer. Then, the CL agent verifies the data. PL agents also respond to the scanned result by doing hardware verification to confirm that the reader only requests the valid data from the RFID system.

The data received from the PL agent follows the CL agent's feature of verification which is important to avoid losing or damaging the data before it is transferred to the RFID middleware. Here, the data is scanned again to perform verification event that has been specified by the CL agent. The scanning process is verified to confirm that the process follows the RFID command protocol. Since the locations of these RFID tags attached to the items are already known, we can easily refer to each set of tags by their serial numbers or UID addresses, according to the location where user or reader scans the tags. This is because UID is unique and the system knows the difference between these RFID tags from the verification of the UID at the communication layer. Here, RFID middleware supports objects checking in and out, reading, writing, filtering, grouping, and routing of data generated by RFID readers. Within this layer, the expected agent application can be developed so that readers can communicate with each other through the interactions of agents.

Finally, at the top layer agent which resides in RFID application, information about the RFID reader that has been scanned is filtered and managed for validating the scanning event at the beginning of the process. Information about the item that is attached with the RFID tag can be identified and tracked by sending it to the system for data management. Then, the result of the identification is submitted to the system and the reader receives the result of the scan stating that the data is successfully read. Interconnection between CL agent and TL agent is required here to confirm that the RFID data is verified by following the actions justified by the agent.

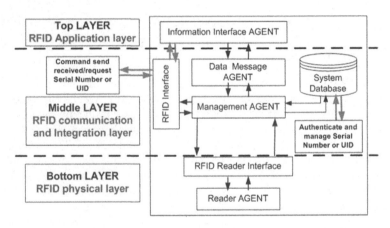

Fig. 6. Multi agent RFID architecture design.

4.3   Agent Shopping System Architecture

As mentioned in the previous section, there are three layer modules developed in the RFID system as stated in Fig 6. The first module which is a physical layer specifies the verification process of physical devices which involved the physical layer agent (PL agent). The PL agent manages hardware information where the tags and reader are handled by the reader agent. The second module is the middle layer (communication and integration layer). This module is responsible to handle and manage interconnection and communication activities using algorithm and protocol to communicate. Data Message agent and Management agent process and handle data or messages at this middle layer. Finally, the last module is the top layer which consists of Interface Agent. This agent is responsible to manage the system integration for data services in which RFID middleware usually supports data management such as filtering, aggregation, routing, and integration to the back end of the system. This module is also responsible to display the result that the system needs. Information about the activity of item identification and tracking is handled by the agent to ensure the application in RFID system is running well.

People who use RFID reader to shop can request information about the item they scan. The information that a user scans includes price, item information, and the number of item that the user has scanned. Reader agent that is requested during the scanning process verifies and validates, for confirmation that the process follows the RFID command protocol. This can be done when an object that is attached with RFID tag is checked into the system by the Management agent and Data agent. Since the locations of these RFID tags attached to items are already known, we can easily refer to each set of tags by their serial numbers or MAC addresses according to the location where the user or reader scans the tags.

Finally, at the Interface agent, the information about the item that user has scanned is filtered and managed. Information about the item that is attached with the RFID tag can be identified and tracked by sending it to the system for data management when the Data or Message agent interacts with the Interface agent at the top layer. Then, the result of the identification is submitted to the system and the user as result or message that contains information about the item they shop.

4.4 Shopping Procedure of Multi-agent RFID System

Fig 7 shows the procedure of the tag scanning event applied in the multi-agent RFID system.

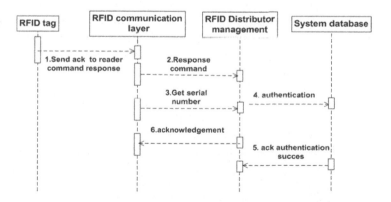

Fig. 7. Shopping procedure for the multi-agent RFID system

The steps involved during the tags scanning process in the multi-agent RFID system are as follows:

Step 1:   Mobile RFID reader obtains RFID tag ID from RFID tag that is attached to the item.

Step 2:   RFID distributor management searches the RFID database management for data repository response command that matches the reader command.

Step 3:   Serial number from the item that is attached with RFID tag is scanned by the RFID distributor management for authentication process by referring to the system database.

Step 4:   Authentication of UID or serial number with private key between the data and the system repository database.

Step 5:   RFID middleware in the RFID system application acknowledges the success respond for the successful validation process.

5   Formal Verification Using CWB-NC New Century Model Checker

During the verification stage, we have used the concurrency workbench of the new century CWB-NC model checker [19] in order to ensure the effectiveness of the agent validations. The CWB-NC model checker performs a depth-first search on the state space and only considers all possible transitions that lead to the states that have not been visited. The CWB-NC model checker also checks the property of the system specification. After the verification process, the formal verification tests the results whether the system is in the correct condition. The CWB-NC model checker will provide identification if the model fails and needs to be modified in order to overcome the uncertainties. But if the errors are not found, the model checker will refine the model and make it reliable to the specification. The verification process will continue until all states are completely checked.

Fig 8 shows RFID system properties specification. This RFID system property is based on RFID system operation environment. It starts when the RFID tags are scanned by the reader to get the UID at the RFID physical layer in which it is managed by the PL agent. Then, when the UID is scanned, the verification and validation process will take place to authenticate and route the information on the tags to the CL agent as stated in Fig 5. The detail information on the tags will be received and the data in a command form of UID will be routed to the next RFID layer managed by the TL agent. Results of the UID commands will be responded by the reader to determine whether the UID reading is successful. If the tags information is successfully read, the tags will be managed by the RFID system where the UID of tags will be verified and validated on other properties belonging to the tags specified in the design requirements. Finally, the system will be ready for the next UID verification and validation.   The specification requirement can be formulated as below:-

Start initial State
  IF NOT scan trace = STOP
    OR *{*   Physical Layer agent (PL agent) receives information UID;
        Communication layer agent (CL agent) receives UID from Physical
        layer agent (PL agent);
        END;
    *}*OR *{*   PL agent sends to CL agent;
        CL agent receives UID from PL agent;
        END;
    *}*THEN
Requirement is violated;
End requirement;

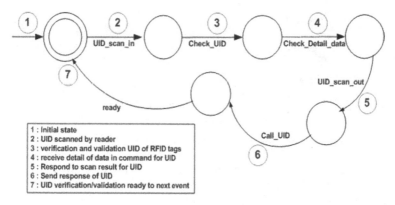

1 : Initial state
2 : UID scanned by reader
3 : verification and validation UID of RFID tags
4 : receive detail of data in command for UID
5 : Respond to scan result for UID
6 : Send response of UID
7 : UID verification/validation ready to next event

Fig. 8. State of RFID system specification for UID checking

 The architecture of the multi-agent and model verification used to check and verify the agent model of the RFID system is shown in Fig 8 and 9, respectively. Based on the interactions of multi-agents in Fig 5, we have further verified the agent's interaction as shown in Fig 9. The physical layer agent (PL agent) will send the input data of the RFID tags with the details containing the tags' serial numbers or UIDs and information to route the data to RFID communication layer where it will be verified by the PL agent. The information will be verified and validated by RFID agents based on the reader-tag check scheme that states the protocol and syntax of commands of the input data as shown in Fig 8.

 The message with the specified syntax received from the PL agent can be accepted by the RFID system. If the information received by the RFID system is invalid, then it will send an acknowledgement to the PL agent to rescan the RFID tags until the data is read in a correct format, otherwise the tags will be blocked. If the information on the RFID tags is valid, then the PL agent will transmit the details of the tags to the communication layer agent (CLA) where formal verification process will be used to check for the suitability of the data in the RFID tags before the authentication process will move to the top layer. At the same time, the CLA will send an acknowledgement to the PL agent informing that the command respond of information is successfully delivered and verified.

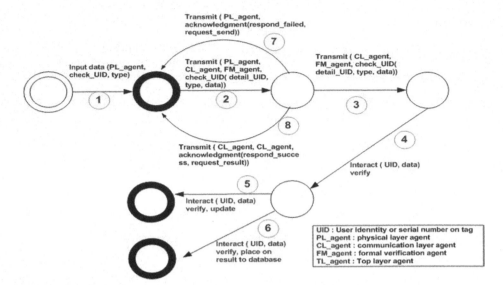

Fig. 9. The interactions of multi-agents in RFID system layer

The RFID system database will be updated after the CL agent completes the verification process. The TL agent plays an important role in ensuring the data transaction is manageable and releasable during the delivery of the tags information. At the early state, the RFID information command message has to be verified by the PL agent in order to ensure the requirement is satisfied according to the specification given by the system. If the specification of the particular state has not been satisfied, the PL agent will generate a counterexample of an error. It decides whether the systems need to do a refinement or an automatic tag information will be sent to the RFID system. The RFID system has used the CWB-NC model checker in order to evaluate the validity of the data and verify the correctness of the data by checking the information contained in the RFID tags.

6 Conclusion

We have proposed a multi-agent architecture for the verification of RFID architecture taxonomy based on tree-node diagrams. It is based on a service, multi-agent RFID layer architecture that integrates the conventional and familiar open system interconnections (OSI) layers with the best practice of enterprise architecture components. The implementation of RFID shopping system has been clearly discussed. Problems and issues of RFID have been handled by understanding the taxonomy of the verification process of the RFID shopping system. Agent based technology is helpful to determine the verification of the RFID architecture on the RFID layer and allows our system to be open, intelligent, and flexible. For the future work, a few RFID network simulations using agent will be studied and used to find better technique and more efficient quality of service (QoS) in RFID system.

Acknowledgement

This work is supported by the Ministry of Science & Technology and Innovation (MOSTI), Malaysia and Research Management Center, Universiti Teknologi Malaysia (UTM) under the Vot 79200.

References

1. Clampitt, H.G., Jones, E.C.: RFID Certification Textbook, 3rd edn., American RFID Solution (2007)
2. Weinstein, R.: RFID, A technical overview and its application to the enterprise. Proceedings of IEEE IT, 27–33 (2005)
3. Paper For Topic: Verification,validation, Certification. Obtain through the internet (accessed 12/6/2008),
   http://www.ece.cmu.edu/~koopman/des_s99/verification
4. Cui, J.F., Chae, H.S.: Agent-based Design of Load Balancing System for RFID Middlewares. In: Proceedings of the 11th IEEE International Workshop on Future Trends of Distributed Computing Systems, FTDCS 2007 (2007)
5. He, H.: What is Service-Oriented Architecture (September 30, 2003),
   http://webservices.xml.com/pub/a/ws/2003/09/30/soa.html
   (accessed 12/4/2008)
6. Le, S., Huang, X., Sharma, D.: A Multi-agent Architecture for RFID Taxonomy. In: Apolloni, B., Howlett, R.J., Jain, L. (eds.) KES 2007, Part III. LNCS (LNAI), vol. 4694, pp. 917–925. Springer, Heidelberg (2007)
7. Foster, I., Jennings, N.R., Kesselman, C.: Brain Meets Brawn: Why Grid and Agents Need Each Other. In: Kudenko, D., Kazakov, D., Alonso, E. (eds.) AAMAS 2004. LNCS (LNAI), vol. 3394, pp. 19–23. Springer, Heidelberg (2005)
8. Pechoucek, M., Marik, V.: Industrial Deployment of Multi-Agent Technologies: Review and Selected Case Studies. To appear in: International Journal on Autonomous Agents and Multi-Agent Systems (2008)
9. Abdel-Naby, S., Giorgini, P.: Locating Agents in RFID Architectures, Technical Report #DIT-06-095 (DEC 2006), Department of Information and Communication Technology, University of Trento, Italy (2006), http://www.dit.unitn.it
10. Leong, K.S., Ng, M.L., Engels, D.: EPC Network Architecture, Auto-ID Center (2004), http://autoid.mit.edu/CS/files/12/download.aspx
11. Traub, K., Allgair, G., Barthel, H., Burstein, L., Garrett, J., Hogan, B., Rodrigues, B., Sarma, S., Schmidt, J., Schramek, C., Stewart, R., Suen, K.K.: The EPCglobal Architecture Framework (2005),
    http://autoid.mit.edu/CS/files/folders/5/download.aspx
12. Floerkemeier, C., Lampe, M.: RFID Middleware Design- Addressing Application Requirements And RFID Constraints. In: Proceedings of the, Joint Conference on Smart Objects and Ambient Intelligent: Innovative Context-aware Services: Usages and Technologies. Grenoble, France, pp. 219–224 (2005)
13. Microsoft Biztalk (2006),
    http://www.microsoft.com/biztalk/biztalk2006r2event.mspx
    (accessed 7/2/2008)
14. IBM, http://www.ibm.com/software/pervasive/
    ws_rfid_premises_server/ (accessed 31/3/2008)

15. Cisco, `http://www.cisco.com/web/strategy/retail/RFID.html` (accessed 15/10/2008)
16. Bea, `http://www.bea.com/framework.jsp?CNT=index.htm&FP=/content/products/weblogic/rfid/` (accessed 31/8/2008)
17. Sybase, `http://www.sybase.com/products/rfidsoftware` (accessed 31/10/2008)
18. Sun, `http://www.sun.com/software/solutions/rfid/index.xml` (accessed 31/8//2008)
19. Concurrency Workbench of the New Century (CWB-NC), `http://www.cs.sunysb.edu/~cwb/` (accessed 15/5//2009)
20. Passive Radio Frequency Identification (RFID)- A Premier for RF regulation by Kevin Powell, Senior Director, Product Development, Matrics Inc., `http://www.rfidjournal.com/whitepapers/download/36` (accessed 15/5//2009)

An Action Selection Architecture for Autonomous Virtual Agents

Lydie Edward, Domitile Lourdeaux, and Jean-Paul Barthès

Laboratoire Heudiasyc, UMR CNRS 6599,
University of Technology of Compiègne

Abstract. Many day-to-day applications involve autonomous agents. When designing autonomous agents, the problem of selecting actions must be considered, as it governs decision making at all times. In this paper we describe how we designed intelligent virtual cognitive agents representing operators performing tasks in a high-risk plant. They must respond to expected as well as unexpected events. The reasoning system controlling the agent behavior is designed to exhibit human behavior. Agents must be able to plan their actions according to their perception, their beliefs and their goals. We developed a planning system that interleaves plan construction and plan execution. The new planner, called AATP[1], produces a plan according to agent goal.

1 Introduction

Autonomous agents should be able to act in the environment following their own decisions. The three main components of autonomy are : perception, planning (decision) and execution (action). In our work, we aim at creating virtual intelligent agents that use all accessible information and knowledge to emulate non trivial behaviors with certain accuracy. The goal is to simulate how humans adapt to deteriorated situations (temporal pressure, physical constraints, stress, tiredness). Our agent planning system addresses the problem of how agents in a particular context are able to choose the appropriate tasks, showing human-like reasoning. We model an agent in charge of managing the world as well as managing other cognitive agents representing operators performing tasks at a high-risk plant. Operators agent must respond to expected and unexpected events.

The abstract problem of how one should proceed from the current state of the world through a sequence of actions to the desired goal state is a planning problem. A planner in IA is a knowledge-based system. Formally the classical planning problem has three inputs according to Hendler [7]:

1. a description of the world in a formal language;
2. a description of the agent goal (i.e. what behavior is desired);
3. a description of the possible actions that can be performed.

The planner output is a sequence of actions that, when executed, satisfy the goal. This classical definition leads to several questions : How do we represent the state of the world ? How do we manage agent goals ? How do we produce the (best) plan to achieve the agent goals ?

---

[1] Autonomous Agent Task Planner.

N.T. Nguyen et al. (Eds.): New Challenges in Compu. Collective Intelligence, SCI 244, pp. 269–280.
springerlink.com

In the paper we first present some previous cognitive and goal-oriented approaches to the problem of action selection. We then introduce our formalism for modeling agent activity. After that, we describe how we represent the world, we present our planner, AATP, and finally discuss some results.

2   Cognitive and Goal-Oriented Approaches

Thomas and Donikian [10] developed a crowd simulator in order to model pedestrian behaviors. They proposed HPTS++ (Hierarchical Parallel Transition Systems) language to model the reactive behavior of their autonomous agents. The language is organized as a hierarchy of automata. One of the interesting characteristics of the model is that the automata allow the description of agent behaviors as well as of their sensors. Information is transmitted to the behavioral module in charge of making a decision according to the behavioral parameters of the pedestrian (prudence, idleness, curiosity, reaction time, distances anticipation).

Grislin [5] proposed a model in order to simulate the behavior of autonomous pedestrians moving in a perturbed environment. Each pedestrian is represented by a configurable autonomous agent (morphology, careful/imprudent behavior, personality). The agent has a perception system and a decision module using a voting mechanism for selecting actions.

Funge [4] proposed a cognitive animation model allowing users to control character knowledge, and the way it is acquired or used for planning. Based on this previous work, Shao and Terzopoulos [9] developed a system representing intelligent virtual pedestrians. Their algorithms aim at emulating appearance, locomotion, and behavior for each character, generating complex collective behaviors. Each pedestrian has an action selection mechanism that determines the action to execute according to its mental states (physiological and psychological social needs).

In the previous approaches only the good procedure are described for an agent activity and tasks are simple. Our goal is different. It is not only to generate expected behaviors but also to generate deviated behaviors and errors linked to agent personality, physical and physiological charcateristics and also to its competencies. The already proposed formalisms aim at describing simple procedures. We need a formalism that allows us to describe complex tasks that can be performed on an industrial plant. We propose a new language, HAWAI-DL[2] [2] that permits such descriptions, and a planner, AATP, that produces agent plans according its personality and supports plan execution or replanning. Unlike the previous approaches, we use a cognitive model to describe the cognitive state of our agents [6].

3   HAWAI-DL Task Representation

In an application proposed by Laird and Rosemblom [8] plans are represented by a graph structure known as a hierarchical task network (HTN). Nodes in the network correspond to tasks, and are represented as STRIPS-style action descriptions [3]:

---

[2] Human Activity and Work Analysis for sImulation-Description Language.

1. Tasks : <Name-of-the-action (arguments)> <Preconditions><Add-list> <Delete-list>
2. Preconditions : <Object><Attribute><Value>

Tasks may be abstract or primitive. Abstract tasks may be decomposed into a partially ordered set of more specific tasks. Primitive tasks are executed without further decomposition. In the network, tasks are connected by a variety of relations. Task relations define the basic hierarchical structure of the network. Ordering relations defines the order in which tasks should be executed.

This formalism is not suitable for describing operator activities in a virtual training environment. To represent activities we add criteria for describing a task more completely. HAWAI-DL consists of a formal description of the task as an arborescent decomposition, starting with abstract tasks (root) decomposed into subtasks to reach actions at the lower level (leafs). Each task is described by several attributes and relations. A task comprises three parts : the task core, the relations with the world and the links with the objects (Fig. 1).

The task core contains all the information needed to describe the task; relations with the world are the world states managing the execution of the task (starting conditions, realization or stop conditions). For example the task "manage the leak" will be realized only if there is a leak.

Scheduling the subtasks of a task is defined by different types of constructors:

1. *Seq : Sequential*
   Subtasks are executed in a given order. The task is finished when all its subtasks are finished or ignored (optional subtasks).
2. *Alt : Alternative*
   Only one subtask is executed. If the execution of the selected subtask fails, then another one is tried.

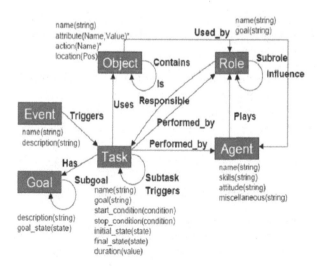

Fig. 1. Representation of HAWAI-DL core

3. *And : Logical And*

Subtasks are executed in any order and do not share data or resources. The task is finished when all subtasks are finished or ignored.

4. *Or : Logical Or*

Subtasks are executed in any order. The task is finished when at least one subtask is finished.

5. *Par : Parallel*

Subtasks are executed in any order and share data and resources. The task is finished when all subtasks are finished or ignored.

6. *Sim : Simultaneous*

Subtasks are executed at the same time. The task is finished when all subtasks are finished or ignored.

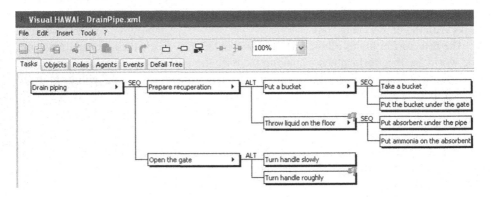

Fig. 2. Extract of the task model

The context of realization of the task is described with precondition attributes. The consequences of the task execution is described with post-condition attributes. For the preconditions we distinguish :

1. Mandatory preconditions: state of the world required to perform the task;
2. Regulatory preconditions: state of the world required by safety regulations;
3. Resources: tools that are required to accomplish the task;
4. Favorable preconditions: the context in which the task is relevant (environment conditions, expertise conditions, time conditions); *Let us take an example. There are two ways to remove a pipe. The environment conditions can be: "bolts are not rusted" or "bolts are rusted". If the bolts are rusted it is better do the task with the following environment condition : "bolts are rusted".*

The description of the activity integrates individual features and situational constraints. A task has conditional and dynamic states. When a precondition is true the task becomes active for this precondition. The conditional states are: mandatory active, regulatory active, favorable active, resource active. The dynamic states are used during plan execution : active/inactive, pending/finished, conflicting, concurrent.

In the remainder of the paper, we will use the following formalism.

$\Omega(\Upsilon)$ = space of tasks.

We represent a task as follows: $\Upsilon = < c_i, Pr\zeta_i, A_i, Ps\zeta_i >$ where :

1. $\Upsilon = Task$
2. $c_i = Task\ constructor$
3. $Pr\zeta_i$: Preconditions (conditions)
   (a) ζ_m :mandatory conditions
   (b) $\zeta_r s$:resources conditions
   (c) ζ_r :regulatory conditions
   (d) $\zeta_f c$:favorable contextual conditions
   (e) $\zeta_f e$:favorable environment conditions
   (f) $\zeta_s c$:safety conditions
   (g) $\zeta_s e$:safety environment conditions
4. A_i : Action
5. $Ps\zeta_i$: Postconditions (conditions)

Preconditions as well as postconditions have the same formalism. Both are conditions. Conditions can be combined with some logical operators (AND, OR). We represent conditions as a quadruple: $\zeta = < op_i, obj_i, prop_i, val_i >$ where :

1. $\zeta = Condition$
2. $op_i = logical operator(=, >, <, and, or)$
3. obj_i : object
4. $prop_i$: object property
5. val_i : value needed for the property

Example : $\zeta_m : (= Pipe_6\ status\ normal)$, $\zeta_f e : (> Pipe_6\ diameter\ 50)$

4   World Representation

Usually the world model is constituted by a set of facts that represent the state of the world, associated with operators (actions). The operators represent the way to transform the world from a state into another state. Such operators can be primitive or complex (that can be decomposed into other operators). Usually the functioning of the system is not explained. One describes only the consequences on the object taking part in an action. This is a problem in particular when objects are linked together by relations.

In our approach, we selected an ontology formalism to describe the world. Objects and actions are concepts. They are made of operations, attributes, is-a links and states (*example: a nut and a screw form a bolt. If the screw is unscrewed and removed then the bolt is unscrewed too*). The ontology representation allows users to have access to various elements of information (*example: what are all the possible actions on an object? What is the link between object i and objet j*). The world state is given by the perception of the environment. Each time the state of the world changes, the model is updated by the manager agent. When an action is done on an object, its state is modified but also the state of the linked objects.

Objects. Objects correspond to all tools, equipment and installation on the modeled industrial site. We establish a hierarchy between objects. Each object is linked to actions regarding this question : what actions can be done with this object? (open, close, wear, knock, unscrew). The main attributes of an object are :

1. State : state of the object according to its role (*open/close, plug/unplug*)
2. Status : functioning state of the object (*normal, broke, jammed*)
3. Position : position in the environment
4. Property : owner, weight, height, diameter
5. Main action : main function of the object (*screwdriver : unscrew*)
6. Secondary actions : other actions that can be done with the object (*screwdriver : knock, open a box*)

Actions. Actions are represented by methods that have preconditions and postconditions to apply when the action is successful or when it is a failure. Actions are modeled as follows :

1. Action code : this code identifies the action when a message is sent to the virtual environment
2. Rules : An if-then rule determines if the action can be applied
3. Target Object : An action belongs but not necessarily to a specific object (*unscrew bolt$_5$*)
4. Main resources : objects/tools needed to accomplish the action
5. Secondary resources : objects that can be used to do the action instead of the main resources.
6. Postconditions : effects on the world

Events. An event indicates an internal or external changing state (bad weather, fire, alarm, leak) of the environment (world). It is not (always) a consequence of the agent actions but it influences the agent activity (goals). Events are linked to a task and tasks can be triggered by an event.

1. Name : Name of the event
2. Auditory : if the event is an audio event (*alarm, explosion*)
3. Visual : if the event is an visual event (*fire, light*)
4. Triggered task : the task triggered by the event
5. Triggering task: the task triggering the event

5   AATP

We developed a planning system which interleaves plan construction and plan execution. AATP produces a hierarchic and ordered plan based on HAWAI-DL. Fig. 3 shows the architecture of our planner. According to the goal of the agent, the planning process produces a plan composed of tasks retrieved from the space of available tasks. Then the execution process executes the plan, it uses the state of the world (to check preconditions) and cognitive rules designed to take into account the state of the agent in the reasoning (Example : a tired agent will not do a task with a high physical load or if it does it, the action will not succeed).

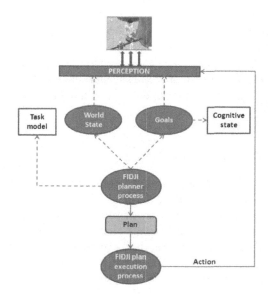

Fig. 3. AATP Architecture

Definition 1. *A goal g is a given state of the world that the agent has to reach. A goal has the same formalism as a condition* $g_i := (op_i, obj_i, prop_i, val_i)$.
For example : the task "unscrew bolt$_5$" will be assigned to the goal
$g_1(= bolt_5\ state\ unscrew)$. *A goal can be a desired state for the agent, for example* $g_2(= agent - thirst\ value\ 0)$.

Definition 2. *A plan is a sequence of actions responding to a goal g by defining a path in the tree of tasks* $\Omega(\Upsilon)$.

5.1 Perception

Each agent has its own representation of the world as a space of concepts with relations and attributes. As fot he world, e use the same ontological formalism. A continuous process informs the agent about what is in its field of vision. The agent perception is what they effectively see, sense, or hear. Consequently, agents do not have a full representation of the environment. If $agent_1$ is doing an action that modifies the state of an object, $agent_2$ will have access to this information only when the object is in its perception field or when it will effectively interact with the object (try do an action). We add constraints linked to the agent state influencing its perception. For example, a stressed agent will not perceive everything in the environment, a tired agent will not perceive all the events or will not pay attention to the perceptive feedbacks (e.g. it will not verify the result of an action after its execution).

5.2 Goals

The goal module retrieves information from the perception module. All events modify the agent goal; the updating goal module creates a new goal that is stored in the agent goal base. This goal will be satisfied according to its priority (a fire for example) or when the agent is available. It could be a goal linked to the task (*example : if the agent decides to do a task and it does not have the appropriated tools a new goal will be to find a tool*). It can also be a goal linked to an agent physiological state. We add three physiological characteristics (thirst, tiredness, hunger) in order to give our virtual agent some more interesting behaviors. If the agent thirst reaches a critical value, its goal will be to "have a drink".

5.3   Cognitive States

We rely on the the COCOM cognitive model proposed by Hollnagel [6]. The model enables to describe, what he called, the control mode of an operator that depends on the time pressure. Hollnagel defines four types of control modes associated to time zones in which an agent can operate: (i) a strategic control mode, in which the agent has a wide time horizon and looks ahead at higher-level goals; (ii) a tactical control mode characterizing situations where performance more or less follows a known procedure or rule, in which the user often chooses the simplest situation and can therefore not respect the safety constraints; (iii) an opportunistic control mode, in which the next action reflects the salient features of the current context (only little planning or anticipation is involved); and (iv) a scrambled control mode, in which the next action is in practice unpredictable or random. Thus depending on the control mode, the operator will plan broadly and choose the actions more adapted to the situation or plan to a more limited degree and compromise on safety aspects to gain productivity. Agent internal states are updated after each action and a new control mode is computed [2].

5.4   Plan Execution Process

According to agent personality (caution, optimism), knowledge level (inexperienced or expert), physiological characteristics (stress, strength, tiredness) and its control mode (strategic, tactic, opportunist or scrambled) the process will choose the task to perform. In the interpretation of the preconditions we add rules linked to the cognitive state of the agent (example : favorable environment conditions are evaluated only by expert agents). Preconditions are tested typically by checking the status (value) of the concerned object. This can be done by sending a query to the manager agent. We describe below how preconditions are evaluated during the execution.

1. Mandatory conditions ζ_m : If they are not true, the task cannot be executed. Agents are trying to make them true by replanning. The state of the current task is settled to pending. The process looks for tasks that respond to the following condition: their postconditions must match the mandatory preconditions of the current task. The founded tasks are executed as a general OR-task.
2. Regulatory conditions ζ_r : If an agent is in a strategic or tactical state, expert and cautious it will check the regulatory conditions. They are not blocking like the mandatory conditions. But in the case of an ALT-task, if there is no tasks to execute, the planner can try to solve the regulatory conditions.

3. Favorable environment conditions $\zeta_f e$: These conditions are interpretable only by expert agent. If they are not true the agent will not execute the task. Example of environment conditions : $\zeta_f e$: $(> Pipe_6 \; diameter \; 50)$

4. Favorable contextual conditions $\zeta_f c$: They are not blocking. They are useful in the case of an OR-task, if all the tasks can be executed then these conditions allow to choose the appropriate task depending on the context. Example of contextual conditions : $\zeta_f c$: $(= weather \; state \; rainy)$

5. Safety conditions $\zeta_s c$: A safety task will be executed by an agent depending on its control mode (strategic, tactical) and its competencies (expert, inexperienced). Same as the $\zeta_f e$, only expert agents will evaluate the safety environment conditions $\zeta_s e$.

6. Resources conditions $\zeta_r s$: First we check if the agent has the resource in its tool box or in its pockets. If the agent does not have the resource, if it is not in a hurry it will look for a plan to get the resource. It has also the possibility to ask another agent (search or lend). If an agent is in opportunistic mode it will reason by analogy. This means it looks for a similar tool that can be used to do the task.

Table 1. Planning Algorithm

Algorithm : Task-Execution-Algorithm(Υ_i)

---

1. Check $\Upsilon(\zeta_m \cap \zeta_r s)$
2. If $\zeta_m \cap \zeta_r s = \text{true}$
   Then If $(agent(cautious) = true) \cap (agent(mode) = strategic \cup tactical)$
   Then Check ζ_r
   Else Solve-mandatory-task-conditions(ζ_m)
   Solve-resource-condition($\zeta_r s$)
3. If $agent(expert) = truc$
   Then Check $\zeta_f e$
4. Check $\zeta_f c$
5. If $\Upsilon(safety) = true$
   Then If $(agent(expert) = true) \cap (agent(mode) = tactical)$
   Then Check $\zeta_s c \cap \zeta_s e$

5.5 Replanning

Replanning may be triggered in response to unexpected events. If the effect of an unexpected event creates a default in the existing plan, the planner modifies the plan. The plan can be repaired by extending it (by adding tasks), or retracting some portion and replanning. However it is not always possible to retain the original plan structure. If the unexpected event is incompatible with the existing plan structure, this structure must be retracted. To find a new plan, the planner will find a corresponding tasks in the space of tasks, especially in what we call hyperonymics tasks, i.e. tasks that do not have links with the main root (independent tasks).

6   Results

6.1   Virtual Environment

We developed a virtual environment corresponding to our working scenario: a pipe substitution operation in a high-risk industrial plant. The operations comprises three phases: (1) prepare and secure the intervention zone, (2) dismantle the mono-pomp group, (3) assemble the new mono-pomp group. The characters were modeled with 3DSMAX and the virtual environment designed with VIRTOOLS.

6.2   Planning Execution

We developed our multi-agent system (manager agent, operators agent) on the OMAS multi-agent platform [1]. Each operator agent has an interface agent called MIT in

Fig. 4. Virtual environment for training

	T	State of the world	Agent state	CM	Actions	Action Selection
A1	τ_0	= handle05 state rusted	Inexperienced	S	/	
A2	τ_0	= handle05 state rusted	Expert	S	/	
A1	τ_1	= handle05 state rusted	Inexperienced	S	Take bucket05	
A2	τ_1	= handle05 state rusted	Expert	S	Put Absorbent2 under Pipe06	
A1	τ_2	= handle05 state rusted	Inexperienced	S	Put Bucket05 under Pipe06	
A2	τ_2	= handle05 state rusted	Expert	S	Put Ammonia2 on Absorbent2	
A1	τ_3	= handle05 state rusted	Inexperienced	S	Turn handle05	
A2	τ_3	= handle05 state rusted	Expert	S	Turn handle05	

Fig. 5. Planning Execution

charge of sending all information about the agent states and the value of its different characteristics to a module interface developed with QT. The values are used to show the evolution in real time of the agent states and plans. We tested our algorithms on a small part of the scenario (Fig. 2).

Agent has to do the task "Drain the pipe". This order matches the goal $g_1 (= pipe_0 6\ state\ drained)$. The planner generates a plan to achieve the goal. Fig. 5 illustrates the action mechanism for four steps. Yellow tasks are *active*, greens are *ended* and red are *in failure*. At step 1, $Agent_1$ chose to prepare the recuperation by putting a bucket under the $Pipe_0 6$. The other choice is a task related to safety, inexperienced agent cannot do this type of task. $Agent_2$ is an expert, and chooses the second way that is to throw absorbent and ammonia on the floor. The second main task of the scenario is to open the gate so that the liquid can flow. There are two ways of doing the task : turn the handle of the gate slowly or roughly. $Agent_1$ chose the first alternative that is to turn slowly. As the handle is rusted the task does not succeed. In the remainder of the simulation, the agent looks for a task for achieving the goal $g_2 (= handle_0 5\ state\ unblock)$. $Agent_2$ chose the same task but it also tried the other alternative related to safety i.e. to turn roughly the handle. It does not succeed and its agitation increases. It will also looks for a way to unblock the handle. Depending on the environment states a leak can appear (this decision is taken by another module).

7   Conclusion and Outlook

In this paper we presented an action selection architecture, called AATP, for virtual autonomous cognitive agents. We described how AATP produces and executes a plan. It takes a goal to achieve in entry, and produces a plan composed of tasks. Plan execution depends on the state of the world and also on state of the agent. We designed a formalism to express actions that can be done in the environment and also tasks. By introducing different categories of preconditions and cognitive rules we are able to generate deviated behaviors according to an agent personality and competencies. The next step of our work will be to test if our model still behaves in an acceptable way with more than 100 tasks to perform. The communication with the virtual environment will be also enriched in order to have a complete interaction between our modules and to validate our system.

In a multi-agent system, agents often perform actions unexpectedly and independently. When each agent is planning on its own without communicating and coordinating with the other agents, each agent has to solve some conflicting problems. This study will be also conducted in the next step of our work.

References

1. Barthès, J.-P.: OMAS - A Flexible Multi-Agent Environment for CSCWD. In: CSCWD 2009, Santiago, Chile, April 23-25 (2008)
2. Edward, L., Lourdeaux, D., Lenne, D., Barthes, J.P., Burkhardt, J.M.: Modelling autonomous virtual agent behaviours in a virtual environment for risk. In: IJVR (2008)
3. Fikes, R.E., Nilsson, N.J.: STRIPS: a new approach to the application of theorem proving to problem solving. Artificial intelligence 2 (1971)

4. Funge, J., Tu, X., Terzopoulos, D.: Cognitive Modeling: Knowledge, Reasoning and Planning for Intelligent Characters. In: Proc. of SIGGRAPH 1999, Los Angeles, CA (1999)
5. Grislin Le Strugeon, E., Hanon, D., Mandiau, R.: Behavioral Self-control of Agent-Based Virtual Pedestrians. In: Ramos, F.F., Larios Rosillo, V., Unger, H. (eds.) ISSADS 2005. LNCS, vol. 3563, pp. 529–537. Springer, Heidelberg (2005)
6. Hollnagel, E.: Time and time again, Theoretical issues in Ergonomics Science 143–158 (2002)
7. Hendler, J., Tate, A., Drummond, M.: AI planning: systems and techniques. AAAI Press, Menlo Park (1990)
8. Laird, J.E., Rosenbloom, P.S.: Integrating execution, planning, and learning in Soar for external environments. AAAI Press, Menlo Park (1990)
9. Shao, W., Terzopoulos, D.: Autonomous Pedestrians. Graphical Models 69(5-6), 246–274 (2007)
10. Thomas, G., Donikian, S.: Virtual humans animation in informed urban environments. In: Computer Animation, pp. 129–136. IEEE Computer Society Press, Philadelphia (2000)

Agent Architecture for Criminal Mobile Devices Identification Systems

Ali Selamat, Choon-Ching Ng, Md. Hafiz Selamat,
and Siti Dianah Abdul Bujang

Universiti Teknologi Malaysia, 81310 UTM Skudai, Johor, Malaysia
aselamat@utm.my, choonching5u@gmail.com
http://www.sps.utm.my/iselab/

Abstract. Current techniques of language identification are based on a method that assigns one or more textual documents into a set of predefined languages that are relevant to page contents. In this paper, we proposed an agent architecture that is used in criminal mobile devices identification systems. It is based on the usage of a software agent to process at least one document by using a dictionary that belong to a set of languages in order to determine the type of language features to be used in the text preprocessing module. Then, the agent will map at least one of the documents with the content of the dictionary in order to identify the languages used in the text by the language identification agent. Finally, the digital forensic agent will check the potential criminal short messages through the predefined keyword repository of corresponding language. Form our experiments, the agent architecture has been able to identify correctly the types of languages written in the short text messaging (SMS) system.[1]

1 Introduction

This chapter discusses how an agent based language identification tool can be used in the real time application [1]. Nowadays, the Short Messaging Services (SMS) is a common technology all around the world. People use it to send a message or information to other users in a simple and fast way. In Malaysia, SMS is used in many languages such as Malay, English, Tamil, Mandarin and others. Moreover, there are many foreign workers from various countries like as Philippine, Myanmar, Bangladesh, Thailand and so on. However, this technology increases several potential criminal activities. For example, threats and spam SMS [2,3]. In this work, agent architecture is developed in order to detect the potential criminal activities of mobile devices. The input to the system will be the text, and this tool will automatically identify the language [4]. The suspicious input data will then be processed for keyword matching detection, based on the corresponding library. The developed tool is able to identify multi languages such as Arabic, Persian, Urdu, Japanese, Indonesian, etc. From the initial experiments, we have found that the agent architecture of identification system can perform the language identification and digital forensic in a good manner.

---

[1] The portion of this research has been certified under Intellectual Property Corporation of Malaysia with the application number **PI 20084793** on 26 November 2008.

N.T. Nguyen et al. (Eds.): New Challenges in Compu. Collective Intelligence, SCI 244, pp. 281–290.
springerlink.com   © Springer-Verlag Berlin Heidelberg 2009

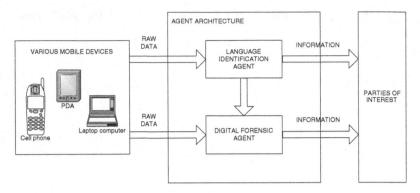

Fig. 1. Agent architecture for criminal mobile devices identification systems

2 Background

In computer science, a software agent is a piece of software that acts for a user or other related programs in an of agency. Such action on behalf of implies the authority to decide which (and if) action is appropriate. The idea is that agents are not strictly invoked for a task, but activate themselves [1]. Currently it is time consuming to do an investigation on mobile devices especially on short messaging services. Currently, it is an important issue to develop a method for efficient information searching and retrieval. In order to achieve this, efforts for reducing tremendous data for investigation should be taken into serious consideration [1]. The practice of digital forensic is new. When computer became common in homes and businesses, the police more and more often came across computers which contained forensic evidence. Thus, police organizations have realized the need for establishing special police units to handle electronic devices. United States of America (USA) was the first to do so, when the FBI established the Computer Analysis and Response Team (CART) on 1984. Later, a similar unit was established in Scotland Yard, United Kingdom (UK).

On criminal activities, languages play an important role. With the rapid growth of computer and network systems in recent years, there has been a corresponding explosive cyber-crime activity [5]. The most commonly seen crime involves: hacking into computer systems, computer viruses, groundless allegation (rumor), threats, etc. Digital forensics can be defined as the practice of scientifically derived and proven technical methods and tools toward the preservation, collection, validation, identification, analysis, interpretation, documentation and presentation of after-the-fact digital data derived from digital sources, for the purpose of facilitating or furthering the reconstruction of events as forensic evidence [6]. Moreover, it can be extended to forensic examination of mobile phones because the proliferation of mobile phones in society has led to a concomitant increase in their use in and connected to criminal activities [7]. Conventional techniques do not provide a further method, which with little expenditure could help identify reliably the language in which the text is composed, even in the case of short texts. Similarity measurements are used for potential threat determination in document. Suspects or relevant data that are identified includes data that are identical to or similar to the extracted unknown data. If there are suspect data, the system transmit an alert to the parties of interest or generates the critical report on a storage device.

Language identification task can be commonly divided into two classes, which are spoken and written language . Spoken language identification method has to choose signal processing, however language identification from text is a particular symbolic or term processing task [8]. In our work, we are focusing on language identification from written words rather than from speech. The main reason for the apparent lack of activity in written language identification is probably that it is not considered as a difficult problem [9,10]. This might be true if the amount of written text available in the identification stage is large enough and computational of very few words such as names and constraints are embedded in the systems [10].

The globalization of the communication industry has resulted with producing various approaches of multilingual system. It is a core technology in many applications such as multilingual conversational system, automatic transliteration system, multilingual speech recognition, text categorization and spoken document retrieval [11,12]. Generally, the phonetic transcription of terms must be obtained online from written text using either rules-based or some other kind of pronunciation models [8]. Most pronunciation models depend on clearly expressed knowledge of the language, and therefore it must be recognized by the system in order to enable the match model. Language identification is often based on only written text, which produces an interesting problem [8]. User intervention is always a possibility, but a completely automatic system would make this phase run easier, and increase the usability of the system. Searching for names in a phonebook using voice input is a good example of a system requiring language identification [8]. Perhaps it can be extended to short messaging services (SMS) or multimedia messaging service (MMS).

Accordingly, there is a need for the suspect data to be identified automatically and precisely, without human intervention.

3   Agent Architecture

The invention relates to a computer software application for enterprise data management. Specifically, it relates to a system and method of automatic language determination for potential tracks in computer systems. Our concern is to identify potential threats of digital textual data from various mobile devices such as PDA, mobile, Bluetooth, laptop, etc. The raw data or text is a combination of a token of letters. N-grams have been applied on raw data for producing features of machine-learning methods due its reliability in language identification. Similarity measurements are used for potential threat determination in document. Suspect or relevant data that are identified includes data that are identical or similar to the extracted unknown data. If there are suspect data, the system transmits an alert to the interested parties, or generates the critical report on a storage device. Parties of interest are local authorities, researchers, news agencies, policy makers, commercial company, etc. In this manner, the suspect data are identified automatically, without any human intervention. Therefore, we have developed the agent architecture as shown in Fig. 1.

3.1 Language Identification Agent

First of all, raw data will be extracted from various mobile devices and then fed into language identification agent. Only digitalized data will be extracted, others will be discarded. The system will check whether the inputs belong to textual data. If the answer is yes, then it proceeds to the following step of code point conversion, otherwise the agent will stop the whole process. The agent then converts the extracted data into the Unicode format. Unicode is a standard encoding method, which has been used worldwide for any language data processing. For those non unicode letters, the agent will remove the irrelevant or unrecognized letters that exist in a particular content. This is done to avoid noisy data from getting mixed with the useful data. The following step is Unicode ranging. Unicode ranging is used to identify the category of particular raw data in Unicode such as Latin, Arabic, Cyrillic, etc. Number of occurrence of each letters in that particular raw data will be summed. The maximum probability is used to select the script language of the particular content. Feature selection takes part by using n- grams for selecting the appropriateness features according to the script language that have been identified. Combination of unigram, bigrams, and trigrams are used for producing the reliable features of input patterns. The following step is to ascertain the language by using machine -learning method, fuzzy ARTMAP. It is based on winner take all concept in both the learning and testing process. Learning is the process of training the model based on the samples have been identified in a particular language. Then, that model is

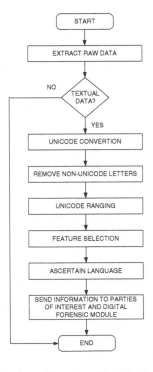

Fig. 2. Action of language identification agent

used to ascertain the language in real time application. When the language of a particular content have been identified, report will be generated and stored in a storage device for further reference by digital forensic module and parties of interest. Finally, the tasks of the language identification agent end.

The algorithm used by language identification agent is stated as follows:

```
Start
Check format of original document
If the original document is not a textual document
    then the document is discarded and the end
Read data from original document
Check the encoding utilized in the original document
Convert the original document to unicode document
Remove all non-unicode letters from unicode document
Segment the document according to the boundary of unicode
Extract features from the segmented document
Identify the language of particular segmented document
The end
```

3.2 Language Identification Agent System Architecture

Referring to Fig. 3, the LIA_M will monitor and extract the raw data from any mobile devices, respectively will identify the textual data to the UCA_{Arabic}, $UCA_{Persian}$, $UCA_{Japanese}$ and until UCA_n. The number of UCA will be based on the multi language of the corresponding system in order to ensure the text data can be handled in parallel. However, if the LIA_M has identified the non textual data in the raw data, the process will be automatically be stopped by the SA. Next, for each parent in UCA will have the number of RA (RA_1, RA_2, RA_3 until RA_n) that will remove the irrelevant or unrecognized letters that exist in a particular content. The number of letters that have been removed by the RA_s will be combined by the existed URA for ranging the unicode. The process will be repeated until the last agent, FSA takes part to select the appropriate features. The total number of agents in LIA system able to create the less number of the total time (t) to retrive the information to the parties of interest.

3.3 Digital Forensic Agent

Digital forensic agent is used to identify the potential threats in a text or message. Similarity measurements are used for potential threat determination in document. Suspect or relevant data that are identified include data that are identical or similar to the extracted unknown data. If there are suspect data, the system transmits an alert to the parties of interest, or generates the critical report on a storage device. In this manner, the suspect data are identified automatically, without intervention by a human being.

For the digital forensic module, it starts with the system receiving textual data and reading its language from the storage device. The tokenization process will divide the document into a number of tokens. Space in the document is the identity for spread out

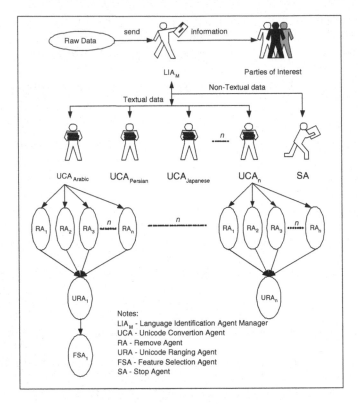

Fig. 3. Architecture of language identification agent approach

the word from the documents. Each token will be held in temporary memory for similarity measurements. Before the process is continued, each word in the token will go through process case folding. All the case of the words will change to small letters. The corresponding language library is read from the storage device for similarity measurements. The content of the language library is already pre-defined based on the analysis that has been done on the potential threat words in the document. Each of the token will be compared with the key words in the library for similarity measurements. If the number of token matched is more than the threshold, then the process will proceed to next step. Otherwise the digital forensic process will be ended. The following step is report generation. The relevant results will be recorded in the storage device or be alerted to the parties of interest. Finally, the process of digital forensic will end.

The algorithm used by digital forensic agent of criminal mobile devices identification system is stated as follows:

```
Start
Read data from original document and its language
Segment the data into token by word
Read corresponding language's semantic library
Calculate the score based on similarity measurement
```

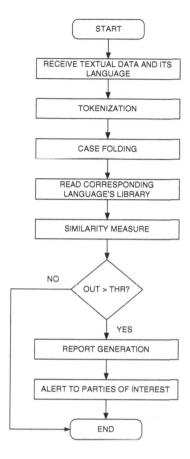

Fig. 4. Action of digital forensic agent

```
If score bigger then threshold
    then generate the report of particular message
    Alert to party of interest
The end
```

3.4   Digital Forensic Agent System Architecture

Referring to the Fig. 5, the DFA will read the language in the raw data or textual data (from LIA) in the storage device. Next, the DFA will send it to TA in order to divide the document into a number of tokens. Before the identification similarity process continued, the process will be segregated by several CFA (CFA$_1$, CFA$_2$, CFA$_3$ and until CFA$_n$). A number of CFA$_s$ will change all the case to small letters and read the corresponding language library in a minimum time. The similarity measurement will be compared by the SA based on the threshold weight. SA will dispatch the result to RA for generate alert reports if the process in identified has potential threats in a text or

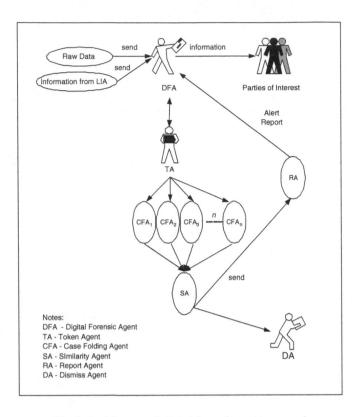

Fig. 5. Architecture of digital forensic agent approach

message. The alert report will be send to DFA back to alert the parties of interest. However, if the process is null, the process will be dismissed by the DA.

4   Result Analysis and Discussion

In general, most of the works are focusing on English SMS spam filtering instead of minority languages. It will lead to highly misclassification of spam or non-spam messages. Some of the researchers suggested doing a language conversion from other languages into English as a preprocessing step in spam filtering. Some of the particular words are hardly to be translated into other language as it cannot represent the real meaning of a particular word. Therefore, language identification module is used as a preprocessing step is important to correctly identify the potential criminal SMS. As a result, we have proposed agent architecture with a combination of language identification agent and digital forensic agent. Once the language of a particular message has been determined, then the digital forensic agent can further process in order to find out the potential SMS spam.

We have selected 500 samples data for testing the language identification agent. The algorithm used to generate the dictionary was the n-grams algorithm proposed by the

[13]. In the initial experiment, we have selected English and Indonesia as the desired languages. For each language dictionary, the *n*-grams generate the particular semantic in the process training. With such dictionary, we have used it to compare with the tested document and find out the argument minimum. The Table 1 shown the result have been done on this experiment. It is noticed that the language identification agent able to determine English and Indonesia precisely. The morphology used on both languages are quite significant different, so it can differentiate the language very well, even on single sentence only. However, we have found that the *n*-grams algorithm hardly to be used to determine too close languages like Malay and Indonesia or the Arabic script languages.

Table 1. The performance of language identification agent using n-grams algorithm

Number of Training Documents	101-500	201-600	301-700	401-800	501-900
Number of Testing Documents	1-100	101-200	201-300	301-400	401-500
Unigram	99.8%	100%	100%	100%	100%
Bigrams	100%	100%	100%	100%	100%

In the future work, we intended to solve the constraints found on the language identification agent and further investigate on how the language identification agent will give an impact on the digital forensic agent. It is assumed that, the correctness of the language identification will directly affect the performance of the digital forensic process.

5   Conclusion

In this work, we have proposed an agent architecture for criminal mobile devices identification systems. It comprises of two agents: language identification and digital forensic. It is time and cost consuming if the investigation of digital raw data involves more than a million units. It is desirable to automate the process by using the agent architecture.

Acknowledgment

This work is supported by the Ministry of Science, Technology & Innovation (MOSTI), Malaysia and Research Management Center, Universiti Teknologi Malaysia (UTM), under the Vot 79200.

References

1. Selamat, A., Selamat, H.: Analysis on the performance of mobile agents for query retrieval. Information Sciences 172(3-4), 281–307 (2005)
2. Cormack, G.V., Hidalgo, J.M.G., Snz, E.P.: Spam filtering for short messages. In: Proceedings of the sixteenth ACM conference on Conference on information and knowledge management, Lisbon, Portugal, pp. 313–320. ACM, New York (2007)

3. Mishne, G., Carmel, D., Lempel, R.: Blocking blog spam with language model disagreement. In: Proceedings of the First International Workshop on Adversarial Information Retrieval on the Web (AIRWeb), Chiba, Japan (2005)
4. Selamat, A., Ng, C.C.: Arabic script language identification using letter frequency neural networks. International Journal of Web Information Systems 4(4), 484–500 (2008)
5. Nykodym, N., Taylor, R., Vilela, J.: Criminal profiling and insider cyber crime. Digital Investigation 2, 261–267 (2005)
6. Enck, W., Traynor, P., McDaniel, P., La Porta, T.: Exploiting open functionality in sms-capable cellular networks. In: Proceedings of the 12th ACM conference on Computer and communications security, pp. 393–404. ACM, New York (2005)
7. Mellars, B.: Forensic examination of mobile phones. Digital Investigation 1, 266–272 (2004)
8. Hakkinen, J., Tian, J.: N-gram and decision tree based language identification for written words. In: Proceedings of the IEEE Workshop on Automatic Speech Recognition and Understanding, ASRU 2001, pp. 335–338 (2001)
9. Schultz, T., Waibel, A.: Language independent and language adaptive large vocabulary speech recognition. In: Proceedings of the Fifth International Conference on Spoken Language Processing, ISCA, pp. 1819–1822 (1998)
10. Schultz, T., Kirchhoff, K.: Multilingual speech processing. Academic Press, London (2006)
11. Li, H.Z., Ma, B., Lee, C.H.: A vector space modeling approach to spoken language identification. IEEE Transactions on Audio, Speech, and Language Processing 15(1), 271–284 (2007)
12. Selamat, A., Omatu, S.: Web page feature selection and classification using neural networks. Information Sciences 158, 69–88 (2004)
13. Cavnar, W.B., Trenkle, J.M.: N-gram-based text categorization. In: Proceedings of the 3rd Annual Symposium on Document Analysis and Information Retrieval, Las Vegas, Nevada, USA, pp. 161–175 (1994)

A Trading Mechanism Based on Interpersonal Relationship in Agent-Based Electronic Commerce

Tokuro Matsuo, Takaaki Narabe, Yoshihito Saito, and Satoshi Takahashi

Graduate School of Science and Engineering, Yamagata University
4-3-16, Jonan, Yonezawa, Yamagata, 992-8510, Japan
matsuo@tokuro.net
http://www.tokuro.net

Abstract. Recently, electronic commerce has been increasing and is also a promising field of applied multiagent technology. In the automated electronic marketplace, agents can trade as buyers and sellers. Buyers and sellers agents evaluate each tradable item with each other, and buy/sell the item. Many electronic commerce sites provide high anonymity for sellers and buyers. Even though there is a special relationship between them such as friends in real world, they generally can not know their individual information in electronic commerce. In actual trade in real world, item's price is sometimes affected by the relationship between them. This paper focuses on the trading in which buyers and sellers have an asserted relationship such as colleague in a company. This paper analyzes items' values which is defined by items' prices and the relationships. Also, we propose a trading protocol including the relationship between sellers and buyers based on the analysis. In the protocol, first, the seller and the buyer evaluate the relationship. Then, the successful trader is determined based on synthesis evaluations. Finally, the seller and successful buyers trade with each other in the calculated price. The Advantage of our protocol is that appropriate evaluations of trading partners can be obtained by various attributes based on Multi-Attribute Utility Theory.

1 Introduction

Recently, the network used electronic commerce has been increasing by the diffusion of the Internet. Multigent-based technique is one of the promising method of automated electronic commerce[5][6]. In this paper, we propose a new electronic trading mechanism design of agent-based commerce.

Existing electronic commerce have various forms of dealing. All participants trade based on the trading rule which is defined by each market. The participants evaluate the items and business partners from various factors, and they negotiates on the item price. However, electronic commerce has high anonymity. For instance, even though a special relationship between buyers and sellers as friendships exists , the relationship is not considered by the negotiation and the price decision in anonymous web-based trading.

An item price is determined by negotiations between buyer and seller. However, we consider a situation in which the item price depends on the relationship between traders.

N.T. Nguyen et al. (Eds.): New Challenges in Compu. Collective Intelligence, SCI 244, pp. 291–302.
springerlink.com   © Springer-Verlag Berlin Heidelberg 2009

Items prices are determined by the relationship and the item's common value. Let τ be an indirect value which is determined by negotiation between seller and buyer. We propose a trading protocol in which we apply τ to the electronic economic mechanism.

In this paper, we also propose a trading method, reflected the relationships between buyers and sellers and satisfied the agreed price. In the protocol, first, the buyer and the seller evaluate the relationship with each other. Second, we use Multi-Attribute Utility Theory to determine τ based on the relationship. Then, the item price is calculated. We develop a new dealing form of the electronic commerce which focuses on the interpersonal relationships between buyers and sellers.

The rest of this paper consists of the following six parts. In Section 2, we explain some terms and the theory used as a preliminary by this research. Also, we describe the result and analysis of questionnaire about their trading experience between acquaintances. In Section 3, we propose a mechanism applying τ to an electronic commerce. In Section 4, we show an example of trading in our mechanism. In Section 5, we discuss the properties and advantages of our mechanism, after that, we show our concluding remarks.

2   Preliminaries

2.1   Multi-attribute Utility Theory

The utility is an index that shows the degree of satisfaction of the player, when he/she selects an item from two or more choices[1]. Philosophers who discovered the concept of "utility" define it as a quantitative standard of the individual happiness[2]. However, they did not argue how to measure it. Today, economists regard it as a method for the descriptions of the preference. The preference is the ranked favors of the individual. Generally, when there are two choices a and b, we describe that a player prefers a to b if he/she chooses a. This is written as $a \succeq b$.

Multi-Attribute Utility Theory is used in the situation where players select the utility value from two or more reasons (or factors). In general, each factor is defined independently. Namely, a certain factor is not correlated with other factors. The utility value of choices are shown by adding the utility value of the factor when Multi-attribute linear utility function is employed. However, such factor may be important for traders, otherwise be not important. Factors can be mapped by a certain function, however that is not just an additive function. We employ the Weight-based Multi-Attribute Utility Theory shown by the utility value of choices with $\sum$ (multiplication of each weight of factor and evaluated value of the factor).

When people choose something such as the case of purchasing, they usually choose it based on an intuitive judgment[3]. The criteria of choices are yielded intuitively in actual decision making. Therefore, we should not cause contradiction between choices by using the Multi-Attribute Utility Theory and choices by a judgment known by intuition. Humphreys proved that there is a high correlation during these seal determination, when the factor was six or less by the experiment. Moreover, it was proven that the more seven or more, that is, the factor increased, the more the correlation decreased[4]. A selector accepts using addition, when the factor is few. However, it is difficult to use addition in the case of a lot of factors, since we have to think about a complex relation between amends between factors. Tversky is described for exclusion by the attribute (factor) to show person's behavior[7]. Exclusion by attribute, first, excludes a factor which is a low utility value about a most important factor. Second, it excludes a factor which is a low utility value about a secondarily important factor. Similarly, a selector excludes some choices and chooses the last one.

Thus, the chosen choices change by only slightly different of selector's condition. Weight and the evaluation of the factor are different since the each person's favor and purpose are different. In this paper, we use the Multi-Attribute Utility Theory with these characteristics to the method of evaluating seller and buyer in commodity exchange. This is written at the following as MAUT.

2.2   Preliminary Research

In this paper, we give a preliminary discussion of correlation between the value of items and traders. For example, if a seller trades with a strange buyer, he/she may deal in the items higher price than the trading with his/her friends. Otherwise, if the buyer declares that he/she handles careful of the bought items, the seller might sell the item at a low price. Namely, the item's value for traders depends on the relationship of them, conditions, and situations. To clarify and investigate such features in trading, we does the questionnaire about trading in actual dealings for traders. The investigations are based on 179 university students. In this questionnaire, the questionnaire includes the experience of buying and selling between their friend, and their items' value that affected from relationships. We also make them imagine the situation of item trading with their friends by the self-role-play. The experiments focus on the influences in the "interpersonal business partner and item's value" and "trading partner's character and item's value" give to decision of negotiation partner and shows the result. The following section explains the result of experiments.

Investigation of experiences as a buyer. As a preliminary research, we investigate that people have ever traded with their friends, colleagues, and acquaintances. Totally, there are one hundred and seventy-nine answerers for our question, that is, answerer has any experience of trading of second-hand goods with your acquaintances. As the result of the question, one hundred and nine people answered that they have had an experience of trading at least one. Forty-seven people answered that they have not had the experiences and the rest is unknown. We asked 109 people the situation in which they traded.

They answered all things that applied as follows for the impression at that time. Table 1 shows the result of concrete impressions in trading with their acquaintance. The followings are options that answerers choose. When the impression did not exist in the options, it answered as "Other impressions".

(**impressionA**) I could buy cheaper price than the trading with strangers.
(**impressionB**) I reassured in the trading.
(**impressionC**) I felt not so good although trading with acquaintances.
(**impressionD**) I had a delight trading.
(**impressionE**) I hope again dealings in the future.

Other impressions include "It was easy to negotiate on the price on tradings with acquaintances", "The dealt item's condition was well because he/she considers the credit as a friend", and etc.

Table 1. Investigation 1

Answer	A	B	C	D	E
Total	92	62	4	27	18

Table 2. Investigation 2

Answer	Yes	No	Other
Total	78	77	24

Table 3. Investigation 3

Answer	F	G	H	I	J
Total	19	41	2	44	11

Investigation of experiences as a seller. Table 2 is a result of answer about experiences as a seller, that is, he/she sold the second-hand goods to his/her acquaintances. Table shows that people more than forty percent have experiences selling the items to their friends, colleagues, and other acquaintances. Seventy-eight people answered their utility at the trading. Table 3 shows the result of concrete impressions in trading with their acquaintance. When the impression that did not exist in choices was obtained, it answered as "Other impressions".

(**impressionA**) Buyer could buy higher price than the trading with strangers.
(**impressionB**) I reassured in the trading.
(**impressionC**) I felt not so good although trading with acquaintances.
(**impressionD**) I had a delight trading.
(**impressionE**) I hope again dealings in the future.

Other impressions include "It was easy to negotiate on the price on tradings with acquaintances", "The dealt item's condition was well because he/she considers the credit as a friend", and etc.

Self-Role-Playing. In the self-role-playing simulation, to investigate the situation of trading, we ask answerers in the following conditions whether they want to trade or do not. Conditions include trader's character, attitude, and several other factors in trade and subsequent treatment.

In the question for answerers, we assume the following three traders exist in trading. When they trade with such traders, we asked how they consider and feel and whether they want to sell/buy or do not.

In any case, we assume that the proposed price is the same and that three people in each case are individuals.

[Case1] When judging based on the person who have a different interpersonal relationship.

(Trader A) He/She knows answerer very well, and intimate.

(Trader B) He/She is not intimate for answerer, but physical distance is short like person who lives near answer's house.

(Trader C) Though it is an acquaintanceship, he/she doesn't have relations at all like staff in same company.

From these options, Table 4 shows the result; trader in which answerers want to trade with. Table 5 also shows the result; trader in which answerers do not want to trade with. Each table includes number of answerers that they do/don't want to trade with a buyer/seller.

Table 4. [Case1] The trader whom answerers want to trade as buyer and seller

	A	B	C
As a seller	103	53	25
As a buyer	110	48	21

Table 5. [Case1] The trader whom answerers do not want to trade as buyer and seller

	A	B	C
As a seller	20	7	152
As a buyer	18	9	154

Next, we discuss the trading where trading partner handles the traded items, like that answerer does not favor: For example, the case that traders resale.

[Case2] When judging based on the purpose of dealing and the different character.

(Trader D) He/She is careful handling his/her bought items.

(Trader E) He/She is not careful handling his/her bought items.

(Trader F) He/She purchases the item to make money by resaling.

From these options, Table 6 shows the result; trader in which answerers want to trade with. Table 7 also shows the result; trader in which answerers do not want to trade with. Each table includes number of answerers that they do/don't want to trade with a buyer/seller.

Table 6. [Case2] The trader whom answerers want to trade as buyer and seller

	D	E	F
As a seller	151	24	6
As a buyer	166	13	0

Table 7. [Case2] The trader whom answerers do not want to trade as buyer and seller

	D	E	F
As a seller	8	86	88
As a buyer	1	13	165

Analysis of Preliminary Research. In the preliminary research, we investigated the trading between the people who have relationships . The investigation includes influences of determination of traders and prices for other-side trader from human relationships between them, trading partner's character and temper, and intention of trading. We consider based on the result of investigations as follows.

1. From Table 2 and Table 3, a lot of people want to sell the item to the acquaintance cheaply.
2. From the result of self-role-play, there is a trend that traders want to deal with chummy person.
3. On contrary, the order of importance of dealings falls down the trading with the unfamiliar person.
4. About 20 percent people have a great value about "Easiness of trading" more than human relationships.
5. When an answerer is the case as a seller, most of answerers answered that they want to sell the item to the other party who has the definite reason and carefully uses it.
6. When the answerer is the case of a buyer, most of them answered that they want to buy the item from the other party whose items are in good preservation.
7. Even though the trade has been finished, most of them consider that they want to trade subsequently with creditable traders.

3 Proposal Technique

In this section, first, we explain the indirect value τ. Then, we show the definition and protocol based on τ.

3.1 The Indirect Value τ

From the result of the preliminary research in Section 2, the relationship between buyers and sellers affects the decision of the business partners and the trading price. For example, some traders may want to trade with their buddy-buddy (good friend). Further, they may sell item to him/her at a low price. In this paper, we propose a trading method in which the business matching and the price determination are taken such relationship between buyers and sellers into consideration. We define the indirect (heterogeneous) value between buyer and seller as τ in which we mentioned previously.

3.2 Model

We give some definitions as follows in our proposed trading mechanism.

- We define a seller set as $S = \{s_1, s_2, \cdots, s_i, \cdots, s_n\}$. s_i is ith seller in the set.
- We define a buyer set as $B = \{b_1, b_2, \cdots, b_j, \cdots, b_m\}$. b_j is jth buyer in the set.
- We define an item set as $G = \{g_1, g_2, \cdots, g_k, \cdots, g_l\}$. g_k is kth buyer in the set.
- We define the price R_{gk}^{si} in which seller s_i evaluates item g_k.
- We also define the price V_{gk}^{bj} that buyer b_j evaluates item g_k.

- We define an evaluation attribute set as $E = \{e_1, e_2, \cdots, e_x, \cdots, e_z\}$. e_x is xth attribute in the set.
- We define $e_x[s_i, b_j]$ about that seller s_i evaluates buyer b_j.
- We define the importance degree of the evaluation for the seller as W_{ex}^{si}. W_{ex}^{si} shows the valuation in which s_i gives a weight on e_x.
- Similarly, we define the evaluation from buyer b_j to seller s_i as $e_x[b_j, s_i]$.
- We define the importance degree of evaluation for the buyer as W_{ex}^{bj}. W_{ex}^{bj} shows the valuation in which b_j gives a weight on e_x.
- Trust value T_{bj}^{si} of buyer b_j for s_i is defined by total sum of evaluation values in which the seller evaluates the buyer. $T_{bj}^{si} = \sum_{x=1}^{z} e_x[s_i, b_j] \cdot W_{ex}^{si}$.
- Trust value T_{si}^{bj} of seller s_i for buyer b_j is defined by total sum of evaluation values in which the buyer evaluates the seller. $T_{si}^{bj} = \sum_{x=1}^{z} e_x[b_j, s_i] \cdot W_{ex}^{bj}$.

Figure 1 shows a visual example of trust value of traders. The graph shows the trader's various attitudes of reliability by nonlinear lines. The horizontal axis represents a value of the trust. The vertical axis represents a magnification rate added to the presentation price. The curves of three patterns are shown in Figure 1. These sorts of patterns are employed in our mechanism.

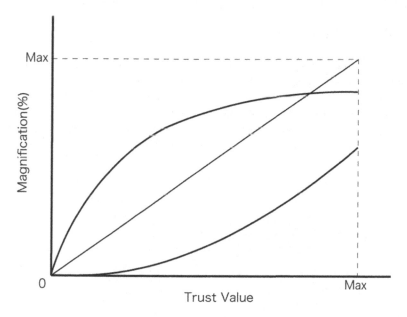

Fig. 1. Example of Trust

- We define the magnification rate obtained in trust value T_{bj}^{si} as $M = f(T_{bj}^{si})$.
- We define the magnification rate obtained by reliability T_{si}^{bj} as $M = f(T_{si}^{bj})$.
- We define the price R_{gk}^{si} in which seller s_i presents buyer b_j in the negotiation. It is multiplied by $M = f(T_{bj}^{si})$ to R_{gk}^{si}.
- We define the price R_{gk}^{bj} that the buyer b_j presents the seller s_i as in the negotiation. It is multiplied by $M = f(T_{si}^{bj})$ to R_{gk}^{bj}.
- We compare with the price between the sellers and the buyers, and decide a successful trader in the negotiation in which the combination of difference of prices $D_{bj}^{si} = |V - R|$ is maximum.
- We define the amount of the final trading price as P, and it is decided by the calculation of $P = R + k(V - R)$. We define k as a magnification rate (trading magnification), which the manager and the traders decide. This magnification rate is decided between $0 \leq k \leq 1$.

3.3 Protocol

In this section, we describe the protocol in our proposed trading mechanism based on τ. In our protocol, the presented price is sealed.

Step1 (The negotiator determination)

First, the seller decides the other party (buyer) who wants to sell items. Second, the buyer who receives the intention answers whether he/she can negotiate with the seller or can not. Then, the buyer and the seller negotiate with each other.

Step2 (Seller's price declaration)

First, seller estimates the price of the item. Second, the seller evaluates the buyer with multi-attribute preferences, and the trust value is put out by the weighted evaluation. Table 8 shows the table of multi-attribute evaluation of buyers. The trust value is dropped to a function shown as the curve in Figure 1. Actually, a lot of patterns of curves exist. To simplify, in this paper, we employ three patterns shown in Figure 1. The magnification rate is multiplied to the price of the item, and the declared price is decided.

Table 8. Multi Attribute Evaluation of Buyer by Seller

$\{s_i, B\}$	b_1	b_2	$\cdots$	b_j	$\cdots$	b_m	W
e_1	$e1_{b1}^{si}$	$e1_{b2}^{si}$	$\cdots$	$e1_{bj}^{si}$	$\cdots$	$e1_{bm}^{si}$	W_{e1}^{si}
e_2	$e2_{b1}^{si}$	$e2_{b2}^{si}$	$\cdots$	$e2_{bj}^{si}$	$\cdots$	$e2_{bm}^{si}$	W_{e2}^{si}
$\cdots$	$\cdots$	$\cdots$	$\cdots$	$\cdots$	$\cdots$	$\cdots$	$\cdots$
e_x	ex_{b1}^{si}	ex_{b2}^{si}	$\cdots$	ex_{bj}^{si}	$\cdots$	ex_{bm}^{si}	W_{ex}^{si}
$\cdots$	$\cdots$	$\cdots$	$\cdots$	$\cdots$	$\cdots$	$\cdots$	$\cdots$
e_z	ez_{b1}^{si}	ez_{b2}^{si}	$\cdots$	ez_{bj}^{si}	$\cdots$	ez_{bm}^{si}	W_{ez}^{si}
T_{bj}^{si}	T_{b1}^{si}	T_{b2}^{si}	$\cdots$	T_{bj}^{si}	$\cdots$	T_{bm}^{si}	

Step3 (Buyer's price declaration)

As same as Step2, first, the buyer estimates the price of the item. Then, the buyer evaluates the seller, and the trust value is put out by the weighted evaluation. Table 9 shows the table of multi-attribute evaluation of sellers. Also, the trust value is dropped to a function shown as the curve in 1. To simplify, as same as the case of seller, we employ three patterns shown in Figure 1. The magnification rate is also multiplied to the price of the item, and the declared price is decided.

Table 9. Seller evaluation tables of buyer

$\{b_j, S\}$	s_1	s_2	$\cdots$	s_i	$\cdots$	s_n	W
e_1	$e1^{bj}_{s1}$	$e1^{bj}_{s2}$	$\cdots$	$e1^{bj}_{si}$	$\cdots$	$e1^{bj}_{sn}$	W^{bj}_{e1}
e_2	$e2^{bj}_{s1}$	$e2^{bj}_{s2}$	$\cdots$	$e2^{bj}_{si}$	$\cdots$	$e2^{bj}_{sn}$	W^{bj}_{e2}
$\cdots$	$\cdots$	$\cdots$	$\cdots$	$\cdots$	$\cdots$	$\cdots$	$\cdots$
e_x	ex^{bj}_{s1}	ex^{bj}_{s2}	$\cdots$	ex^{bj}_{si}	$\cdots$	ex^{bj}_{sn}	W^{bj}_{ex}
$\cdots$	$\cdots$	$\cdots$	$\cdots$	$\cdots$	$\cdots$	$\cdots$	$\cdots$
e_z	ez^{bj}_{s1}	ez^{bj}_{s2}	$\cdots$	ez^{bj}_{si}	$\cdots$	ez^{bj}_{sn}	W^{bj}_{ez}
T^{bj}_{si}	T^{bj}_{s1}	T^{bj}_{s2}	$\cdots$	T^{bj}_{si}	$\cdots$	T^{bj}_{sn}	

Step4 (Decision of final trading partner and the actual trading price)

The difference of declared price are computed. The successful traders are determined based on the maximum difference. The actual trading price P is decided based on the trading magnification rate.

4 Example

In this section, we explain an example when a seller and three buyers exist in the market. Tradable item g is one (single item). We assume the evaluation criteria as "Friendship degree", "Easiness of trading", and "Expectation degree for item". The evaluation value and priority are assumed to be five level rating evaluation from 1 to 5. For example, the total value is calculated as $\sum s \cdot W$. Through Table 10 to 12 show the examples of multiplication of evaluation values and weights. We give a concrete situation of negotiation process using these tables. The maximum value of magnification rate M obtained from the trust value is set as 40%. When the actual trading price is decided, trading magnification rate k is set as 50%.

1. Seller s applies a negotiation to three buyers b_1, b_2, b_3. As the result, buyer b_1, b_2 agrees with the seller's proposal. Seller s starts negotiating with buyer b_1, b_2.
2. Seller s estimates that item g is $ 1000 yen or more than . Seller s evaluates buyer b_1, b_2 by multiple attribute evaluations.
3. Buyer b_1, b_2's valuations of item g are each $ 900 and $ 1200 for item. Each buyer b_1, b_2 evaluates seller s based on multiple attribute evaluations.

Table 10. Buyers Evaluation by Seller s

Evaluation of seller s	b_1	b_2	W
Friendship degree	4	3	4
Easiness of trading	3	2	5
Expectation level to item	4	5	4
Trust value T	47	42	

Table 11. Seller Evaluation by Buyer b_1

Evaluation of buyer b_1	s	W
Friendship degree	4	4
Easiness of trading	3	5
Expectation level to item	5	4
Trust value T	51	

4. After the calculation of trust value, we apply this value to the curves as shown in Figure 1. We calculate the magnification rate of the item. The magnification rate is multiplied to the item price. As an example, we use nonlinear curves in Figures 2 and 3.

5. Seller s chooses the curve in the graph (Figure 2) regarding reliability of buyer b_1, b_2. The magnification rate of the rating regarding trust value is multiplied with the item price. Buyer b_1, b_2 also choose the curve in the graph (Figure 3) regarding reliability of seller s.

6. From the determined magnification rate, the each price in which seller s presents to buyer b_1, b_2 is decided as $R_{b1}^s = 1000 \times 0.75 = 750$ dollars and $R_{b2}^s = 1000 \times 0.78 = 780$ dollars. Also, the price in which buyer b_1 presents to seller s is $V_s^{b1} = 900 \times 1.33 = 1197$ dollars, and the price in which the buyer b_2 presents to seller s becomes $V_s^{b2} = 1200 \times 1.25 = 1500$ dollars.

7. To decide the combination of dealings, like a combinatorial matching, we calculate a price difference D. When seller s trades with buyer b_1, $D = 1197 - 750 = 447$ dollars. When seller s trades with buyer b_2, $D = 1500 - 780 = 720$ dollars. Thus, the matching result is s, b_2.

8. In this example, since trading magnification rate k is 50%, the actual trading price becomes $780 + 0.5(1500 - 780) = 780 + 360 = 1140$ dollars.

Table 12. Seller Evaluation of Buyer b_2

Evaluation of buyer b_2	s	W
Friendship degree	3	4
Easiness of trading	3	5
Expectation level to item	5	4
Trust value T	47	

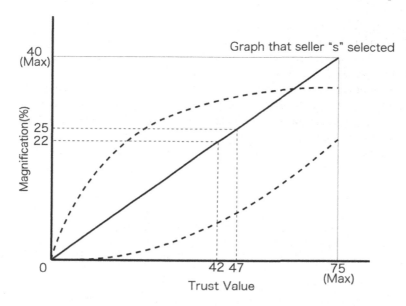

Fig. 2. A Curve in which Seller s Selects

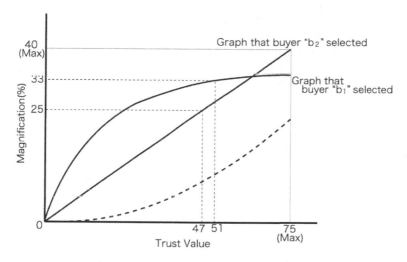

Fig. 3. A Curve in which Each Buyer Selects

The actual trading price is between each trader's valuation. Thus, it is insured individual rationality.

5 Conclusion

This paper focuses on the relationship between the buyer and seller in trading. The proposed mechanism is a trading form where traders relationships are considered. The advantage of our mechanism is what an appropriate trading price are calculated based on multiple attribute evaluations including both business partner in himself/herself and items. The relations between individuals are reflected in the price by applying the effect value obtained from multiple attribute evaluation.

As the development of our work, our mechanism can be reputation system in the Internet auctions. While the existing Internet auction reputation system employs only rating method, our mechanism separates trader's humanity from item condition. Thus, new comer traders can know appropriately the potential business partner's information. Further, our trading price determination mechanism help traders' decision making.

References

1. Nishimura, K.: Contemporary Economics Intermediate Microeconomics A Modern Approach, 2nd edn. Iwanami-Shoten (2001)
2. Varian, H.R.: Intermediate Microeconomics A Modern Approach, 7th edn. Keiso-Shobo (2007)
3. Kobashi, Y.: The decision is supported. University of Tokyo Press (1998)
4. Humphreys, P.: Application of multi-attribute utility theory. In: Humphreys, P., Zeeuw, G. (eds.) Decision making and change in human affairs. Reidel, Dordrecht (1977)
5. Takahashi, S., Matsuo, T.: An Approach to Efficient Trading Model of Hybrid Traders based on Volume Discount. In: The 21st International Conference on Industrial, Engineering & Other Applications of Applied Intelligent Systems, IEA/AIE (2008)
6. Matsuo, T., Ito, T.: A Bidders Cooperation Support System for Agent-based Electronic Commerce. In: Chung, P.W.H., Hinde, C.J., Ali, M. (eds.) IEA/AIE 2003. LNCS (LNAI), vol. 2718, pp. 369–378. Springer, Heidelberg (2003)
7. Tversky, A.: Elimination by aspects: a theory of choice. Psychological Review 79, 281–299 (1972)

Agent Based Architecture for Pricing the Services in Dynamic User/Network Environment

Drago Žagar, Slavko Rupčić, and Snježana Rimac-Drlje

University of Osijek
Faculty of Electrical Engineering
Osijek, Croatia
{Drago.Zagar,Slavko.Rupcic,Snjezana.Rimac}etfos.hr

Abstract. The main goals of Internet service providers are to maximize profit and maintain a negotiated quality of service. However, achieving these objectives could become very complex if we know that Internet service users might during the session become highly dynamic and proactive. This connotes changes in user profile or network provider/s profile caused by high level of user mobility or variable level of user demands. This paper proposes a new pricing architecture for serving the highly dynamic customers in context of dynamic user/network environment. The most convenient concept that could comprise highly adaptive and proactive architecture is agent based architecture. The proposed agent based architecture places in context some basic parameters that will enable objective and transparent assessment of the costs for the service those Internet users receive while dynamically change QoS demands and cost profile.

Keywords: QoS, Price Optimization, Agent architecture, Proactive.

1 Introduction

In parallel with rise of a confidence in Internet network, also rise the expectations to its reliability which connotes satisfactory level of *Quality of Service* (QoS). These expectations force the Internet users to negotiate with its *Internet Service Providers – ISPs*, about the guaranties required for specific levels of QoS. The price for Internet access is generally the main (but also the easiest) criterion for new customers to compare different ISPs. Very important questions for customers are how to estimate the received quality of service and how to get the optimal service value. This process is obviously very complex and connotes dynamic analyze and adaptation of many different parameters. These parameters can be used to compare the services offered by different ISPs and to inform a price/quality trade-off decision for a consumer. Parameters should be designed to be sufficiently flexible to be used across a mixture of connection technologies. Some elements of QoS can't always be clearly measurable in a numerical and objective way (security, reliability and ease of billing). Usually, these parameters can be measured only subjectively, and may also not ever be a part of price/quality trade-off [2][13].

N.T. Nguyen et al. (Eds.): New Challenges in Compu. Collective Intelligence, SCI 244, pp. 303–311.

A development of recent complex and demanding applications continuously increases the requirements on a quality of the service, provided by communication's infrastructure. It is very possible that the same complex applications and the services will be executed by different level of quality that would satisfy every single user. This is much more important if we know that applications and services could, during the communication, require different level of QoS. The variable QoS requirements could become from variable demands of a user and an application and from the variable network characteristics, respectively. Furthermore, variations of QoS as well as the QoS degradations can also be a consequence of using on user's devices with different possibilities, in different networks, or on mobile devices. A proper relation between the quality and the cost of the service will result in adequate user satisfaction.

The variable requirements have to be negotiated between the end system and the network, every time when the user or the network changes the demands on QoS [18].

At the moment there exist several pricing models that could be used for Internet accounting, but not all are generally applicable. This is especially emphasized in dynamic user/network environment in which user and network could change their QoS characteristics and profiles. The dynamic accounting model could be a benefit for the customer, because he will find the appropriate and optimal model of network usage.

This paper presents a new pricing architecture for serving the dynamic users in context of dynamic user/application/network environment. The most convenient concept that could comprise highly adaptive and proactive architecture is agent based architecture. The proposed agent based architecture places in context some basic parameters that will enable objective and transparent assessment of the costs for the service those Internet users receive while dynamically change QoS demands and cost profile.

The paper is organized as follows: second section presents different aspects of service price estimation and optimization as a motivation for new proposal, the third section proposes a new agent based architecture for dynamic pricing the services in distributed network environment, the fourth section presents price optimization procedure for the services with dynamic QoS demands and last section concludes the paper.

2 Aspects of Service Price Estimation and Optimization

The service price on the market is established in close connection between service user and service provider. The service providers set and adapt the service price in order to maximize the profit. The service users try to optimize the ratio of received service value and the service price.

Revenue from services can be basically calculated in accordance to different factors (e.g. time, cost, quantity of services). Consumers select the class of service in accordance to their preferences and habits, and compare prices of similar services enabled by the various service providers. The main goal is to maximize the benefits (level of satisfaction) within limited budget [3][4][5].

There exist many different strategies for optimization of service price from different aspects. They could include different pricing architectures, pricing strategies and service models. Flat rate pricing model is stimulant to over use the network, which, especially in connection with the increasing number of subscribers, can lead to network congestion. The result is a reduction of the quality of service and QoS management. In contrast, usage-based pricing scheme can be a good step to general price/resources optimisation. The QoS architectures that are mostly used and relatively largely implemented (e.g. Diffserv model) deal with this problem in a user oriented manner i.e. a user always chooses the service class providing the best QoS [6] [11][12].

Classical pricing schemes are becoming outdated because of three major reasons:

i. Existing simple schemes are not stimulant to improve utilization.
ii. Some users might wish to pay more than others to avoid congestion and get better QoS.
iii. The current Internet and, even much more, the next-generation Internet has to deal with numbers of applications with very diverse quality of service requirements. If we comprise these requirements in a pricing model, we could improve users' level of satisfaction [8].

Additionally to these reasons we might add the fourth, which is obvious: the users could during the session dynamically change their demands in accordance to variable requirements and network condition. In such a way the user can get the improved (optimal) cost per session (service).

The study of dynamic pricing a QoS-enabled network environment has become one of the hottest research areas in recent years. Many optimal pricing schemes have been proposed in the past few years. QoS pricing schemes proposed so far often entail either congestion control or admission control or even both. Most of them either assume a well-known user *utility function* or *user demand function* and establish an optimization model to maximize either the social welfare or provider's revenue [9].

Some early versions of congestion-dependent pricing also considered charges per packet. Such approaches often use an auction to determine the optimal price per packet, resulting in *prices that vary with demand*. Several research studies investigate the issue of user behavior upon price change [7][12].

Users' willingness to pay will simply depend on the type of services offered and on the price charged which could be described by the utility function. This function gives a measure of the user's sensitivity to the perceived QoS level that is to the variations of the network resources allocated to the relevant flow. From an economic point of view, the utility function is strictly related to the users' demand curve, which describes users' willingness to pay as a function of the perceived QoS level [10].

Some researchers presume that optimal pricing schemes require a centralized architecture to collect all the required information and perform an optimal price setting or resource allocation. From this, very probable point of view, it is obvious that the optimal pricing schemes do not scale well. Also, it is not feasible to adopt a centralized approach when multiple domains/service providers are involved. If the centralized approach is used only within a domain, then it is not clear if the locally optimal solution will lead to a globally optimal one.

An ideal model for network access would be the continuous accessibility. In every access point we could have defined a price for Internet access, and user willing to pay can approach the network. The price should be set as high as optimal network usability will be achieved. One mechanism for access price estimation could be

"Walrasian tatonnement", in which a trial price could be set and the users can decide whether to accept or refuse the offer. If the total amounts of users' requests exceed the network capacity then the price should be set to a higher level. The main problem by this scheme is that user must cautiously monitor the price. If a time of day traffic sample is enough predictive than this scheme could work, but it is well known that Internet traffic will be highly penetrating and unpredictable [1].

To provide a better QoS guarantee in accordance to service price a proactive approach is preferred to a reactive approach in many cases. This fits dynamic environment much better, especially if we know that *users are free to adjust their QoS demands during a service session although this should not be mandatory.*

As well as the centralized QoS pricing approach obviously doesn't lead to a general solution our attempt introduces distributed agent based architecture for QoS pricing. It is especially appropriate for dynamic network environment and can proactively lead a system to state which is dynamically optimized.

3 An Agent Based Architecture

A process of QoS negotiation could be a very complex procedure, time wasting, and burden by huge number of the measures to define the quality of service and by shortage of proper definition of applied measures. Furthermore, we often do not have the appropriate methodology to measure and observe the negotiated QoS. If these problems were addressed on an ad-hoc basis, between every single customer and its ISP, the implementation of QoS will lead to big effort and large traffic measurements that don't scale well in general network model. These problems are much more emphasized if we introduce dynamic network environment, which could be lunched by different variations in network surroundings.   The dynamic distributed environment could be shown as tight connection between network, user and application (Figure 1.). Variations of QoS as well as QoS degradations are not necessarily caused by overall traffic in the network, but could also be a consequence of using devices with different possibilities, different networks, or mobile devices. We can assume that during the communication new and complex applications and services would require a different level of service quality and would be executed with different level of quality of service. Furthermore, characteristics of services offered in distributed environments are exposed to high variations, from the light quality degradation to a complete loss of communication link. The end system has to compensate these variations and provide the user with acceptable level of service quality. The proactive applications are predestined for distributed and dynamic environment with limited resources and unpredictable variations in quality of communications [14][15].

The processing quality of service in distributed dynamic environment includes several coherent activities:

♦ definition of demands on service quality in the form of subjective user wishes or acceptable applications quality; performances, synchronization, costs, etc.;

♦ mapping of obtained results into parameters of service quality for different system components or layers;

♦ negotiation between system components or layers to insure that all components can satisfy the desirable quality of service.

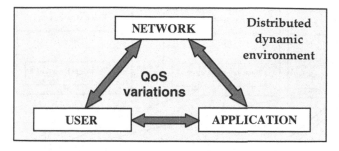

Fig. 1. Distributed dynamic environment

The main motivation for proposing new agent based architecture was a need to improve QoS, user costs and user satisfaction. This is especially request of the applications that could require a different level of QoS during the session. Examples of such applications are distributed multimedia, virtual reality, mobile applications, telemedicine, etc. They all have in common that in order to reduce the costs could during the session change their requests on QoS and could be executed with variable (acceptable) quality of service.

The architecture proposal should be based on the following three underlying objectives:

- Economic efficiency;
- Simplicity and scalability and
- Optimal revenue from the user point of view.

The first goal is to achieve the optimal overall customer value of received services. The second goal is to design a simple and scalable architecture/scheme. The last goal is to provide optimal level of user QoS guarantees and to minimize the cost/service ratio.

These objectives are sometimes conflicting or even contradictory. Most researchers believe that optimal pricing solutions are very hard or even impossible to achieve as the scale of the problem is becoming large. Taking into account all measures that should be taken to improve different aspects of service pricing it must be pointed out that in practice, as in many other fields, maximal simplicity is often more important than maximal efficiency [9][16][17].

In the Figure 2. is shown the proposed architecture for pricing the services in dynamic user/network environment. The architecture consists of different service providers competing for users. Every service provider should have its own price strategy that will attract the users to choose it. Every user also should have its own price strategy that will obtain the optimal QoS/price value. If the user changes the ISP during the communication (mobility, QoS request change, device change...) this process should be as transparent as possible. The best way to employ dynamic price system is implementation of price agents by users' and ISPs'. By using predefined price strategies price agents have to dynamically negotiate QoS pricing. A very important question is how to implement a billing system in dynamic environment in efficient and secure manner. One obvious solution could be a centralized approach by independent agency that could provide service of cost management. Every user

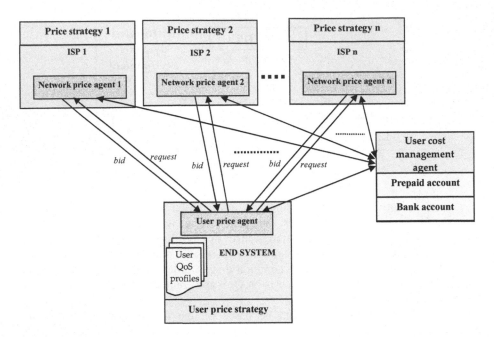

Fig. 2. The architecture for pricing the services in dynamic user/network environment

should open a user account which could be managed dynamically by User cost management agent. User cost management agent could communicate with user price agent and network price agents as well as user bank account or credit system to arrange independent and transparent billing process.

4   Price Optimization for the Services with Dynamic QoS Requirements

The proposed architecture enables efficient User/Network QoS/price management. In order to optimize a level of QoS service the applications and services could, during the communication, require different level of QoS. The variable requirements have to be negotiated between the end system and the network, every time when the user or the network changes the QoS demands. In Figure 3. is shown a process of QoS/price adaptation. The price should be adapted when external factors change. Some of external factors could be user request, application request and changes in network environment. The negotiation strategy is based on price adaptation/optimisation that is implemented in end system and service providers. In order to improve negotiation the user and ISP (ISPs) could change QoS/price policy. The dynamic result will be QoS/Price working point that will worth till the next QoS change.

An oftenness of price change is also very important detail. For transparent and continuous QoS providing it is necessary to define the limits (QoS, price…) that will trigger a new negotiation procedure. As a result the negotiation process between

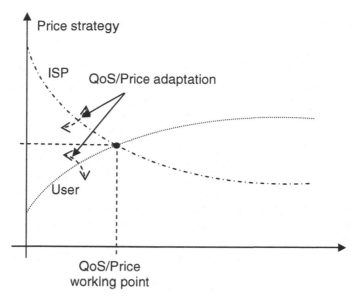

Fig. 3. QoS/Price adaptation policy

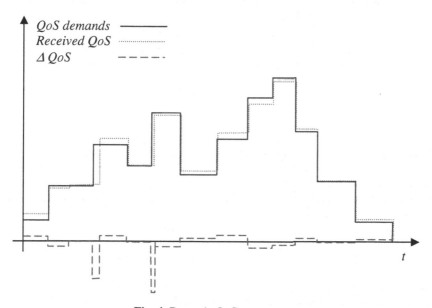

Fig. 4. Dynamic QoS management

network and user could become more efficient. In the Figure 4. is shown the process of QoS adaptation in which the received QoS should be as close as requested QoS. To best meet the QoS requirements the QoS difference ΔQoS should be as small as possible.

5 Conclusions

In response to fluctuations in either end system or network resources the dynamic heterogeneous network environment is desirable to dynamically adapt QoS/Price. Variations that are comprised could be caused by the limited/changed system resources or by variations in the application demands. The main goal of QoS/Price control is to maximise user's satisfaction by the best use of limited resources.

This paper proposes pricing architecture for serving the highly dynamic customers in dynamic user/network environment. The most convenient concept that could comprise highly adaptive and proactive architecture is agent based architecture. The proposed agent based architecture incorporates some basic parameters that will enable objective and transparent assessment of the costs for the service those Internet users receive while dynamically change QoS demands and cost profile.

In the proposed agent based QoS/Price architecture model, the main improvements could be summarised as follows:

- Proactive agent based approach insures transparent pricing in dynamic QoS environment.
- proposed model insures the best level of QoS/Price ratio that end user receives and an optimal level of user's satisfaction;
- centralised accounting served by independent User cost management agent enables transparent dynamic accounting;

Further work will include a formal verification and experimental validation of the proposed agent based architecture for pricing in dynamic user/network environment. The proposed architecture will be a base for further investigation related to heterogeneous and dynamic QoS environment.

References

1. Fulp, E.W., Reeves, D.S.: Optimal Provisioning and Pricing of Internet Differentiated Services in Hierarchical Markets. In: Proceedings of the IEEE International Conference on Networking (2001)
2. Gupta, A., Stahl, D.O., Whinston, A.B.: The Economics of network management. Communications of the ACM 42(9) (September 1999)
3. Dutta-Roy, A.: The cost of quality in Internet-style networks. IEEE spectrum, 57–62 (September 2000)
4. Pras, A., Beijnum, B.J., Sprenkels, R., Parhonyi, R.: Internet Accounting. IEEE Communications Magazine, 108–113 (May 2001)
5. Dube, P., Liu, Z., Wyner, L., Xia, C.: Competitive equilibrium in e-commerce: Pricing and outsourcing. In: Computers & Operations Research. Elsevier, Amsterdam (2006)
6. Hayel, Y., Tuffin, B.: A mathematical analysis of the cumulus pricing scheme. Computer Networks (2004)
7. Jin, N., Jordan, S.: The effect of bandwidth and buffer pricing on resource allocation and QoS. Computer Networks 46 (2004)
8. Ros, D., Tuffin, B.: A mathematical model of the Paris Metro Pricing scheme for charging packet networks. Computer Networks 46 (2004)

9. Li, T., Iraqi, Y., Boutaba, R.: Pricing and admission control for QoS enabled Internet. Computer Networks 46 (2004)
10. Blefari-Melazzi, N., Di Sorte, D., Reali, G.: Accounting and pricing: a forecast of the scenario of the next generation Internet. Computer Communications 26, 2037–2051 (2003)
11. Wang, X., Schulzrinne, H.: Comparative study of two congestion pricing schemes: auction and tatonement. Computer Networks 46, 111–131 (2004)
12. Di Sorte, D., Reali, G.: Pricing and brokering services over interconnected networks. Journal of Network and Computer Applications 28, 249–283 (2005)
13. Žagar, D., Vilendečić, D., Mlinarić, K.: Network Resource Management for the Services with Variable QoS Requirements. In: Conference on Telecommunications, ConTel 2003, Zagreb (2003)
14. Turk, T., Blažič, B.J.: User's responsivness in the price-controlled best-effort QoS model. Computer Comunication 24, 1637–1647 (2001)
15. Wang, M., Liu, J., Wang, H., Cheung, W.K., Xie, X.: On-demand e-supply chain integration: A multi-agent constraint-based approach. Expert Systems with Applications 34, 2683–2692 (2008)
16. Wang, M., Wang, H., Vogel, D., Kumar, K., Chiu, D.K.W.: Agent -based negotiation and decision making for dynamic supply chain formation. Engineering Applications of Artificial Intelligence (2008)
17. Comuzzi, M., Francalanci, C., Giacomazzi, P.: Pricing the quality of differentiated services for media-oriented real-time applications: A multi-attribute negotiation approach. Computer Networks 52, 3373–3391 (2008)
18. Ferguson, P., Huston, G.: Quality of Service on the Internet: Fact, Fiction, or Compromise? In: Proceedings of the INET 1998, Geneva (1998)

Part V
Other Applications

Accuracy in Predicting Secondary Structure of Ionic Channels

Bogumil Konopka[1], Witold Dyrka[1], Jean-Christophe Nebel[2],
and Malgorzata Kotulska[1,*]

[1] Wroclaw University of Technology, Institute of Biomedical Engineering and
Instrumentation, ul. Wybrzeze Wyspianskiego 27, 50-370 Wroclaw, Poland
`malgorzata.kotulska@pwr.wroc.pl`
[2] Kingston University, Faculty of Computing, Information Systems & Mathematics, Kingston
Upon Thames, Surrey, KT1 2EE, United Kingdom

Abstract. Ionic channels are among the most difficult proteins for experimental
structure determining, very few of them has been resolved. Bioinformatical
tools has not been tested for this specific protein group. In the paper, prediction
quality of ionic channel secondary structure is evaluated. The tests were carried
out with general protein predictors and predictors only for transmembrane
segments. The predictor performance was measured by the accuracy per residue
Q and per segment SOV. The evaluation comparing ionic channels and other
transmembrane proteins shows that ionic channels are only slightly more
difficult objects for modeling than transmembrane proteins; the modeling
quality is comparable with a general set of all proteins. Prediction quality
showed dependence on the ratio of secondary structures in the ionic channel.
Surprisingly, general purpose PSIPRED predictor outperformed other general
but also dedicated transmembrane predictors under evaluation.

Keywords: ionic channel, secondary structure, prediction.

1 Introduction

Ionic channels are transmembrane proteins with a gated pore that control passive
influx and outflow of ions and molecules through the cellular membrane. Channels
play a crucial role in the living processes of cells and mutations that alter their three-
dimensional structure lead to very serious health consequences. Although
conductivity of ionic channels is usually very selective, they have successfully been
used for intracellular drug delivery. Therefore, the knowledge on their structure and
functionality is of great importance. Since ionic channels are water insoluble
transmembrane proteins, anchored deeply in the lipid bilayer, and are usually very
large molecules, their extraction, crystallization and analysis are very difficult tasks.
Consequently, very few structures of ionic channels have been resolved. Hence,
possibility of computational structure prediction would be very valuable.

Predicting secondary structure of an ion channel is an essential step towards
obtaining a complete three-dimensional architecture of the molecule. This process

---

* Corresponding author.

N.T. Nguyen et al. (Eds.): New Challenges in Compu. Collective Intelligence, SCI 244, pp. 315–326.
springerlink.com   © Springer-Verlag Berlin Heidelberg 2009

involves predicting secondary structure of the overall protein and location of the transmembrane region. Many specialized tools, including PHDhtm, HMMTOP, DAS, PRED-TMR, SOSUI, CONPRED, address the problem of transmembrane region detection. Their reliability was reported as close to 70% (for a review see [1-3]). Since they only predict residues from transmembrane segments, they do not provide information about the secondary structure of the overall protein. For this purpose, general secondary structure prediction (SSP) tools could be employed. Evaluation of the most contemporary SSP tools show that their efficiency on soluble proteins reaches 80% of the correctly predicted helices, beta structures and coils [4].

Although methods and tools dedicated to SSP have been available for more than thirty years, since the Chou & Fasman algorithm was published [5], they have not been thoroughly tested on an extensive ion channel data set since, until recently, only a few channel structures were available in the Protein Data Bank (PDB) [6] in the form of pdb files. Due to technology advances, the number of resolved proteins is growing steadily. Therefore, it is now possible to evaluate SSP tools based on a significant number of ionic channels, for example using database published by S. White's group [7,8]. To our knowledge, it has not been tested whether the efficiency of general purpose SSP tools is equally high for ionic channels as for soluble proteins and how it compares to a set of other transmembrane proteins. We would also like to know what the shortcomings of their predictions are in case of ionic channels. Moreover, we want to identify which tools should be recommended for such a specific subset of transmembrane proteins, especially for the more complex channels which are made of several chains and/or contain beta strands.

In the presented work, we test the performance of SSP tools, comparing their outcome for a set of ionic channels and other transmembrane (TM) proteins. The results are related to evaluation on all proteins provided by Eva benchmark [4,9]. The analysis concerns prediction quality for three standard secondary structures: α-helix (H), β-strand (E), and coil (C). The evaluation is expressed by Q_i value reflecting per residue accuracy and SOV_i value showing the accuracy per segment. For the general purpose tools the analysis is performed on the whole protein sequence, while it is limited to transmembrane regions for the tools dedicated to transmembrane proteins. Furthermore, we investigate what is the relation between the predictor's quality and the ratio of each secondary structure in the channel, derived for each of the predictors.

2 Materials and Methods

2.1 Protein Data Sets

The set of ionic channels consisted of proteins from the redundant database published by S. White's Laboratory at UC Irvine [8]. From this database (of 9.04.2009) we excluded the channels that did not meet some quality criteria, i.e. resolution < 5 A, factors evaluating quality of the pdb model: R < 0.3 and $|R - R_{free}| < 0.05$ (R_{free} defined in [10]). The set included 76 chains from 19 ionic channels, and the set with recognized transmembrane region had 57 chains from 18 ionic channels. For control, a set with resolution < 2.5 A was also tested. The results with this very small set (only 18 chains available in PDB) did not differ significantly from those obtained with the larger set (data not shown). Hence, we decided that currently the larger, lower-resolution set can be used as more reliable, due to the larger representation.

In the case of other transmembrane proteins, where a larger number of structures is available, the quality criteria were like of the strict channel set: similarity cutoff 40%, resolution < 2.5 A, R > 0.3, $|R - R_{free}| < 0.05$. The set of other transmembrane proteins was generated based on the database PDBTM maintained by Institute of Enzymology in Budapest [11,12] as the set of α-helical proteins (of 9.04.2009) excluding ionic channels. Finally the set included 64 chains from 47 proteins, while the subset with recognized transmembrane region consisted of 55 chains from 39 proteins.

2.2 Evaluation

The evaluation was based on PDB files where α-helices and β-strands are defined. Other residues were assumed to be within a coil. Prediction quality was evaluated by per residue accuracy Q_i, and per segment accuracy SOV_i, $i \in \{H, E, C\}$ [13]. Q_i is the percentage of correctly predicted residues in the i-th state related to the real number of aminoacids in the relevant secondary structure; Q_3 is the percentage of correctly predicted residues in all three states related to the chain length. SOV and SOV_i show a segment overlap for all three states and the i-th state, respectively [14].

2.3 Tools

2.3.1 Predictors
The efficiency and applicability of SSP predictors for deriving a channel protein secondary structure was tested on three highly effective general purpose predictors: PSIPRED, JNET and Proteus Consensus methods. All the methods utilize Position Specific Scoring Matrices (PSSM) profiles generated by PSI-BLAST [15] and implement feed-forward back-propagation neural networks, however their details differ significantly. The evaluation of the methods is presented after EVA [4] as of April 2008 [9].

PSIPRED algorithm [16] consists of two neural networks, each with a single hidden layer. First of the nets receives a PSSM with window length of 15 as an input and outputs 3-state secondary structure prediction. Second neural network is designed to filter outputs from the main net. Hidden layer sizes were 75 and 60 respectively. Interestingly, transmembrane fragments were excluded from PSIPRED training set. The EVA evaluation of PSIPRED showed AVE_Q3: 77.8 and AVE_SOV: 75.3 (see Fig.1). JNET [17] also utilizes the architecture based on two neural networks: one for sequence to structure prediction and one for structure to structure filtering. In addition to a PSSM based module, it consists of three other pairs of nets fed with multiple sequence alignments, PSI-BLAST frequency profiles and HMM profiles generated by HMMER [18]. Then, if there is a disagreement between predictions from different modules, a separate neural network is trained to make the final decision. The window lengths for the basic nets were 17 and 19 respectively. Hidden layers in JNET v.1.0 used for the tests were built from 9 neurons. The EVA evaluation of JNET showed AVE_Q3: 72.3 and AVE_SOV: 67.7.

PROTEUS CONSENSUS [19] method combines the three stand-alone secondary structure predictors (JNET, PSIPRED and their own TRANSSEC). Their results are used as an input to a standard feed-forward network containing a single hidden layer, which acts as a jury-of-experts. While TRANSSEC follows the two-tier design, its first neural network includes two hidden layers (sizes: 160 and 20) and uses a 19 residues window. The second net has one 44-neuron large hidden layer and operates on window of size 9. Moreover second network input is based on secondary structure multiple alignment.

Accuracy in predicting secondary structure of transmembrane region in ionic channels was tested by specialized transmembrane SSP tools: TMHMM, DAS, HMMTOP, ConPred and PHDhtm. These tools only predict the structure of transmembrane protein regions.

HMMTOP [20,21] and TMHMM [22,23] use probabilistic framework of the Hidden Markov Model to predict transmembrane helixces. In TMHMM, various regions of a membrane protein are represented by submodels, which consist of several HMM states. This architecture allows modeling lengths of helices. HMMTOP applies similar concepts.

PHDhtm is a two-level feed-forward neural network method developed by Rost and coworkers [24,25]. First level network is fed with the data from local sequence alignment for a window size of 13 and with global statistics for the whole protein. The output units code whether the central residue of the window is a transmembrane one or not. This data together with global statistics become an input of the second level network, which is used to elongate predicted helices.

Dense Alignment Surface (DAS, [26,27]) predictor consists on a protein sequence comparison by using a dot-plot based on scoring matrices which emphasize polarity properties of sequence regions. Therefore, when two transmembrane proteins are processed, a characteristic chess-board pattern is produced. A query sequence is compared with known transmembrane proteins from the library and an average cross-weighted cumulative score profile is calculated.

CONPRED2 [28,29] is a method that combines the results of several TM predictors. It offers two similar algorithms to localize the consensus position of the center of a given helix. Then, the TM region is extended by 10 residues towards N and C termini from the center.

Transmembrane region from the pdb file was recognized based on the data provided by TMDET [30] by transmembrane database PDBTM.

2.3.2 Evaluation Analysis
The analysis was carried out by SSPE, home made software developed by B.K. in National Instruments CVI environment, with a major use of ANSI C functions. The input streams contain PDB files of tested proteins and relevant files with predicted secondary structures. The software produces a set of parameters and the superposition of the predicted and the original sequences.

3 Results

The general purpose SSP tools were tested on the ionic channels and other transmembrane proteins data sets. Table 1 shows a comparison of their results. The overall accuracy Q_3 for the whole molecule prediction was typically around 80%, and per segment accuracy SOV above 70%. The highest accuracy for all protein sets was shown by PSIPRED (Fig.1), slightly outperforming the CONSENSUS method. The poorest results were obtained by JNET, especially in both groups of transmembrane proteins (Table 1A,B)). However, since JNET tends to underpredict the limits of helices and β-strands, it showed a very good result in the coil prediction for all protein groups.

The overall prediction accuracy of ionic channels was lower than for other transmembrane proteins (Table 1A,B). However, it was comparable with the overall prediction accuracy of soluble proteins evaluated by EVA (Fig.1) [9]. Interestingly, two best predictors performed very unequally for different secondary structures of transmembrane proteins (Table 1A,B). For example, in the set of ionic channels β-strands were very accurately predicted ($Q_E = 88\%$, $SOV_E = 83\%$), similarly good result were obtained for α-helices ($Q_H = 79\%$, $SOV_H = 82\%$), coils were predicted very poorly ($SOV_C = 59\%$, for the best result by CONSENSUS algorithm). In contrast, prediction quality of β-strands for other transmembrane proteins was surprisingly low ($Q_E = 49\%$, $SOV_E = 45\%$, Table 1B). Therefore, we tested the hypothesis whether these results are related to the ratio of each secondary structure within a molecule. In prediction of α-helices in ionic channels, PSIPRED performance showed dependence on the ratio of helices (Fig.2); JNET did not show any relation, however its prediction of α-helices was very dispersed (Fig.3). This phenomenon was not observed in other transmembrane proteins (Fig.4 shows the results for PSIPRED).

Table 1. Prediction quality of the general purpose SSP tools applied for the set of (A) ionic channels, (B) other transmembrane proteins, (C) soluble proteins. The whole molecule was modeled. The best results are underlined, the worst in italics – if discernible.

(A) ionic channels

SSP	Statistic	Q_3	Q_H	Q_E	Q_C	SOV	SOV_H	SOV_E	SOV_C
PSIPRED	Mean	77	79	88	65	70	82	83	57
	SD	8	14	5	14	16	13	4	19
JNET	Mean	*61*	*47*	73	72	55	*49*	72	*57*
	SD	13	21	8	15	15	19	8	17
CONSENSUS	Mean	76	75	86	66	70	79	81	59
	SD	8	17	6	14	15	16	6	18

(B) other TM proteins

SSP	Statistic	Q_3	Q_H	Q_E	Q_C	SOV	SOV_H	SOV_E	SOV_C
PSIPRED	Mean	81	84	49	76	76	83	45	62
	SD	9	12	38	17	13	13	39	20
JNET	Mean	*61*	52	44	82	55	52	42	61
	SD	16	23	32	14	17	23	33	19
CONSENSUS	Mean	80	81	48	78	76	81	43	62
	SD	10	15	37	17	14	14	36	20

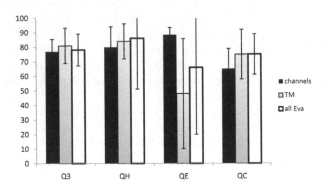

Fig. 1. Performance of PSIPRED, a general purpose SSP tool that showed best overall reliability on ionic channels in our tests. The evaluation shows that this predictor does not perform worse for ionic channels than other transmembrane proteins and an average (results from EVA benchmark [9]).

Prediction quality of β-strands in ionic channels did not show any dependence on their rate (Fig.5). However, the analysis on other transmembrane proteins showed that β-structures had a higher chance to be unrecognized if their residue number is very low (below 5%), Fig.6; at higher representation the prediction quality was constant. The very low representation of β-residues in non-channel transmembrane proteins adversely affected the results of Q_E and SOV_E (Table 1B).

We also tested the predictors performance with regard to the most difficult ionic channel, i.e. the channel which was evaluated the least accurately. In the set of all ionic channels, the worst prediction for all SSP tools was obtained for the channel 3beh. This is the transmembrane regions of a bacterial cyclic nucleotide-regulated channel, solved in 2008, at the resolution of 3.2 A [31]. All 4 chains of this protein showed very poor performance − slightly above 60%, the worst prediction was obtained for chain A. Comparison of the general purpose SSP tools (Table 2) confirms that PSIPRED outperforms other tested tools, even in such a difficult case. Interestingly, JNET confirmed its superiority in coil prediction also in this case (Table 2). Another example of a poorly predicted ionic channel (below 70%) with more balanced rate of all three structures is presented in Table 3 − 2vl0, chain A. Here, all predictors obtained similar performance with JNET very significantly outperforming the others in coil prediction.

The results of transmembrane predictors applied on TM regions of ionic channels are presented in Table 4 and compared with the prediction accuracy by general purpose SSP tools only applied to the transmembrane parts of the ionic channels and other transmembrane proteins (Table 5). According to the tests of 2002 by Chen et al. [1], they obtained the following predictors performance on transmembrane helices: TMHMM ($Q_H = 66\%$), DAS ($Q_H = 46\%$), HMMTOP ($Q_H = 67\%$), ConPred ($Q_H = 65\%$), PHDhtm ($Q_H = 74\%$). Our results obtained in 2009 on the set of ionic channels are much more promising (Table 4) − typically $Q_H \approx 80\%$ with PHDhtm obtaining $Q_H = 89\%$. Even better results are achieved in segment prediction, where HMMTOP obtained $SOV_H = 85\%$ (Table 4).

Surprisingly, better prediction was provided by general purpose SSP tools, where PSIPRED obtained $Q_H = 92\%$ and was the best in segment prediction, obtaining

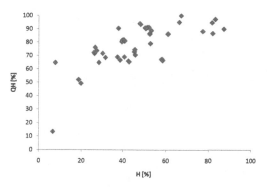

Fig. 2. For ionic channel proteins modeled by PSIPRED, prediction quality of α-helices QH depends on the rate H of α-helical residues (dots). Similar dependence was observed for CONSENSUS.

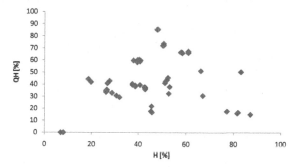

Fig. 3. For ionic channel proteins modeled by JNET, prediction quality of α-helices QH is very dispersed with no evident trend.

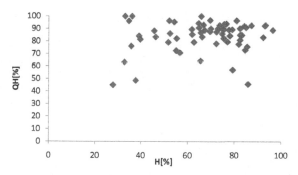

Fig. 4. For other transmembrane proteins, prediction quality of α-helices does not depend on the rate H of α-helical residues although very low rate of H adversely affects the prediction. Data shown for PSIPRED, similar results found for CONSENSUS.

$SOV_H = 86\%$ (Table 5). The summary of the SSP tool prediction on transmembrane helices in ionic channels is presented in Fig.7. Note that the prediction rate was also affected by the quality of the transmembrane region recognition.

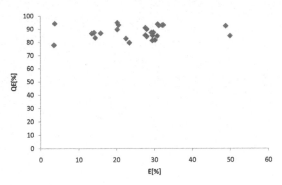

Fig. 5. For ionic channel proteins, prediction quality of α-structures QB does not depend on their rate in any general purpose SSP predictor (data shown for PSIPRED).

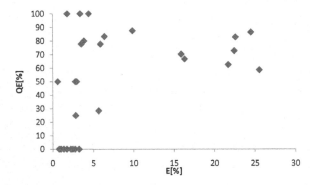

Fig. 6. For other TM proteins, Q_E is low with high standard devation since many chains have less than 3% of α-structures. Data for PSIPRED, similar results for all general SSP tools.

Table 2. The worst case study of ionic channel prediction. Data for 3beh, chain A.

channel	3beh, chain A, H = 53%, E = 0%, C = 47%, 3beh - 4 chains							
	Q_3	Q_H	Q_E	Q_C	SOV	SOV_H	SOV_E	SOV_C
PSIPRED	64	89	no	35	24	89	no	9
JNET	39	38	no	40	19	48	no	11
CONSENSUS	63	86	no	38	28	85	no	11

Table 3. Good performance of JNET in coil prediction is an asset in case of some ionic channels with a high coil rate. Data for 2vl0, chain A.

channel	2vl0, chain A, H = 26%, E = 30%, C = 44%							
	Q3	QH	QE	QC	SOV	SOV_H	SOV_E	SOV_C
PSIPRED	67	71	85	52	61	87	83	40
JNET	63	36	67	76	57	42	67	60
CONSENSUS	66	69	83	54	66	84	82	49

Table 4. Prediction of transmembrane α-helices in ionic channels and other transmembrane proteins by SSP-TM predictors.

SSP	Statistic	ionic channels		other TM proteins	
		QH	SOV_H	QH	SOV_H
TMHMM	Mean	78	77	85	88
	SD	15	18	15	16
DAS	Mean	77	84	74	80
	SD	8	9	17	16
HMMTOP	Mean	80	85	81	85
	SD	9	14	22	23
CONPRED	Mean	81	83	85	87
	SD	10	14	20	20
PHDhtm	Mean	89	72	84	80
	SD	8	16	20	19

Table 5. Prediction of transmembrane α-helices in ionic channels and other transmembrane proteins by general purpose SSP predictors.

SSP	Statistic	ionic channels		other TM proteins	
		QH	SOV_H	QH	SOV_H
PSIPRED	Mean	92	86	97	87
	SD	6	11	4	13
JNET	Mean	56	50	57	54
	SD	27	22	29	27
CONSENSUS	Mean	92	82	95	84
	SD	7	16	7	15

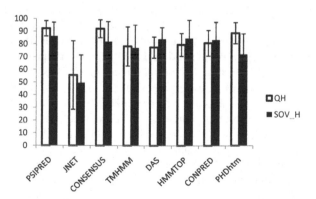

Fig. 7. Performance of all tested SSP tools in predicting transmembrane α-helices of ionic channels. PSIPRED holds the highest result.

4 Conclusions

The applicability of currently available secondary structure predictors for ionic channels was evaluated. The experiments included three general SSP tools (PSIPRED, JNET, and Proteus CONSENSUS method) and five dedicated transmembrane predictors (TMMHMM, DAS, HMMTOP, CONPRED, PHDhtm). All of the tools were applied to the whole molecule of ionic channel and to molecules of other transmembrane proteins. The results were related to their outcome for all proteins (populated mostly by soluble proteins) published by Eva. The predictor performance was measured by the accuracy per residue Q and per segment SOV. Correlation between predictor performance and protein structure was studied and the worst case analysis for low-performance ionic channels provided.

The study shows that the overall result for general SSP tools is very satisfactory. The best prediction quality was obtained by PSIPRED: $Q_3 = 77\%$, $Q_H = 79\%$, $Q_E = 88\%$, SOV = 70%, $SOV_H = 82\%$, $SOV_E = 83\%$. Very similar results were obtained by Proteus CONSENSUS, the worst accuracy came from JNET, especially in predicting α-helices. However, since JNET had a good accuracy in coil prediction, it could be utilized in some proteins. PSIPRED maintained its superiority also for the most complex ionic channel according to prediction performance. The experiments also reveal that a high helix content improves helix prediction quality of the best tools. However, if the content of β-sheets exceeds at least 5%, their amount does not influence predictor accuracy of this secondary structure. In general, it was shown that ionic channels are predicted at a similar level of accuracy as soluble proteins, and only slightly worse than other transmembrane proteins.

The analysis of the prediction of transmembrane helices of ionic channels by transmembrane predictors showed that, although the best predictor could reach almost 90% of accuracy per residue (PHDhtm), they are outperformed by the general SSP tool, i.e. PSIPRED ($Q_H = 93\%$). The same SSP tools also show superiority in the prediction per-segment, obtaining $SOV_H = 86$.

The superiority of PSIPRED in both groups of predictors is surprising for two reasons. First, PSIPRED is a general purpose tool. More than that – all transmembrane proteins were excluded from the training set of PSIPRED, contrary to JNET. Secondly, the recognition method of PSIPRED is simpler than that of JNET. We propose the hypothesis that, in protein recognition, less complex heuristic methods work better in situations when the final testing set comes from a different group that that for which the solution was optimized (all general purpose prediction tools are biased towards soluble proteins, as the most numerous in PDB). Very complex methods seem over-sensitive for such changes. The TM-devoted tools have been optimized towards transmembrane regions, therefore their performance in the structure prediction is influence by the correct prediction of the TM residues.

The study revealed that the whole molecule of ionic channel could be modeled by a general purpose SSP tool with a sufficiently good accuracy. It indicated PSIPRED is the predictor of the best quality among tested and showed the relation between the ratio of each state in the channel and the prediction accuracy. Moreover, predictions could be refined or confirmed by transmembrane predictors that are capable of indicating the location of transmembrane regions.

Acknowledgments. This work was in part supported by Erasmus grant for B.K. and Young Scientist grants of Wroclaw University of Technology and of British Council Programme (WAR/342/108) for W.D.

References

1. Chen, C.P., Kernytsky, A., Rost, B.: Transmembrane helix predictions revisited. Protein Sci. 11(12), 2774–2791 (2002)
2. Cuthbertson, J.M., Doyle, D.A., Sansom, M.S.: Transmembrane helix prediction: a comparative evaluation and analysis. Protein Eng. Des. Sel. 18(6), 295–308 (2005)
3. Punta, M., Forrest, L.R., Bigelow, H., Kernytsky, A., Liu, J., Rost, B.: Membrane protein prediction methods. Methods 41(4), 460–474 (2007)
4. Koh, I.Y., Eyrich, V.A., Marti-Renom, M.A., Przybylski, D., Madhusudhan, M.S., Eswar, N., Graña, O., Pazos, F., Valencia, A., Sali, A., Rost, B.E.: Evaluation of protein structure prediction servers. Nucleic Acids Res. 31(13), 3311–3315 (2003)
5. Chou, P.Y., Fasman, G.D.: Prediction of protein conformation. Biochemistry 13(2), 222–245 (1974)
6. Berman, H.M., Westbrook, J., Feng, Z., Gilliland, G., Bhat, T.N., Weissig, H., Shindyalov, I.N., Bourne, P.E.: The Protein Data Bank Nucleic Acids Research, vol. 28, pp. 235–242 (2002)
7. Jayasinghe, S., Hristova, K., White, S.H.,, M.: A database of membrane protein topology. Protein Sci. 10(2), 455–458 (2001)
8. Membrane Proteins of Known Structure, http://blanco.biomol.uci.edu/Membrane_Proteins_xtal.html
9. EVA: Evaluation of automatic structure prediction servers, http://cubic.bioc.columbia.edu/eva/
10. Brünger, A.T., Free, R.: value: cross-validation in crystallography. Methods Enzymol. 277, 366–396 (1997)
11. Tusnady, G.E., Dosztanyi, Z., Simon, I.: PDB_TM: selection and membrane localization of transmembrane proteins in the protein data bank. Nucleic Acids Res. 33(Database issue), 275–278 (2005)
12. PDBTM: Protein Data Bank of Transmembrane Proteins, http://pdbtm.enzim.hu/
13. Rost, B., Sander, C.: Prediction of protein secondary structure at better than 70% accuracy. J. Mol. Biol. 232(2), 584–599 (1993)
14. Zemla, A., Venclovas, C., Fidelis, K., Rost, B.: A modified definition of Sov, a segment-based measure for protein secondary structure prediction assessment. Proteins 34(2), 220–223 (1999)
15. Altschul, S.F., Madden, T.L., Schaffer, A.A., Zhang, J., Zhang, Z., Miller, W., Lipman, D.J.: Gapped BLAST and PSI-BLAST: a new generation of protein database search programs. Nucleic Acids Res. 25(17), 3389–3402 (1997)
16. Jones, D.T.: Protein secondary structure prediction based on position-specific scoring matrices. J. Mol. Biol. 292, 195–202 (1999)
17. Cuff, J.A., Barton, G.J.: Application of multiple sequence alignment profiles to improve protein secondary structure prediction. Proteins 40(3), 502–511 (2000)
18. Durbin, R., Eddy, S., Krogh, A., Mitchison, G.: Biological sequence analysis: probabilistic models of proteins and nucleic acids. Cambridge University Press, Cambridge (1998)

19. Montgomerie, S., Sundararaj, S., Gallin, W.J., Wishart, D.S.: Improving the accuracy of protein secondary structure prediction using structural alignment. BMC Bioinformatics 7, 301 (2006)
20. Tusnady, G.E., Simon, I.: Principles governing amino acid composition of integral membrane proteins: application to topology prediction. J. Mol. Biol. 283(2), 489–506 (1998)
21. HMMTOP, http://www.enzim.hu/hmmtop/html/submit.html
22. Krogh, A., Larsson, B., von Heijne, G., Sonnhammer, E.L.: Predicting transmembrane protein topology with a hidden Markov model: application to complete genomes. J. Mol. Biol. 305(3), 567–580 (2001)
23. TMHMM server, v. 2.0, http://www.cbs.dtu.dk/services/TMHMM/
24. Rost, B., Casadio, R., Fariselli, P., Sander, C.: Transmembrane helices predicted at 95% accuracy. Protein Sci. 4(3), 521–533 (1995)
25. PredictProtein - Structure Prediction and Sequence Analysis, http://www.predictprotein.org/
26. Cserzo, M., Wallin, E., Simon, I., von Heijne, G., Elofsson, A.: Prediction of transmembrane alpha-helices in procariotic membrane proteins: the Dense Alignment Surface method. Prot. Eng. 10, 673–676 (1997)
27. DAS-TMfilter server, http://mendel.imp.ac.at/sat/DAS/DAS.html
28. Arai, M., Mitsuke, H., Ikeda, M., Xia, J.X., Kikuchi, T., Satake, M., Shimizu, T.: ConPred II: a consensus prediction method for obtaining transmembrane topology models with high reliability. Nucleic Acids Res 32(Web Server issue), W390–W393 (2004)
29. Conpred II, http://bioinfo.si.hirosaki-u.ac.jp/~ConPred2/
30. Tusnady, G.E., Dosztanyi, Z., Simon, I.: TMDET: web server for detecting transmembrane regions of proteins by using their 3D coordinates. Bioinformatics 21(7), 1276–1277 (2005)
31. Clayton, G.M., Altieri, S., Heginbotham, L., Unger, V.M., Morais-Cabral, J.H.: Structure of the transmembrane regions of a bacterial cyclic nucleotide-regulated channel. Proc. Natl. Acad. Sci. USA 105, 1511–1515 (2008)

Secure Information Splitting Using Grammar Schemes

Marek R. Ogiela and Urszula Ogiela

AGH University of Science and Technology
Al. Mickiewicza 30, PL-30-059 Krakow, Poland
{mogiela,ogiela}@agh.edu.pl

Abstract. Information splitting is used in many tasks of the intelligent sharing of secrets and key data in business organisations. The significance of information splitting depends on its nature, while the significance of information sharing may depend on its importance and the meaning it has for the organisation or institution concerned. This study presents and characterises models for multi-level information splitting and information management with the use of the linguistic approach and formal grammars. The appropriate methods for secret sharing to be chosen for the specific type of an organisational structure will be identified depending on this structure.

Keywords: Secret sharing, threshold schemes, intelligent information management.

1 Introduction

Multi-level information splitting algorithms are named after the type of splitting they perform. This split can, for example, be hierarchical or by layers. The principal difference between the presented types of splits concerns the method of introducing the split itself. When a split is made within homogenous, uniform groups of layers, then it is a layer split, whereas if the split is made regardless of the homogeneity of the group or layer but by reference to several groups ordered hierarchically, this is a hierarchical split.

Information can be divided both within the entire structure in which some hierarchical dependency is identified, or within a given group as well as within any homogenous layer. This is why, depending on the type of information split, it makes sense to identify correctly selected information splitting algorithms.

Algorithms for the multi-level splitting of confidential or secret information are designed using structural analysis and the linguistic recording of data. The structural analysis used for this kind of task is based on the analysis of the structure of the business organisation and can be designed for a specific organisation, or splitting and sharing algorithms can be designed for a broader group of organisations, which proves that the method is universal. However, one must be aware that the group should be homogenous in terms of the structure of organisations forming part of it.

Another important component of information splitting algorithms is the use of linguistic data recording methods [13]. This type of information recording and presentation

N.T. Nguyen et al. (Eds.): New Challenges in Compu. Collective Intelligence, SCI 244, pp. 327–336.
springerlink.com   © Springer-Verlag Berlin Heidelberg 2009

refers to a syntactic data analysis. The key in this approach is that it uses mathematical/ informatics formalisms and linguistic methods which allow an additional stage to be introduced which enhances the functionality of classical threshold schemes of information sharing [3, 4, 10, 12, 16]. Such enhanced linguistic schemes can be used for information sharing in various data management structures.

2 Information Splitting in Layered Structures

The essence of the presented approach is that within a given layer it is possible to divide secret information in such a way that every person involved in the process of encrypting the information becomes the owner of a certain part of the secret. Even though such persons are equal owners of parts of the secret from the perspective of the information splitting process, the secret can be recreated omitting some of them. If the secret is split between the members of a given group in equal parts, this means that every member will receive the same amount of the secret, and then all of them have to reveal their parts to recreate the original message. There is obviously no absolute requirement for all owners of parts of the secret to reveal their parts, because, for example, threshold schemes for information splitting (like the Tango algorithm [16]) guarantee that secret information can be recreated with the involvement of a smaller number of participants than the number between which the shares were distributed.

Since every participant of the information splitting and also the information reconstruction process is treated as an equal process participant, there is no person in the

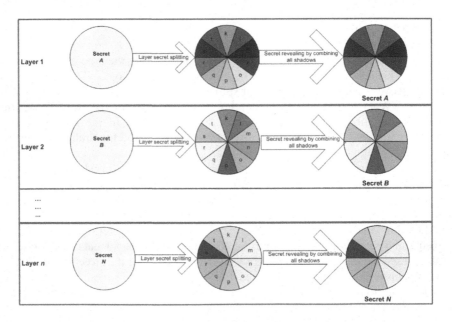

Fig. 1. Splitting information constituting a secret within a selected layer

group who could reconstruct the information without involving others. This situation is shown in Fig. 1 which presents the method of splitting a secret within a given layer.

Such a split of information between the members of a given group in which every one has the same privileges is a layer split. It is worth noting that the layer split may refer to the following types of splits:

- **Of various secrets split in various layers in the same (similar) way** - this situation means that the secret is split in the same way (in the sense of the method), regardless of the layer dealing with this secret (Fig. 1). Obviously, the number of participants of the secret split in various layers is determined by the instance supervising the split (the decision-maker), and in addition it is unchanged in the remaining layers. What does change is the information constituting the secret being split in the specific layer.

- **Of the same secret split in different ways depending on the layer** - if we take information A, which can be a secret for several layers within which it is split, then, for instance, this secret can be split among n participants in the first layer, the same secret can be split in the superior (second) layer between $n-k$ participants, which is a number smaller than in the subordinate layer, and in the third layer the same secret can be split among $n-k-p$ participants. The values n, k, p can be defined freely depending on the size of the group from which the selected persons - secret trustees - are chosen (Fig. 2).

- **Various secrets in different layers** - this type of a split concerns a situation in which different pieces of information can be split between different groups of persons (Fig. 3). So for a business organisation this situation may mean that at the decision-making level the secret split comprises specific strategic information of the organisation, but at the executive stage marketing and promotion information of the organisation may be split.

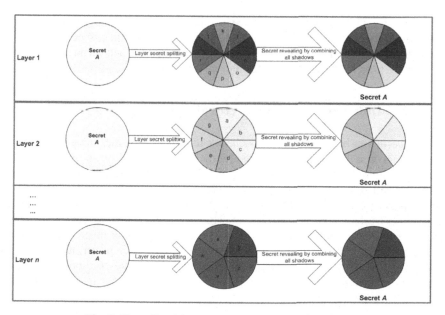

Fig. 2. The split of the same secret within various layers

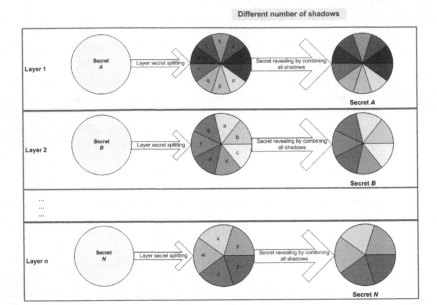

Fig. 3. Various splits of secret information within different layers

The above layer splits of secrets can apply to splitting information (constituting a secret) at various levels - the operational, tactical and strategic levels of a given organisation. Of course, the selection of the appropriate splitting method depends on the type of the organisational structure and the materiality (importance) of the shared information.

A layer split is characteristic for flat structures, as it is very easy to split information within a given layer. Such a split makes sense only within a specific layer of a flat structure, as the split of information between the director and managers is already characteristic for a layered split of information.

3   Information Flow in a Hierarchical Split

The essence of the hierarchical approach lies in considering the hierarchy operating within the business organisation. It is the hierarchical nature of business organisations that allows hierarchical secret splits to be introduced. Such a split may have the form of a split of varied information (secret) within a given hierarchy, taking into consideration that higher up in the hierarchy this secret can be reconstructed by other trustees (or a single other trustee) of parts of the secret. This situation is illustrated in Fig. 4.

Hierarchical information splits are much more frequent than layered splits, as the hierarchical nature of the structure is much more commonplace in various types of organisations. This is why a hierarchical information split can be used both in lean and flat structures, taking into account the superiority of persons managing the organisation and the subordination of particular departments and their managers.

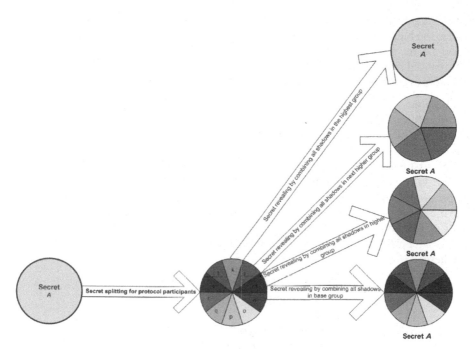

Fig. 4. Hierarchical secret split

In the case of a hierarchical information split, it is noticeable that secret splits are very varied and the ways of splitting and sharing information are very numerous and depend on the individual situation of the organisation and the materiality of shared information. This is why the methods of secret splitting presented in this publication, concerning both the hierarchical and the layered split, can be used in various types of organisational structures.

4   Proposed Splitting Algorithm Based on the Mathematical Linguistic Concept

This chapter proposes an original algorithm enhancing the operation of information splitting and sharing schemes which generates an additional information component (called a shadow) in the form of the linguistic information necessary to reconstruct the entire secret. The general methodology of using formal languages to enhance a traditional threshold scheme is as follows:

1. one of the traditional secret splitting schemes (e.g. Blakley's, Shamir's or Tang's algorithm [6, 14, 16]) is used in the given organisational structure;
2. the split data is transformed into a bit sequence;
3. a grammar is defined which generates bit positions for the shared data;
4. the bit sequence is parsed with an analyser defined for the introduced grammar;

5. the parsing generates a sequence of production numbers (grammar rules) which allow the bit representation of the shared secret to be generated;
6. the secret represented by a sequence of production numbers is split using the threshold scheme selected (in step 1);
7. shadows are distributed to particular participants of the protocol.

These stages determine the basic actions necessary to generate the components of shared information which can be communicated to the participants of the entire procedure of allocating the split data. A proposal for splitting information using context-free grammars is presented below.

The first type of a multi-level information split algorithm is based on single-bit information recording, which is converted into a linguistic record of information using the proposed grammar when the algorithm is executed.

An example of this type of grammar is the G_{BIT1} grammar presented below which looks as follows:

$$G_{BIT1}=(V_N, V_T, SP, STS)$$

where:

V_N = {BIT, Z, O} – a set of non-terminal symbols;
V_T = {0, 1, λ} – a set of terminal symbols;
{λ} – an empty symbol;
STS = BIT – a grammar start symbol;
SP – a set of productions, defined as follows:

1.   BIT $\rightarrow$ Z BIT | O BIT | λ
2.   Z $\rightarrow$ 0
3.   O $\rightarrow$ 1

The presented grammar is a context-free grammar which makes a bit conversion of sequences of zeroes and ones into sequences of numbers of the grammar productions. In practice, this operation means that the resultant sequence will contain numbers of the derivation rules of the grammar, i.e. integers from the [1,..,5] interval. The representation is converted by a syntax analyser which changes the bit sequence into numbers of the linguistic rules of the grammar within a time of square complexity. Its operation is exemplified in Figure 5.

The solution presented in Fig. 5 shows how selected information is converted into its bit representation which is coded using the proposed grammar. The coded form of information can be split in the way presented in Figure 5. This is an (m, n)-threshold split in which just the main part of the secret, that is m or n-m of secrets is necessary to reconstruct the split secret. Every one of these main split parts allows the split secret to be successfully reconstructed. However, combining these components yields only the contents of the secret, which allows the input information to be decoded using grammatical reasoning methods (i.e. meaning analysis methods).

The proposed modification of a threshold algorithm for information splitting and sharing consists in using a grammar at the stage of converting the bit representation into sequences of numbers of linguistic rules in the grammar.

After this transformation is completed, any secret splitting scheme can be used, and the components can be distributed among any number n of protocol participants. If the allocation of grammatical rules remains a secret, then this is an arbitration proto-col in which the reconstruction of a secret by the authorised group of shadow owners requires the involvement of a trusted arbitrator who has information on grammar rules.

If the grammar is disclosed, the secret can be reconstructed without the involve-ment of a trusted person just on the basis of the secret components possessed by the authorised group of participants of the information splitting algorithm.

The proposed information sharing algorithm may apply to the execution of any classical (m, n)-threshold secret sharing algorithm. In the case of data splitting and sharing algorithms, the split secret is not the bit sequence itself, but the sequence composed of numbers of syntactic rules of the grammar introduced for the splitting. Depending on its structure and type, it can contain values of two or more bits. This is why the stage of converting the bit representation of the shared secret can also be generalised from the version coding single bits to the coding of bit blocks of various lengths. However, to avoid too many generation rules in the defined grammar, it is worth imposing a restriction on the length of coded bit blocks in the proposed scheme. It seems easy and natural to consider bit blocks no longer than 4-5 bits. Information theory says that all representations of values coded with such lengths of machine words will fall within the range of 16 or 32 values, which, when combined with a few additional grammatical rules, allows us to estimate the total number of productions of this grammar as not exceeding 20 for 4-bit words and 40 for 5-bit words.

To illustrate the idea of an enhanced linguistic coding, a generalised version of a linguistic information splitting algorithm will be presented for a grammar that con-verts blocks of several bits. The graphic representation of using grammar expansion in classical threshold schemes is presented in Fig. 5.

$$G=(V_N, V_T, SP, STS),$$

where:

V_N = {SECRET, BIT_BLOCK, 1_BIT, 2_BIT, 3_BIT, 4_BIT} – a set of non-terminal symbols

V_T= {ONE BIT VALUE, TWO BITS VALUE, THREE BITS VALUE,

FOUR BITS VALUE, λ} – a set of terminal symbols which define each bit block value.

{λ} – defines an empty symbol.

STS = SECRET - the grammar start symbol.

A production set SP is defined in following way.

1. SECRET → BIT_BLOCK
2. BIT_BLOCK → BIT_BLOCK BIT_BLOCK
3. BIT_BLOCK → 1_BIT | 2_BIT | 3_BIT | 4_BIT
4. BIT_BLOCK → λ
5. 1_BIT → ONE BIT VALUE
6. 2_BIT → TWO BITS VALUE
7. 3_BIT → THREE BITS VALUE
8. 4_BIT → FOUR BITS VALUE

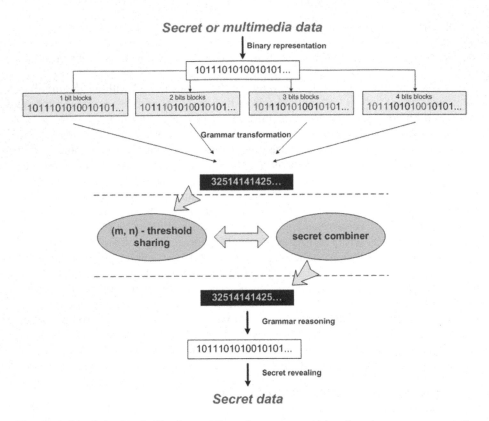

Fig. 5. A linguistic threshold scheme. The enhancement consists in using a grammar at the stage of changing the bit representation into sequences of numbers of grammar rules

This type of grammar allows more complex information coding tasks to be executed, as the information is converted into the bit representation and in the next step is converted into a record of 2, 3 or 4-bit clusters which become the basis for coding the original information. The rest of the information splitting process is the same as in the previous case, but the presentation itself of the coded information is completely different, as it is created using a completely different grammar.

With regard to the proposals of the linguistic enhancement of threshold schemes presented here it is notable that the level of security achieved is independent of the length of blocks subjected to conversion with the use of rules of the introduced grammar.

The methods of multi-level information splitting or sharing presented in this chapter, which use bit blocks of various lengths, show how information splitting algorithms can be significantly enhanced by adding elements of linguistic and grammatical data analysis. This is a novel solution. The length of bit blocks has a major impact on the speed and length of the stage of coding the input information representation, which is the stage that prepares information to be coded as a secret.

5   Conclusion

This publication presents new methods of developing linguistic threshold schemes used to split and share information. Such methods constitute a secure enhancement of traditional secret splitting algorithms and utilise a stage at which information is coded using the appropriately defined regular or context-free grammar [13]. The introduction of such formalisms yields more robust threshold schemes and also and additional component of the secret, which can be assigned to one of the splitting protocol participants. The many possible applications of such methods include their use for the intelligent management of important or confidential information in government institutions or businesses. Algorithms of multi-level information splitting allow information which is not available to all employees of a given organisation or its environment to be securely split or shared. The security of such algorithms is due to using cryptographic information encryption protocols at the stage of developing these algorithms, which, combined with linguistic methods of describing and interpreting data (information in this case) means that these protocols ensure the security of the entire splitting process, that is information encryption, its splitting and reconstruction. An important characteristic of the proposed method is its suitability for use in hierarchical structures and layered management models.

Acknowledgements. This work has been supported by the AGH University of Science and Technology under Grant No. 10.10.120.783.

References

1. Asmuth, C.A., Bloom, J.: A modular approach to key safeguarding. IEEE Transactions on Information Theory 29, 208–210 (1983)
2. Ateniese, G., Blundo, C., De Santis, A., Stinson, D.R.: Constructions and bounds for visual cryptography. In: Meyer auf der Heide, F., Monien, B. (eds.) ICALP 1996. LNCS, vol. 1099, pp. 416–428. Springer, Heidelberg (1996)
3. Beguin, P., Cresti, A.: General short computational secret sharing schemes. In: Guillou, L.C., Quisquater, J.-J. (eds.) EUROCRYPT 1995. LNCS, vol. 921, pp. 194–208. Springer, Heidelberg (1995)
4. Beimel, A., Chor, B.: Universally ideal secret sharing schemes. IEEE Transactions on Information Theory 40, 786–794 (1994)
5. Blakley, G.R.: Safeguarding Cryptographic Keys. In: Proceedings of the National Computer Conference, pp. 313–317 (1979)
6. Blakley, B., Blakley, G.R., Chan, A.H., Massey, J.: Threshold schemes with disenrollment. In: Brickell, E.F. (ed.) CRYPTO 1992. LNCS, vol. 740, pp. 540–548. Springer, Heidelberg (1993)
7. Blundo, C., De Santis, A.: Lower bounds for robust secret sharing schemes. Inform. Process. Lett. 63, 317–321 (1997)
8. Charnes, C., Pieprzyk, J.: Generalised cumulative arrays and their application to secret sharing schemes. Australian Computer Science Communications 17, 61–65 (1995)
9. Desmedt, Y., Frankel, Y.: Threshold Cryptosystems. In: Brassard, G. (ed.) CRYPTO 1989. LNCS, vol. 435, pp. 307–315. Springer, Heidelberg (1990)
10. van Dijk, M.: On the information rate of perfect secret sharing schemes. Designs, Codes and Cryptography 6, 143–169 (1995)

11. Hang, N., Zhao, W.: Privacy-preserving data mining Systems. Computer 40(4), 52–58 (2007)
12. Jackson, W.-A., Martin, K.M., O'Keefe, C.M.: Ideal secret sharing schemes with multiple secrets. Journal of Cryptology 9, 233–250 (1996)
13. Ogiela, M.R., Ogiela, U.: Linguistic Extension for Secret Sharing (m, n)-threshold Schemes. In: SecTech 2008 - 2008, International Conference on Security Technology, Hainan Island, Sanya, China, December 13-15, pp. 125–128 (2008) ISBN: 978-0-7695-3486-2; doi:10.1109/SecTech.2008.15
14. Shamir, A.: How to Share a Secret. Communications of the ACM, 612–613 (1979)
15. Simmons, G.J.: An Introduction to Shared Secret and/or Shared Control Schemes and Their Application in Contemporary Cryptology. In: The Science of Information Integrity, pp. 441–497. IEEE Computer Society Press, Los Alamitos (1992)
16. Tang, S.: Simple Secret Sharing and Threshold RSA Signature Schemes. Journal of Information and Computational Science 1, 259–262 (2004)
17. Wu, T.-C., He, W.-H.: A geometric approach for sharing secrets. Computers and Security 14, 135–146 (1995)
18. Zheng, Y., Hardjono, T., Seberry, J.: Reusing shares in secret sharing schemes. The Computer Journal 37, 199–205 (1994)

Comparative Analysis of Neural Network Models for Premises Valuation Using SAS Enterprise Miner

Tadeusz Lasota[1], Michał Makos[2], and Bogdan Trawiński[2]

[1] Wrocław University of Environmental and Life Sciences, Dept. of Spatial Management
Ul. Norwida 25/27, 50-375 Wroclaw, Poland
[2] Wrocław University of Technology, Institute of Informatics,
Wybrzeże Wyspiańskiego 27, 50-370 Wrocław, Poland
Tadeusz.Lasota@wp.pl, Michal.Makos@gmail.com,
Bogdan.Trawinski@pwr.wroc.pl

Abstract. The experiments aimed to compare machine learning algorithms to create models for the valuation of residential premises were conducted using the SAS Enterprise Miner 5.3. Eight different algorithms were used including artificial neural networks, statistical regression and decision trees. All models were applied to actual data sets derived from the cadastral system and the registry of real estate transactions. A dozen of predictive accuracy measures were employed. The results proved the usefulness of majority of algorithms to build the real estate valuation models.

Keywords: neural networks, real estate appraisal, AVM, SAS Enterprise Miner

1 Introduction

Automated valuation models (AVMs) are computer programs that enhance the process of real estate value appraisal. AVMs are currently based on methodologies from multiple regression analysis to neural networks and expert systems [11]. The quality of AVMs may vary depending on data preparation, sample size and their design, that is why they must be reviewed to determine if their outputs are accurate and reliable. A lot need to be done to create a good AVM. Professional appraisers instead of seeing in AVMs a threat should use them to enhance services they provide.

Artificial neural networks are commonly used to evaluate real estate values. Some studies described their superiority over other methods [1], [7]. Other studies pointed out that ANN was not the "state-of-the-art" tool in that matter [13] or that results depended on data sample size [9]. Studies showed that there is no perfect methodology for real estate value estimation and mean absolute percentage error ranged from almost 4% to 15% in better tuned models [4] .

In our pervious works [3], [5], [6] we investigated different machine learning algorithms, among others genetic fuzzy systems devoted to build data driven models to assist with real estate appraisals using MATLAB and KEEL tools. In this paper we report the results of experiments conducted with SAS Enterprise Miner aimed at the

N.T. Nguyen et al. (Eds.): New Challenges in Compu. Collective Intelligence, SCI 244, pp. 337–348.
springerlink.com
© Springer-Verlag Berlin Heidelberg 2009

comparison of several artificial neural network and regression methods with respect to a dozen performance measures, using actual data taken from cadastral system in order to assess their appropriateness to an internet expert system assisting appraisers' work.

2 Cadastral Systems as the Source Base for Model Generation

The concept of a data driven models for premises valuation, presented in the paper, was developed on the basis of the sales comparison method. It was assumed that whole appraisal area, which means the area of a city or a district, is split into sections (e.g. clusters) of comparable property attributes. The architecture of the proposed system is shown in Fig. 1. The appraiser accesses the system through the internet and chooses an appropriate section and input the values of the attributes of the premises being evaluated into the system, which calculates the output using a given model. The final result as a suggested value of the property is sent back to the appraiser.

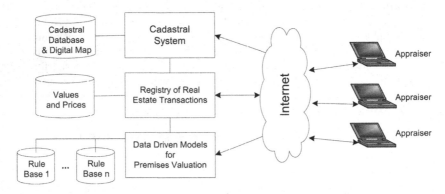

Fig. 1. Information systems to assist with real estate appraisals

Actual data used to generate and learn appraisal models came from the cadastral system and the registry of real estate transactions referring to residential premises sold in one of big Polish cities at market prices within two years 2001 and 2002. They constituted original data set of 1098 cases of sales/purchase transactions. Four attributes were pointed out as price drivers: usable area of premises, floor on which premises were located, year of building construction, number of storeys in the building, in turn, price of premises was the output variable.

3 SAS Enterprise Miner as the Tool for Model Exploration

SAS Enterprise Miner 5.3 is a part of SAS Software, a very powerful information delivery system for accessing, managing, analyzing, and presenting data [8], [10]. SAS Enterprise Miner has been developed to support the entire data mining process from data manipulation to classifications and predictions. It provides tools divided into five fundamental categories:

Sample. Allows to manipulate large data sets: filter observation, divide and merge data etc.

Explore. Provides different tools that can be very useful to get to know the data by viewing observations, their distribution and density, performing clustering and variable selection, checking variable associations etc.

Modify. That set of capabilities can modify observations, transform variables, add new variables and handle missing values etc.

Model. Provides several classification and prediction algorithms such as neural network, regression, decision tree etc.

Assess. Allows to compare created models.

SAS Enterprise Miner allows for batch processing which is a SAS macro-based interface to the Enterprise Miner client/server environment. Batch processing supports the building, running, and reporting of Enterprise Miner 5 process flow diagrams. It can be run without the Enterprise Miner graphical user interface (GUI). User-friendly GUI enables designing experiments by joining individual nodes with specific functionality into whole process flow diagram. The graph of the experiments reported in the present paper is shown in Fig. 2.

Following SAS Enterprise Miner algorithms for building, learning and tuning data driven models were used to carry out the experiments (see Table 1):

Table 1. SAS Enterprise Miner machine learning algorithms used in study.

Type	Code	Descritption
Neural networks	MLP	Multilayer Perceptron
	AUT	AutoNeural
	RBF	Ordinary Radial Basis Function with Equal Widths
	DMN	DMNeural
Statistical regression	REG	Regression
	DMR	Dmine Regression
	PLS	Partial Least Square
Decision trees	DTR	Decision Tree (CHAID, CART, and C4.5)

MLP - Neural Network. Multilayer Perceptron is the most popular form of neural network architecture. Typical MLP consists of input layer, any number of hidden layers with any number of units and output layer. It usually uses sigmoid activation function in the hidden layers and linear combination function in the hidden and output layers. MLP has connections between the input layer and the first hidden layer, between the hidden layers, and between the last hidden layer and the output layer.

AUT - AutoNeural. AutoNeural node performs automatic configuration of neural network Multilayer Perceptron model. It conducts limited searches for a better network configuration.

RBF - Neural Network with ORBFEQ i.e. ordinary radial basis function with equal widths. Typical radial neural network consists of: input layer, one hidden layer and output layer. RBF uses radial combination function in the hidden layer, based on the squared Euclidean distance between the input vector and the weight vector. It uses the exponential activation function and instead of MLP's bias, RBFs have a width associated with each hidden unit or with the entire hidden layer. RBF has connections between the input layer and the hidden layer, and between the hidden layer and the output layer.

DMN – DMNeural. The DMNeural node enables to fit an additive nonlinear model that uses the bucketed principal components as inputs. The algorithm was developed to overcome the problems of the common neural networks that are likely to occur especially when the data set contains highly collinear variables. In each stage of the DMNeural training process, the training data set is fitted with eight separate activation functions. The algorithm selects the one that yields the best results. The optimization with each of these activation functions is processed independently.

REG – Regression: Linear or Logistic. Linear regression method is a standard statistical approach to build a linear model predicting a value of the variable while knowing the values of the other variables. It uses least mean square in order to adjust the parameters of the linear model/function. Enables usage of the stepwise, forward, and backward selection methods. Logistic regression is a standard statistical approach to build a logistic model predicting a value of the variable while knowing the values of the other variables. It uses least mean square in order to adjust the parameters of the quadratic model/function. Enables usage of the stepwise, forward, and backward selection methods.

PLS - Partial Least Squares. Partial least squares regression is an extension of the multiple linear regression model. It is not bound by the restrictions of discriminant analysis, canonical correlation, or principal components analysis. It uses prediction functions that are comprised of factors that are extracted from the Y'XX'Y matrix.

DMR - Dmine Regression. Computes a forward stepwise least-squares regression. In each step, an independent variable is selected that contributes maximally to the model R-square value.

DTR - Decision Tree. An empirical tree represents a segmentation of the data that is created by applying a series of simple rules. Each rule assigns an observation to a segment based on the value of one input. One rule is applied after another, resulting in a hierarchy of segments within segments. It uses popular decision tree algorithms such as CHAID, CART, and C4.5. The node supports both automatic and interactive training. Automatic mode automatically ranks the input variables, based on the strength of their contribution to the tree. This ranking can be used to select variables for use in subsequent modeling.

4   Experiment Description

The main goal of the study was to carry out comparative analysis of different neural network algorithms implemented in SAS Enterprise Miner and use them to create and learn data driven models for premises property valuation with respect to a dozen of performance measures. Predictive accuracy of neural network models was also compared with a few other machine learning methods including linear regression, decision trees and partial least squares regression. Schema of the experiments comprising algorithms with preselected parameters is depicted in Fig. 2.

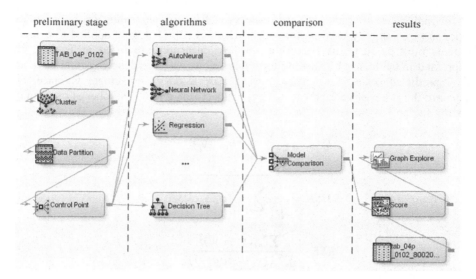

Fig. 2. Schema of the experiments with SAS Enterprise Miner

The set of observations, i.e. the set of actual sales/purchase transactions containing 1098 cases, was clustered and then partitioned into training and testing sets. The training set contained 80% of data from each cluster, and training one – 20%, that is 871 and 227 observations respectively. As fitness function the mean squared error (*MSE*) was applied to train models. A series of initial tests in order to find optimal parameters of individual algorithms was accomplished. Due to lacking mechanism of cross validation in any Enterprise Miner modeling nodes, each test was performed 4 times (with different data partition seed) and the average MSE was calculated.

Table 2. Performance measures used in study

Denot.	Description	Dimen-sion	Min value	Max value	Desirable outcome	No. of form.
MSE	Mean squared error	d^2	0	∞	min	1
RMSE	Root mean squared error	d	0	∞	min	2
RSE	Relative squared error	no	0	∞	min	3
RRSE	Root relative squared error	no	0	∞	min	4
MAE	Mean absolute error	d	0	∞	min	5
RAE	Relative absolute error	no	0	∞	min	6
MAPE	Mean absolute percentage error	%	0	∞	min	7
NDEI	Non-dimensional error index	no	0	∞	min	8
r	Linear correlation coefficient	no	-1	1	close to 1	9
R^2	Coefficient of determination	%	0	∞	close to 100%	10
var(AE)	Variance of absolute errors	d^2	0	∞	min	11
var(APE)	Variance of absolute percentage errors	no	0	∞	min	12

In final stage, after the best parameters set for each algorithm were determined, a dozen of commonly used performance measures [2], [12] were applied to evaluate models built by respective algorithms. These measures are listed in Table 2 and expressed in the form of following formulas below, where y_i denotes actual price and $\hat{y}_i$ – predicted price of i-th case, $avg(v)$, $var(v)$, $std(v)$ – average, variance, and standard deviation of variables $v_1, v_2, ..., v_N$, respectively and N – number of cases in the testing set.

$$MSE = \frac{1}{N} \sum_{i=1}^{N} (y_i - \hat{y}_i)^2 \tag{1}$$

$$RMSE = \sqrt{\frac{1}{N} \sum_{i=1}^{N} (y_i - \hat{y}_i)^2} \tag{2}$$

$$RSE = \frac{\sum_{i=1}^{N}(y_i - \hat{y}_i)^2}{\sum_{i=1}^{N}(y_i - avg(y))^2} \tag{3}$$

$$RRSE = \sqrt{\frac{\sum_{i=1}^{N}(y_i - \hat{y}_i)^2}{\sum_{i=1}^{N}(y_i - avg(y))^2}} \tag{4}$$

$$MAE = \frac{1}{N} \sum_{i=1}^{N} |y_i - \hat{y}_i| \tag{5}$$

$$RAE = \frac{\sum_{i=1}^{N}|y_i - \hat{y}_i|}{\sum_{i=1}^{N}|y_i - avg(y)|} \tag{6}$$

$$MAPE = \frac{1}{N} \sum_{i=1}^{N} \frac{|y_i - \hat{y}_i|}{y_i} * 100\% \tag{7}$$

$$NDEI = \frac{RMSE}{std(y)} \tag{8}$$

$$r = \frac{\sum_{i=1}^{N}(y_i - avg(y))\,(\hat{y}_i - avg(\hat{y}))}{\sqrt{\sum_{i=1}^{N}(y_i - avg(y))^2}\,\sqrt{\sum_{i=1}^{N}(\hat{y}_i - avg(\hat{y}))^2}} \tag{9}$$

$$R^2 = \frac{\sum_{i=1}^{N}(\hat{y}_i - avg(y))^2}{\sum_{i=1}^{N}(y_i - avg(y))^2} * 100\% \tag{10}$$

$$var(AE) = var(|y - \hat{y}|) \tag{11}$$

$$var(APE) = var(\frac{|y - \hat{y}|}{y}) \tag{12}$$

5   Results of the Study

5.1   Preliminary Model Selection

Within each modeling class, i.e. neural networks, statistical regression, and decision trees, preliminary parameter tuning was performed using trial and error method in order to choose the algorithm producing the best models. All measures were calculated for normalized values of output variables except for MAPE, where in order to avoid the division by zero, actual and predicted prices had to be denormalized. The nonparametric Wilcoxon signed-rank tests were used for three measures: MSE, MAE, and MAPE. The algorithms with the best parameters chosen are listed in Table 3.

Table 3. Best algorithms within each algorithm class

Name	Code	Description
Neural Network: MLP	MLP	Training Technique: Trust-Region; Target Layer Combination Fun: Linear; Target Layer Activation Fun: Linear; Target Layer Error Fun: Normal; Hidden Units: 3
AutoNeural	AUT	Architecture: Block Layers; Termination: Overfitting; Number of hidden units: 4
Neural Network: ORBFEQ	RBF	Training Technique: Trust-Region; Target Layer Combination Fun: Linear; Target Layer Activation Fun: Identity; Target Layer Error Fun: Entropy; Hidden Units: 30
DMNeural	DMN	Train activation function (selection): ACC; Train objective function (optimization): SSE
Regression	REG	Regression Type: Linear; Selection Model: None
Dmine Regression	DMR	Stop R-Square: 0.001
Partial Least Square	PLS	Regression model: PCR
Decision Tree	DTR	Subtree Method: N; Number of leaves: 16; p-value Inputs: Yes; Number of Inputs: 3

5.2   Final Results

Final stage of the study contained comparison of algorithms listed in Table 3, using all 12 performance measures enumerated in pervious section. The results of respective measures for all models are shown in Fig. 3-14, it can be easily noticed that relationship among individual models are very similar for some groups of measures.

Fig. 9 depicts that the values of MAPE range from 16.2% to 17.4%, except for PLS with 18.9%, what can be regarded as fairly good, especially when you take into account, that no all price drivers were available in our sources of experimental data.

Fig. 11 shows there is high correlation, i.e. greater than 0.8, between actual and predicted prices for each model. In turn, Fig.12 illustrating the coefficients of determination indicates that above 85% of total variation in the dependent variable (prices) is explained by the model in the case of AUT and REG models and less than 60% for DMS and PLS models.

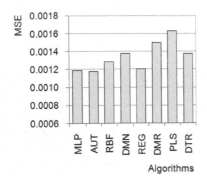

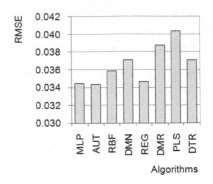

Fig. 3. Comparison of MSE values

Fig. 4. Comparison of RMSE values

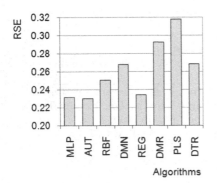

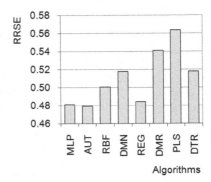

Fig. 5. Comparison of RSE values

Fig. 6. Comparison of RSSE values

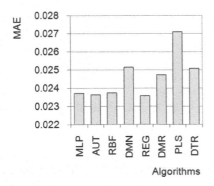

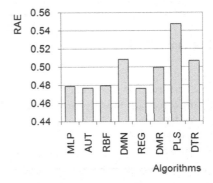

Fig. 7. Comparison of MAE values

Fig. 8. Comparison of RAE values

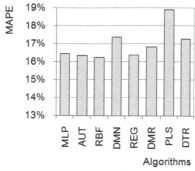

Fig. 9. Comparison of MAPE values

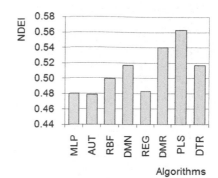

Fig. 10. Comparison of NDEI values

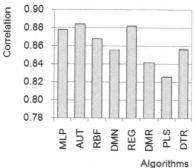

Fig. 11. Comparison of correlation coefficient (r) values

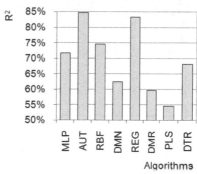

Fig. 12. Comparison of determination coefficient (R^2) values

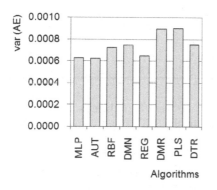

Fig. 13. Comparison of var(AE) values

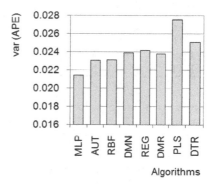

Fig. 14. Comparison of var(APE) values

The nonparametric Wilcoxon signed-rank tests were carried out for three measures: MSE, MAPE, and MAE. The results are shown in Tables 4, 5, and 6 in each cell results for a given pair of models were placed, in upper halves of the tables – p-values, and in lower halves final outcome, where N denotes that there are no

Table 4. Results of Wilcoxon signed-rank test for squared errors comprised by MSE

MSE	MLP	AUT	RBF	DMN	REG	DMR	PLS	DTR
MLP		0.962	0.494	0.022	0.470	0.218	0.000	0.064
AUT	N		0.758	0.180	0.650	0.244	0.001	0.114
RBF	N	N		0.284	0.924	0.271	0.001	0.035
DMN	Y	N	N		0.426	0.942	0.000	0.780
REG	N	N	N	N		0.152	0.007	0.023
DMR	N	N	N	N	N		0.002	0.176
PLS	Y	Y	Y	Y	Y	Y		0.097
DTR	N	N	Y	N	Y	N	N	

Table 5. Results of Wilcoxon test for absolute percentage errors comprised by MAPE

MAPE	MLP	AUT	RBF	DMN	REG	DMR	PLS	DTR
MLP		0.957	0.557	0.031	0.000	0.178	0.000	0.106
AUT	N		0.854	0.102	0.000	0.372	0.001	0.143
RBF	N	N		0.088	0.000	0.180	0.000	0.067
DMN	Y	N	N		0.000	0.392	0.001	0.977
REG	Y	Y	Y	Y		0.000	0.000	0.000
DMR	N	N	N	N	Y		0.001	0.286
PLS	Y	Y	Y	Y	Y	Y		0.097
DTR	N	N	N	N	Y	N	N	

Table 6. Results of Wilcoxon signed-rank test for absolute errors comprised by MAE

MAE	MLP	AUT	RBF	DMN	REG	DMR	PLS	DTR
MLP		0.946	0.619	0.050	0.739	0.260	0.002	0.134
AUT	N		0.692	0.145	0.934	0.306	0.001	0.135
RBF	N	N		0.178	0.799	0.195	0.001	0.091
DMN	N	N	N		0.212	0.646	0.000	0.887
REG	N	N	N	N		0.166	0.004	0.059
DMR	N	N	N	N	N		0.006	0.350
PLS	Y	Y	Y	Y	Y	Y		0.111
DTR	N	N	N	N	N	N	N	

differences in mean values of respective errors, and Y indicates that there are statistically significant differences between particular performance measures. In the majority of cases Wilcoxon tests did not provide any significant difference between models, except the models built by PLS algorithms, and REG with respect to MAPE.

Due to the non-decisive results of majority of statistical tests, rank positions of individual algorithms were determined for each measure (see Table 7). Observing median, average, minimal and maximal ranks it can be noticed that highest rank positions gained AUT, MLP, REG, RBF algorithms and the lowest PLS and DMR. Table 7 indicates also that some performance measures provide the same rank positions, and two groups of those measures can be distinguished. First one based on mean square errors contains MSE, RMSE, RSE, RRSE, NDEI, and the second one based on mean absolute errors comprises MAE and RAE.

Table 7. Rank positions of algorithms with respect to performance measures (1 means the best)

	MLP	AUT	RBF	DMN	REG	DMR	PLS	DTR
MSE	2	1	4	5	3	7	8	6
RMSE	2	1	4	5	3	7	8	6
RSE	2	1	4	5	3	7	8	6
RRSE	2	1	4	5	3	7	8	6
MAE	3	2	4	7	1	5	8	6
RAE	3	2	4	7	1	5	8	6
MAPE	4	2	1	7	3	5	8	6
NDEI	2	1	4	5	3	7	8	6
r	3	1	4	6	2	7	8	5
R^2	4	1	3	6	2	7	8	5
var(AE)	2	1	4	5	3	7	8	6
var(APE)	1	2	3	5	6	4	8	7
median	2.00	1.00	4.00	5.00	3.00	7.00	8.00	6.00
average	2.50	1.33	3.58	5.67	2.75	6.42	8.00	5.92
min	1	1	1	5	1	4	8	5
max	4	2	4	7	6	7	8	7

6 Conclusions and Future Work

The goal of experiments was to compare machine learning algorithms to create models for the valuation of residential premises, implemented in SAS Enterprise Miner 5.3. Four methods based on artificial neural networks, three different regression techniques and decision tree were applied to actual data set derived from cadastral system and the registry of real estate transactions.

The overall conclusion is that multilayer perceptron neural networks seem to provide best results in estimating real estate value. Configuration of AutoNeural node (which is actually implementation of MLP) gave a bit better results than MLP itself almost in every error/statistical measure. The analysis of charts leads to a conclusion that these eight algorithms can be divided into two groups with respect to their performance. To the first group with better results belong: AutoNerual, Neural Network: MLP, Linear Regression and Neural Network: ORBFEQ. In turn, to the second group with worse outcome belong: Decision Tree, DMNeural, Partial Least Squares and Dmine Regression.

Some performance measures provide the same distinction abilities of respective models, thus it can be concluded that in order to compare a number of models it is not necessary to employ all measures, but the representatives of different groups. Of course the measures within groups differ in their interpretation, because some are non-dimensional as well as in their sensitivity understood as the ability to show the differences between algorithms more or less distinctly.

High correlation between actual and predicted prices was observed for each model and the coefficients of determination ranged from 55% to 85% .

MAPE obtained in all tests ranged from 16% do 19%. This can be explained that data derived from the cadastral system and the register of property values and prices can cover only some part of potential price drivers. Physical condition of the premises and their building, their equipment and facilities, the neighbourhood of the building, the location in a given part of a city should also be taken into account, moreover overall subjective assessment after inspection in site should be done. Therefore we intend to test data obtained from public registers and then supplemented by experts conducting on-site inspections and evaluating more aspects of properties being appraised. Moreover further investigations of multiple models comprising ensembles of different neural networks using bagging and boosting techniques is planned.

References

1. Do, Q., Grudnitski, G.: A Neural Network Approach to Residential Property Appraisal. Real Estate Appraiser, 38–45 (December 1992)
2. Hagquist, C., Stenbeck, M.: Goodness of Fit in Regression Analysis – R^2 and G^2 Reconsidered. Quality & Quantity 32, 229–245 (1998)
3. Król, D., Lasota, T., Trawiński, B., Trawiński, K.: Investigation of Evolutionary Optimization Methods of TSK Fuzzy Model for Real Estate Appraisal. International Journal of Hybrid Intelligent Systems 5(3), 111–128 (2008)
4. Lai, P.P.: Applying the Artificial Neural Network in Computer-assisted Mass Appraisal. Journal of Housing Studies 16(2), 43–65 (2007)
5. Lasota, T., Mazurkiewicz, J., Trawiński, B., Trawiński, K.: Comparison of Data Driven Models for the Validation of Residential Premises using KEEL. International Journal of Hybrid Intelligent Systems (in press, 2009)
6. Lasota, T., Pronobis, E., Trawiński, B., Trawiński, K.: Exploration of Soft Computing Models for the Valuation of Residential Premises using the KEEL Tool. In: Nguyen, N.T., et al. (eds.) 1st Asian Conference on Intelligent Information and Database Systems (ACIIDS 2009), pp. 253–258. IEEE, Los Alamitos (2009)
7. Limsombunchai, V., Gan, C., Lee, M.: House Price Prediction: Hedonic Price Model vs. Artificial Neural Network. American J. of Applied Science 1(3), 193–201 (2004)
8. Matignon, R.: Data Mining Using SAS Enterprise Miner. Wiley Interscience, Hoboken (2007)
9. Nguyen, N., Cripps, A.: Predicting Housing Value: A Comparison of Multiple Regression Analysis and Artificial Neural Networks. Journal of Real Estate Research 22(3), 3131–3336 (2001)
10. Sarma, K.: Predictive Modeling with SAS Enterprise Miner: Practical Solutions for Business Applications. SAS Press (2007)
11. Waller, B.D., Greer, T.H., Riley, N.F.: An Appraisal Tool for the 21st Century: Automated Valuation Models. Australian Property Journal 36(7), 636–641 (2001)
12. Witten, I.H., Frank, E.: Data Mining: Practical Machine Learning Tools and Techniques. Elsevier, Morgan Kaufmann, San Francisco (2005)
13. Worzala, E., Lenk, M., Silva, A.: An Exploration of Neural Networks and Its Application to Real Estate Valuation. J. of Real Estate Research 10(2), 18–201 (1995)

Author Index